Compendium of Bedding Plant Diseases and Pests

A. R. Chase
Chase Agricultural Consulting
Cottonwood, Arizona, U.S.A.

Margery L. Daughtrey
Cornell University
Section of Plant Pathology and Plant–Microbe Biology
Long Island Horticultural Research and Extension Center
Riverhead, New York, U.S.A.

Raymond A. Cloyd
Kansas State University
Department of Entomology
Manhattan, Kansas, U.S.A.

APS PRESS
The American Phytopathological Society
St. Paul, Minnesota, U.S.A.

Front cover image:
Bedding plants in the landscape: celosia, gomphrena, New Guinea impatiens, and ageratum. (Courtesy M. L. Daughtrey—© APS)

Back cover images:
Background: coleus plants. (Courtesy M. L. Daughtrey—© APS)
Inset: Top left, zinnia, and lower left, osteospermum. (Courtesy P. Ek—© APS)
Top right, dahlia, and lower right, geranium. (Courtesy M. L. Daughtrey—© APS)

Library of Congress Control Number: 2017961885
International Standard Book Numbers:
Print: 978-0-89054-601-7
Online: 978-0-89054-602-4
Mobi: 978-0-89054-603-1
ePub: 978-0-89054-604-8

Printed in the United States of America on acid-free paper.

The American Phytopathological Society
3340 Pilot Knob Road
St. Paul, Minnesota 55121, U.S.A.

Preface

Compendium of Bedding Plant Diseases and Pests summarizes the information currently available about diseases, disorders, and arthropod (insect and mite) pests of annual ornamental bedding plants. This new compendium provides valuable information that will allow producers to deal more effectively with the challenges of growing bedding plants in both indoor and outdoor settings.

Plant materials have changed significantly over the years with the introductions of many new species and cultivars. In addition, changes in propagation procedures of ornamental bedding plants have resulted in increasing numbers of outbreaks of bacterial, fungal, and viral diseases, as well as many infestations of insect pests, such as fungus gnats, thrips, and whiteflies. Greenhouse operations routinely exchange plant materials, thus facilitating the movement of pesticide-resistant pathogens and arthropod pests worldwide. Strategies such as clean stock production, sanitation, and applications of traditional pesticides (fungicides, bactericides, and insecticides/miticides) and biological controls are often implemented to manage diseases and arthropod pests of annual bedding plants.

The prevention of diseases and disorders, as well as damage caused by arthropod pests, is an important concern for producers of annual bedding plants. These plants have multiple exposures to disease-causing organisms and arthropod pests throughout the production cycle. During production from seed or cuttings to finished plants, bedding plants are often moved from one greenhouse operation to another—sometimes, to a different state or region. The latest technology allows moving rooted cuttings from offshore production sites, and unrooted cuttings and seed are frequently moved between countries. Final incorporation into the landscape brings bedding plants into yet another environment, with additional sources of arthropod pests and disease inoculum.

The nature of bedding plant production is such that hundreds of species are grown within one operation, often in the same greenhouse range, which makes it challenging to simultaneously manage all the potential pathogens and arthropod pests. In addition, greenhouse producers' options for pest management may be limited because of the rapid production cycles and relatively low value of bedding plant crops. Therefore, regular monitoring should be conducted, with full awareness of the susceptibility of each crop to specific diseases and arthropod pests. Anticipating problems will help greenhouse producers initiate preventive approaches associated with pest management. Prevention strategies are always much more effective than curative strategies.

Compendium of Bedding Plant Diseases and Pests is organized into these major parts:

- The Introduction provides an overview of the industry and issues specific to bedding plant production.
- Part I, Infectious Diseases, is divided into sections on diseases caused by bacteria, fungi, oomycetes, nematodes, viroids, and viruses. More than 75 individual diseases are discussed in terms of symptoms, causal agent or organism, disease cycle and epidemiology, and management.
- Part II, Abiotic Diseases and Disorders, addresses stress-induced conditions caused by a number of factors, including air pollution, excess and insufficient light, nutritional imbalances, pesticide toxicity (phytotoxicity), and water imbalances.
- Part III, Arthropod Pests, provides an overview of production and management topics specific to insect and mite pests and then discusses the life cycles of, damage caused by, and management options for 22 arthropod pests.
- Three appendixes provide supporting information, including scientific and common names of more than 100 host plants, common names of diseases along with their pathogens, and scientific and common names of insect, mite, and mollusk pests. A glossary and an index provide helpful tools for readers of all interests and levels of knowledge.

This new compendium contains 188 color photographs and will be a useful resource for bedding plant producers, helping them to identify problems and to implement effective disease and arthropod pest management programs (preventive and responsive). The information used to develop this compendium has been compiled from both the scientific literature and our own extensive experience.

Some of the sections for given diseases, disorders, and pests are not as comprehensive as we would like because of a lack of research. We have included as much helpful information as was available at the time of publication. Several sections include information from *Compendium of Flowering Potted Plant Diseases* (by M. L. Daughtrey, R. L. Wick, and J. L. Peterson); the authors of those sections are acknowledged within this book. We would like to acknowledge the authors of the section "Diseases Caused by Viroids," written for this new compendium, and thank them for their contribution:

Rosemarie W. Hammond (Retired), U.S. Department of Agriculture–Agricultural Research Service (USDA–ARS), Molecular Plant Pathology Laboratory, Beltsville, Maryland
Giuseppe Stancanelli, European Food Safety Authority, Plant Health Unit, Parma, Italy
Jacobus Th. J. Verhoeven, National Plant Protection Organization, National Reference Centre, Wageningen, the Netherlands

We also thank the following individuals for their assistance with development of this compendium. Their generosity in reviewing portions of the content and in supplying illustrations is very much appreciated:

Scott Adkins, USDA–ARS, U.S. Horticultural Research Laboratory, Fort Pierce, Florida
Caitlyn Allen, Department of Plant Pathology, University of Wisconsin–Madison
Margot C. Becktell, Department of Biological Sciences, Colorado Mesa University, Grand Junction
D. Michael Benson (Retired), Department of Plant Pathology, North Carolina State University, Raleigh

Carolee Bull, Department of Plant Pathology and Environmental Microbiology, Pennsylvania State University, State College
John Hammond, USDA–ARS, United States National Arboretum, Beltsville, Maryland
Rosemarie W. Hammond (Retired), USDA–ARS, Molecular Plant Pathology Laboratory, Beltsville, Maryland
James LaMondia, The Connecticut Agricultural Experiment Station Valley Laboratory, Windsor
Benham E. L. Lockhart, Department of Plant Pathology, University of Minnesota, St. Paul
Theresa Meers, Department of Business and Agri-Industries, Parkland College, Champaign, Illinois
Gary W. Moorman (Retired), Department of Plant Pathology and Environmental Microbiology, Pennsylvania State University, State College
Hanu R. Pappu, Department of Plant Pathology, Washington State University, Pullman
Leanne Pundt, College of Agriculture, Health, and Natural Resources, Extension, University of Connecticut, Torrington
Karen K. Rane, Department of Plant Pathology, Plant Diagnostics Laboratory, University of Maryland, College Park
Naidu Rayapati, Department of Plant Pathology, Washington State University, Prosser
Katja Richert-Pöggeler, Julius Kühn-Institut, Braunschweig, Germany
Amy Rossman (Retired), USDA–ARS, Systematic Botany and Mycology Laboratory, Beltsville, Maryland
Nina Shishkoff, USDA–ARS, Foreign Disease–Weed Science Research, Frederick, Maryland

Readers should note that the information provided in this book on chemical controls is intended for educational purposes only. Legal restrictions and regulations for pesticide use vary among countries, and restrictions and regulations within one country may change over time. In the United States, even counties within a given state may interpret product labels in different ways. Therefore, readers should not consider the discussions of pesticides in this compendium as recommendations or endorsements. We are reporting information from research findings and other sources, and readers must use this information only within the limitations of local and federal labeling to avoid making illegal pesticide applications. In addition, readers should be sure to follow product labels for plant protection materials. In the United States, the Cooperative Extension Service and agencies in state departments of agriculture should be consulted for advice on the appropriate and legal use of pesticides.

A. R. Chase
Margery L. Daughtrey
Raymond A. Cloyd

Contents

Introduction

At one point in the 1970s, bedding plants were in such favor that a very active grower–researcher–educator group was formed called "Bedding Plants, Inc." (later, the "Professional Plant Growers Association"). This was the heyday of research on these crops, and many innovative production systems were developed, including so-called plug production. Now, it is hard to envision a time in which growers planted their own seeds and transplanted seedlings into a series of increasingly larger pots.

Bedding plants are employed to create a colorful, temporary planting of fast-growing plants with low maintenance demands. This display is often seasonal, with defined windows during the spring, summer, fall, and winter months. The crops chosen for each season are usually able to tolerate the special weather conditions of that season for a specific climate, from the warmest to the coldest. The plants used for bedding are generally annuals, biennials, and tender perennials. Succulents, bulbs, and tropicals have become more popular, especially in areas with cold climates where a lush summer garden is sought. Sometimes, bedding plants are called "patio plants," because they are easily adapted to large pots and other containers positioned on patios, terraces, decks, and other areas around public buildings and private homes.

The modern bedding plant industry breeds and produces plants that have a neat, dwarf habit and that flower uniformly and reliably. Each year, more genera and species are used as bedding plants, and many new cultivars are available of familiar staples such as pansy and salvia. Many of the plants covered in this compendium are identified in Tables 1 and 2, and a more formal and comprehensive list is provided in Appendix I.

Bedding plants are bred primarily for their beauty and suitability to large-scale production schemes—and often with little attention to sustainability in the landscape. Even less attention is paid to beddings plants' tolerance of common diseases and arthropod pests; newly introduced genera are sometimes accompanied by new pests.

An additional complication is the highly mobile nature of modern ornamentals production. Unrooted cuttings may be produced from stock plants that are grown in Central or South America, moved to a rooting facility in Africa, and then sent all over the globe—all within a few weeks. In the face of these challenges, researchers are struggling to keep up with the identification of new problems and the development of improved methods of pest prevention and management.

(Prepared by A. R. Chase and M. L. Daughtrey)

TABLE 1. Common Bedding Plants: Latin Genus Names to Common Names[a]

Genus	Common Name(s)
Ageratum	Ageratum, floss flower
Amaranthus	Amaranth
Anemone	Anemone, wind flower
Angelonia	Angelonia, summer snapdragon
Antirrhinum	Snapdragon
Argyranthemum	Marguerite daisy, argyranthemum
Aster	Aster
Begonia	Begonia (wax begonia and others)
Bellis	English daisy
Bidens	Bidens
Brassica	Ornamental kale, ornamental cabbage
Browallia	Amethyst flower, browallia
Calendula	Calendula, pot marigold
Calibrachoa	Calibrachoa
Callistephus	China aster
Campanula	Campanula, bellflower
Capsicum	Ornamental pepper
Catharanthus	Annual vinca, Madagascar periwinkle
Celosia	Celosia, cockscomb
Centaurea	Bachelor's button, cornflower
Chaenostoma	Bacopa
Chrysanthemum	Chrysanthemum
Clarkia	Godetia, farewell-to-spring
Cleome	Spider flower
Coreopsis	Tickseed
Cosmos	Cosmos, Mexican aster
Cuphea	Cigar flower
Cyclamen	Cyclamen
Dahlia	Dahlia
Delphinium	Larkspur
Dianthus	Pink, Sweet William
Diascia	Twinspur, diascia
Dimorphotheca	Dimorphotheca, African daisy
Eustoma	Lisianthus
Gaillardia	Blanket flower, gaillardia
Gerbera	Gerber daisy, gerbera daisy, Transvaal daisy
Gomphrena	Common globe amaranth, gomphrena
Helichrysum	Strawflower
Impatiens	Impatiens (garden, balsam, New Guinea)
Ipomoea	Sweetpotato vine
Linaria	Toadflax
Lobelia	Lobelia
Lobularia	Sweet alyssum
Matthiola	Garden stock, ten-weeks stock, hoary stock
Nemesia	Nemesia
Nicotiana	Flowering tobacco
Ocimum	Basil
Osteospermum	Cape Marguerite, trailing African daisy, osteospermum
Pelargonium	Geranium (ivy, zonal, regal)
Pericallis	Cineraria
Petunia	Petunia
Phlox	Phlox (annual)
Platycodon	Balloon flower
Plectranthus (syns. *Coleus, Solenostemon*)	Coleus
Portulaca	Moss rose, portulaca, purslane
Primula	Primrose, true oxlip, false cowslip
Salvia	Salvia, sage
Scaevola	Fan flower
Senecio	Dusty miller
Tagetes	Marigold
Torenia	Torenia, wishbone flower
Tropaeolum	Nasturtium
Verbena	Verbena
Viola	Pansy, viola
Zinnia	Zinnia

[a] Courtesy A. R. Chase and M. L. Daughtrey—© APS.

TABLE 2. Common Bedding Plants: Common Names to Latin Genus Names[a]

Common Name(s)	Genus
Ageratum, floss flower	*Ageratum*
Alyssum	*Lobularia*
Amaranth	*Amaranthus*
Amethyst flower, browallia	*Browallia*
Anemone, wind flower	*Anemone*
Angelonia, summer snapdragon	*Angelonia*
Aster	*Aster*
Bachelor's button, cornflower	*Centaurea*
Bacopa	*Chaenostoma*
Balloon flower	*Platycodon*
Basil	*Ocimum*
Begonia (wax begonia and others)	*Begonia*
Bidens	*Bidens*
Blanket flower, gaillardia	*Gaillardia*
Cabbage, ornamental	*Brassica*
Calendula, pot marigold	*Calendula*
Calibrachoa	*Calibrachoa*
Campanula, bellflower	*Campanula*
Cape Marguerite, trailing African daisy, osteospermum	*Osteospermum*
Celosia	*Celosia*
China aster	*Callistephus*
Chrysanthemum	*Chrysanthemum*
Cigar flower	*Cuphea*
Cineraria	*Pericallis*
Cockscomb	*Celosia*
Coleus	*Plectranthus* (syns. *Coleus, Solenostemon*)
Common globe amaranth, gomphrena	*Gomphrena*
Cosmos, Mexican aster	*Cosmos*
Cyclamen	*Cyclamen*
Dahlia	*Dahlia*
Dimorphotheca, African daisy	*Dimorphotheca*
Dusty miller	*Senecio*
English daisy	*Bellis*
Fan flower	*Scaevola*
Flowering tobacco	*Nicotiana*
Geranium (ivy, zonal, regal)	*Pelargonium*
Gerber daisy, gerbera daisy, Transvaal daisy	*Gerbera*
Godetia, farewell-to-spring	*Clarkia*
Impatiens (garden, balsam, New Guinea)	*Impatiens*
Kale, ornamental	*Brassica*
Larkspur	*Delphinium*
Lisianthus	*Eustoma*
Lobelia	*Lobelia*
Marguerite daisy, argyranthemum	*Argyranthemum*
Marigold	*Tagetes*
Moss rose, portulaca, purslane	*Portulaca*
Nasturtium	*Tropaeolum*
Nemesia	*Nemesia*
Pansy	*Viola*
Pepper, ornamental	*Capsicum*
Petunia	*Petunia*
Phlox (annual)	*Phlox*
Pink	*Dianthus*
Primrose, true oxlip, false cowslip	*Primula*
Salvia, sage	*Salvia*
Snapdragon	*Antirrhinum*
Spider flower	*Cleome*
Stock	*Matthiola*
Strawflower	*Helichrysum*
Sweet William	*Dianthus*
Sweetpotato vine	*Ipomoea*
Tickseed	*Coreopsis*
Toadflax	*Linaria*
Torenia, wishbone flower	*Torenia*
Verbena	*Verbena*
Vinca (annual vinca, Madagascar periwinkle)	*Catharanthus*
Viola	*Viola*
Zinnia	*Zinnia*

[a] Courtesy A. R. Chase and M. L. Daughtrey—© APS.

Part I. Infectious Diseases

Diseases Caused by Bacteria

Bacteria are microscopic, prokaryotic organisms that are ubiquitous and involved in degradation of organic matter, nitrogen fixation, transformation and release of nutrients in soil, and pathogenesis of plants and animals. Bedding plants are subject to a number of devastating bacterial diseases, most commonly caused by *Dickeya* (syn. *Erwinia*), *Pectobacterium* (syn. *Erwinia*), *Pseudomonas*, and *Xanthomonas* species. Bacteria in the genera *Agrobacterium*, *Curtobacterium*, *Ralstonia*, and *Rhodococcus* (syn. *Corynebacterium*) are less commonly encountered but can result in economic losses.

Selected References

Bradbury, J. F. 1986. Guide to Plant Pathogenic Bacteria. CABI, Wallingford, Oxfordshire, UK.

Dowson, W. J. 1957. Plant Diseases Due to Bacteria, 2nd ed. Cambridge University Press, London.

Goto, M. 1992. Fundamentals of Bacterial Plant Pathology. Academic Press, New York.

Kado, C. I. 2010. Plant Bacteriology. American Phytopathological Society, St. Paul, MN.

Lelliott, R. A., and Stead, D. E. 1987. Methods for the Diagnosis of Bacterial Diseases of Plants. Blackwell Scientific, Oxford.

Schaad, N. W., Jones, J. B., and Chun, W., eds. 2001. Laboratory Guide for Identification of Plant Pathogenic Bacteria, 3rd ed. American Phytopathological Society, St. Paul, MN.

(Prepared by A. R. Chase and M. L. Daughtrey)

Crown Gall

Crown gall is caused by *Agrobacterium tumefaciens*. Because the pathogen is soilborne, this disease is particularly problematic for certain fruit and nursery crops but rarely occurs on bedding plants, which are grown primarily in peat-based growing mixes. Bedding plants produced from seed have a short production period in soilless mix and thus are not likely to become contaminated with the bacterium that causes crown gall. Because of an industry trend to use vegetative propagation for bedding plants (growing them from cuttings to allow faster adoption of new cultivars), the incidence of crown gall has increased but is still quite rare.

Symptoms

Galls usually form at the root crown of the plant but occasionally occur on the roots or on a stem or branch. On bedding plants propagated from cuttings, galls can appear at cutting wounds and at other injured sites on stems and leaves. Galls begin as small, whitish masses of callus tissue but may enlarge to several centimeters in diameter (Fig. 1). They are firm, and their surfaces are rough and irregular (Fig. 2).

The causal bacterium can be very difficult to detect even when galls are present. Screening healthy-looking plant materials within clean stock programs is challenging, because the bacterium is known to be present in tiny numbers in unevenly distributed pockets prior to the formation of conspicuous galls.

Causal Organism

A. tumefaciens (syn. *Rhizobium radiobacter*) is a gram-negative, aerobic, rod-shaped bacterium with peritrichous

Fig. 1. Crown gall (*Agrobacterium tumefaciens*) on an osteospermum cutting. (Courtesy A. R. Chase—© APS)

Fig. 2. Crown gall (*Agrobacterium tumefaciens*) on Marguerite daisy (argyranthemum). (Courtesy M. L. Daughtrey—© APS)

flagella. Three biovars have been proposed to account for differences in genotypic traits. *A. tumefaciens* transfers bacterial oncogenes to the plant from a circular piece of DNA called a "plasmid," which resides in the bacterial cell and is moved into the host plant cell. This natural gene transfer results in an overproduction of auxin and cytokinin in the plant, which in turn results in an unregulated and disorganized growth of cells to the extent that galls develop.

Many isolates of the bacterium from soil and plants are not pathogenic and cannot be differentiated from pathogenic isolates on the basis of standard physiological and biochemical tests. The soil-dwelling *A. tumefaciens* is considered ubiquitous.

Disease Cycle and Epidemiology

The host range of *A. tumefaciens* is very large, including about 600 species of angiosperms. The following bedding plants are known to be hosts (primarily through inoculation studies): *Anemone* spp., *Argyranthemum frutescens, Begonia semperflorens* (wax begonia) and *B. rex-cultorum* (rex begonia), a *Campanula* sp., *Catharanthus roseus* (annual vinca), *Chrysanthemum maximum* (Shasta daisy), *Dahlia* hybrids, a *Lobelia* sp., *Osteospermum ecklonis, Pelargonium* × *hortorum* (florist's geranium), and *Salvia splendens.*

Management

Management practices employed for clean propagation of vegetative cuttings are appropriate for management of crown gall. No bactericides are recognized as effective in preventing gall formation on any bedding plant. All plants with symptoms of crown gall should be discarded, along with their containers, and work and growing surfaces should be disinfested. Given the broad host range of *A. tumefaciens,* it cannot be tolerated in the greenhouse.

Biological control of crown gall is possible on some woody ornamentals with the bacterium *A. radiobacter,* strain K84. However, crown gall occurs too infrequently to warrant preventive treatments on bedding plants; no studies of the feasibility of such treatments have been reported.

Selected Reference

Kado, C. I. 2010. Plant Bacteriology. American Phytopathological Society, St. Paul, MN.

(Prepared by M. L. Daughtrey, R. L. Wick, and J. L. Peterson; revised by A. R. Chase and M. L. Daughtrey)

Leaf Spot and Blight Caused by *Pseudomonas cichorii*

Leaf spots and blights on ornamentals can be caused by a variety of bacteria, including species of *Pseudomonas.* The diseases incited by *P. cichorii* are reported worldwide but do not appear to cause continuous or significant losses on most crops. The possible exception is chrysanthemum, for which leaf spot and blight appears almost annually in landscape beds, primarily in areas in which the climate is warm and summer rains are common. *P. cichorii* can be a problem during production of certain cultivars of chrysanthemum, as well.

Symptoms

P. cichorii causes lesions similar to those caused by *P. syringae* and by many pathovars of *Xanthomonas.* Lesions are initially water soaked but expand rapidly and turn black. During dry periods, the lesions may turn tan and stop growing temporarily.

Lesions on *Pelargonium* × *hortorum* (florist's geranium) caused by *P. cichorii* may vary depending on environmental conditions. Plants irrigated from overhead and those subjected to rainfall can develop large (5–10 mm), irregularly shaped, dark-brown to black lesions. In contrast, when plants are produced in the greenhouse, where leaf surfaces can be kept drier, lesions may be smaller and have tan centers and dark margins. Yellowing of leaves invariably occurs, and flower buds may also become infected.

On *Gerbera jamesonii* (gerber daisy), tan to black spots develop along leaf veins and margins and often are quite large (Fig. 3). On some bedding plants, stem infections also occur.

Causal Organism

P. cichorii is a gram-negative rod that is fluorescent on King's medium B; it is oxidase positive, arginine dihydrolase negative, levan negative, and potato rot negative. It causes a hypersensitive reaction when infiltrated into tobacco leaves. Other characters useful for identification can be found in publications by Bradbury, Lelliott, and Schaad (see Selected References).

P. syringae causes symptoms similar to those caused by *P. cichorii* and can be differentiated from *P. cichorii* by PCR amplification of diagnostic regions, such as the 16S–23S rDNA intergenic spacer region, followed by DNA sequencing of the amplicon and alignment to the sequences of known isolates.

Host Range and Epidemiology

P. cichorii causes diseases of a variety of ornamentals used as bedding plants: *Catharanthus roseus* (annual vinca), *Chrysanthemum morifolium, Coreopsis* spp. (tickseed), *Cosmos bipinnatus, Cyclamen persicum,* gerber daisy, *Lobelia erinus, Pelargonium* spp. (geranium), and *Primula elatior* (primrose). Host specificity is not known to exist.

The optimal temperature for *P. cichorii* is 20–28°C. As is generally true for other bacterial diseases of foliage, periods of high humidity and leaf wetness are necessary for infection and disease development. At an optimal temperature, moisture for a period of 48–72 h favors infection and results in a high incidence of disease and lesion expansion. Chrysanthemum is known to carry epiphytic populations of *P. cichorii,* and this is probably the case with many other hosts. Thus, exchange of propagative materials provides for long-distance distribution

Fig. 3. Leaf spots (*Pseudomonas cichorii*) on gerber daisy. (Courtesy A. R. Chase—© APS)

of *P. cichorii,* whereas irrigation and rain result in dispersal within the crop and infection periods when the temperature is favorable and leaf surfaces remain wet.

Management

Plants known to be carriers of *P. cichorii,* such as chrysanthemum, should be kept separate from other known hosts. Bactericides including copper, quaternary ammonium compounds, and an extract from *Reynoutria sachalinensis* can be very effective when used preventively. Streptomycin sulfate is also somewhat effective.

Selected References

Bradbury, J. F. 1986. Guide to Plant Pathogenic Bacteria. CABI, Wallingford, Oxfordshire, UK.

Engelhard, A., Mellinger, H. C., Ploetz, R. C., and Miller, J. W. 1983. A leaf spot of the florists' geranium incited by *Pseudomonas cichorii.* Plant Dis. 67:541-544.

Garibaldi, A., Bertetti, D., Gilardi, G., and Saracco, P. 2009. Two new bacterial pathogens on ornamentals in Italy. Protez. delle Colt. (2):60.

Gilardi, G., Gullino, M. L., and Garibaldi, A. 2009. *Coreopsis lanceolata* new host of *Pseudomonas cichorii* in Piedmont (Northern Italy). Protez. delle Colt. (4):50-51.

Jones, J. B., and Engelhard, A. W. 1983. Outbreak of a stem necrosis on chrysanthemum incited by *Pseudomonas cichorii* in Florida. Plant Dis. 67:431-433.

Jones, J. B., Chase, A. R., Harbaugh, B. K., and Raju, B. C. 1985. Effect of leaf wetness, fertilizer rate, leaf age, and light intensity before inoculation on bacterial leaf spot of chrysanthemum. Plant Dis. 69:782-784.

Jones, J. B., Raju, B. C., and Engelhard, A. W. 1984. Effects of temperature and leaf wetness on development of bacterial spot of geraniums and chrysanthemums incited by *Pseudomonas cichorii.* Plant Dis. 68:248-251.

Kitazawa, Y., Netsu, O., Nijo, T., Yoshida, T., Miyazaki, A., Hara, S., Okana, Y., Maejima, K., and Namba, S. 2014. First report of bacterial leaf blight on cosmos (*Cosmos bipinnatus* Cav.) caused by *Pseudomonas cichorii* in Japan. J. Gen. Plant Pathol. 80:499-503.

Lelliott, R. A., and Stead, D. E. 1987. Methods for the Diagnosis of Bacterial Diseases of Plants. Blackwell Scientific, Oxford.

Miller, J. W., and Knauss, J. F. 1973. Bacterial blight of *Gerbera jamesonii* incited by *Pseudomonas cichorii.* Plant Dis. Rep. 57:504-505.

Putnam, M. L. 1999. Bacterial blight, a new disease of *Lobelia ricardii* caused by *Pseudomonas cichorii.* Plant Dis. 83:966.

Schaad, N. W., Jones, J. B., and Chun, W., eds. 2001. Laboratory Guide for Identification of Plant Pathogenic Bacteria, 3rd ed. American Phytopathological Society, St. Paul, MN.

(Prepared by M. L. Daughtrey, R. L. Wick, and J. L. Peterson; revised by A. R. Chase and M. L. Daughtrey)

Leaf Spots Caused by *Pseudomonas syringae* Pathovars

Pseudomonas syringae was originally described as a pathogen of *Syringa vulgaris* (lilac), for which it was named. The bacterium has a wide host range, including woody species, vegetable crops, grasses, and herbaceous ornamentals. Historically, considerable differences in host specificity have been reported, and *P. syringae* has been assigned more than 50 pathovar designations.

Of all the pathovars, *P. syringae* pv. *syringae* has the widest reported host range and is the most important to bedding plants. *P. syringae* pv. *antirrhini* is known to affect *Antirrhinum majus* (snapdragon) (Fig. 4) and has also been shown to be a pathogen of a *Calceolaria* sp. (slipperwort) by inoculation. *P. syringae* pv. *primulae* has been reported occasionally on *Primula* spp. (primrose), and *Tagetes* spp. (marigold) are susceptible to *P. syringae* pv. *tagetis.*

Symptoms

Pathovars of *P. syringae* typically cause leaf spots characterized by water-soaked lesions, which may become dark brown to black, gray, or tan. Desiccation of infected tissue often results in formation of a thin, papery lesion, which cracks as the leaf expands. When partially expanded leaves become infected, they develop spots and become distorted (Fig. 5).

On *Impatiens walleriana* (garden impatiens), small spots with purple margins typically form, often originating at the leaf margins (Fig. 6). *I. hawkeri* (New Guinea impatiens) usually develops fairly large, water-soaked lesions, which turn dark brown. *Pelargonium* × *hortorum* (florist's geranium) develops lesions that are indistinguishable from those caused by *P. cichorii,* which is more commonly found on this plant. Small, brown to black spots develop, which may coalesce into large, necrotic areas on the leaf. Yellowing of adjacent tissues occurs within a few days after lesions appear. Affected leaves die and remain attached to the plant without wilting.

Similar symptoms occur on marigold infected with *P. syringae* pv. *tagetis,* which was first reported from the United States in 1980. In addition, apical chlorosis, flecks, and large, necrotic spots were seen alone or in combination on marigold.

Fig. 4. Leaf spots (*Pseudomonas syringae*) on snapdragon. (Courtesy A. R. Chase—© APS)

Fig. 5. Leaf spots (*Pseudomonas syringae*) on salvia. (Courtesy A. R. Chase—© APS)

Causal Organisms

P. syringae pathovars are gram-negative rods, are fluorescent on King's medium B, and cause a hypersensitive reaction when infiltrated into tobacco leaves. Studies using RFLP analysis have shown that strains from snapdragon aligned with those from tomato (*P. syringae* pv. *tomato*); this finding could impact management strategies, because tomato transplants and flower crops are often produced in the same greenhouse during the spring.

Host Range and Epidemiology

P. syringae pathovars collectively affect many hosts, including *Dahlia* hybrids, garden impatiens, New Guinea impatiens, snapdragon, marigold, and florist's geranium among crops used as bedding plants. In general, *P. syringae* pathovars are capable of causing disease at low temperatures (15–20°C) and thus often cause disease in the northern United States early in the bedding plant production season. Disease is exacerbated by high humidity and extended periods of leaf wetness. *P. syringae* has been reported to be seedborne in several crop plants.

P. syringae pv. *tagetis* has been reported from a number of composite crops and weeds in addition to marigold, including *Helianthus annuus* (sunflower), *Ambrosia artemisiifolia* (common ragweed), *H. tuberosus* (Jerusalem artichoke), and *Taraxacum officinale* (dandelion).

Management

Sanitation is an important disease management principle and is particularly pertinent to bacterial problems. Plants in plug trays should be inspected carefully on receipt and again a few days later to check for water-soaked spots, which could indicate bacterial disease. Entire trays and individual plants in larger pots that develop symptoms should be discarded, as treatments are not curative. Workers should wash their hands after handling diseased plants and soil, and diseased plant debris should be removed from the growing area as promptly as possible. Bacteria are easily splashed from plant to plant by irrigation water. Splashing should be minimized and leaf wetness duration reduced as much as practical by irrigating early in the day or by subirrigating. Handling foliage when it is wet should be avoided.

Nutrition may affect disease susceptibility. High levels of nitrogen have been reported to increase the susceptibility of garden impatiens to *P. syringae* pv. *syringae*. Plant tissue age is an important factor in infection on marigold. Only tissue that is immature can be infected, so protection of new tissue is critical.

Copper sulfate pentahydrate and copper hydroxide are registered in the United States for managing *Pseudomonas* spp. However, resistance to copper was detected in a strain of *P. syringae* causing disease on impatiens in California. The bacterium contained what was apparently the same 47-kilobase (-kb) plasmid (pPSI1) that had been characterized earlier from *P. syringae* pv. *tomato* strains with copper resistance in California, suggesting that the plasmid might have been exchanged between the two different but closely related bacteria. Even without the extra complication of pesticide resistance, bactericides are only marginally effective in controlling bacterial diseases on bedding plants, making sanitation and environmental controls extremely important.

Fig. 6. Leaf spots (*Pseudomonas syringae*) on garden impatiens. (Courtesy A. R. Chase—© APS)

Selected References

Cooksey, D. A. 1990. Plasmid-determined copper resistance in *Pseudomonas syringae* from impatiens. Appl. Environ. Microbiol. 56:13-16.

Cooksey, D. A., and Koike, S. T. 1990. A new foliar blight of *Impatiens* caused by *Pseudomonas syringae*. Plant Dis. 74:180-182.

Manceau, C., and Horvais, A. 1997. Assessment of genetic diversity among strains of *Pseudomonas syringae* by PCR-restriction fragment length polymorphism analysis of rRNA operons with special emphasis on *P. syringae* pv. *tomato*. Appl. Environ. Microbiol. 63:498-505.

Styer, D. J., and Durbin, R. D. 1981. Influence of growth stage and cultivar on symptom expression in marigold, *Tagetes* sp., infected with *Pseudomonas syringae* pv. *tagetis*. HortScience 16:768-769.

Styer, D. J., Worf, G. L., and Durbin, R. D. 1980. Occurrence in the United States of a marigold leaf spot incited by *Pseudomonas tagetis*. Plant Dis. 64:101-102.

Wick, R. L., and Rane, K. K. 1987. *Pseudomonas syringae* leaf spot of *Pelargonium* × *hortorum*. (Abstr.) Phytopathology 77:1620.

(Prepared by M. L. Daughtrey, R. L. Wick, and J. L. Peterson; revised by A. R. Chase and M. L. Daughtrey)

Leafy Gall (Bacterial Fasciation)

Bacterial fasciation is not prevalent in the bedding plant industry. The disease occurs sporadically, and losses are usually minimal. It is most commonly observed on *Pelargonium* × *hortorum* (florist's geranium) but has not been prevalent since growers switched to soilless mixes. Worldwide, bacterial fasciation has been reported from at least 19 states in the United States and in Canada, Mexico, northern Europe, Asia, Australia, New Zealand, the Middle East, Egypt, and Colombia. The term "leafy gall" was first proposed in 1933 and is the name most often used.

Symptoms

Fasciation may take several forms. Cylindrical organs, such as stems and peduncles, become flat and bandlike. Buds may proliferate, resulting in leafy, cauliflower-like galls, especially at the plant's root crown (Figs. 7 and 8). Witches'-brooms may also occur, but necrosis of tissue is rare. On florist's geranium, stubby, stunted shoots are seen. Leafy gall often appears as a number of very short, hypertrophied shoots that develop at the base of the cutting. They may be just barely visible or hidden beneath the soil line.

Symptoms of leafy gall are sometimes confused with the toxic effects of an herbicide drift or overspraying or misapplication of a plant growth regulator (uncommon on bedding plants). In addition, infections by other organisms—including *Agrobacterium tumefaciens* (which causes crown gall), '*Candidatus* Phytoplasma asteris' (which causes aster yellows), and viruses—are sometimes confused with leafy gall.

Causal Organism

Rhodococcus fascians (syn. *Corynebacterium fascians*) is a gram-positive, nonmotile, pleomorphic rod. On various culture media, it is cream to orange or yellow, depending on the isolate. Virulent strains possess a plasmid that confers pathogenicity.

Host Range and Epidemiology

R. fascians has a wide host range that includes both monocots and dicots. Reports of distinctive symptoms and the results of PCR testing have identified the following as hosts of *R. fascians:* anemone, annual vinca, aster, chrysanthemum, coreopsis, cosmos, dahlia, florist's geranium, garden impatiens, marigold, osteospermum, petunia, phlox, primrose, snapdragon, verbena, and viola.

The ecology of *R. fascians* is not well known. The bacterium can be seedborne, and some researchers have speculated that it is soilborne. It colonizes the cotyledonary buds as the plant emerges from the soil and is usually found on the surfaces of symptomatic meristematic tissues, where it produces growth regulators. The optimal temperature for growth is 24–27°C.

Management

Hot-water treatment of seed effectively controls fasciation of nasturtium; however, seed is not an important source of this disease in bedding plants. Leafy gall occurs too infrequently to warrant attention to preventive management practices beyond excellent sanitation and scouting. Use of bactericides is not effective.

When fasciation occurs, plants should be discarded immediately. Growers should never propagate from a plant that shows fasciation or galling.

Fig. 7. Leafy gall (*Rhodococcus fascians*) on sweetpotato vine. (Courtesy A. R. Chase—© APS)

Fig. 8. Leafy gall (*Rhodococcus fascians*) on an *Erodium* sp. (Courtesy A. R. Chase—© APS)

Selected References

Cooksey, D. A., and Keim, R. 1983. Association of *Corynebacterium fascians* with fasciation disease of *Impatiens* and *Hebe* in California. Plant Dis. 67:1389.

Davis, M. J. 1986. Taxonomy of plant-pathogenic coryneform bacteria. Annu. Rev. Phytopathol. 24:115-140.

Nikolaeva, E. V., Kang, S., Olson, T. N., and Kim, S. H. 2012. Real-time PCR detection of *Rhodococcus fascians* and discovery of new plants associated with *R. fascians* in Pennsylvania. Online. Plant Health Progress. doi:10.1094/PHP-2012-0227-02-RS

Putnam, M. L., and Miller, M. L. 2007. *Rhodococcus fascians* in herbaceous perennials. Plant Dis. 91:1064-1076.

(Prepared by M. L. Daughtrey, R. L. Wick, and J. L. Peterson; revised by A. R. Chase and M. L. Daughtrey)

Phytoplasma Diseases

The "yellows" diseases of plants were originally thought to be caused by viruses because of their unique symptoms and researchers' inability to culture any pathogens from diseased plants. In 1967, researchers discovered that certain nonculturable, wall-less bacteria in the phloem caused these viruslike symptoms. These bacteria resembled animal pathogens called "mycoplasmas," and thus, the new plant pathogens were called "mycoplasmalike organisms (MLOs)" until 1994. These plant pathogens were later referred to as "phytoplasmas," and the various members of this group are classified as belonging to a new candidate genus, "Phytoplasma."

Typical symptoms of phytoplasma infection include yellowing or bronzing of leaves, virescence (green coloration of ordinarily colored flower petals), phyllody (leaflike flower petals), stunting, sterile flowers, abnormal fruit and seed, proliferation of roots, and witches'-brooms (Figs. 9 and 10). Phytoplasmas are vectored by psyllids, leafhoppers, and planthoppers. They are restricted to the phloem of an infected plant and can be detected by electron microscopy, but their pleomorphic shape makes them hard to distinguish from other plant cell components. Serological methods and DNA-based techniques have been utilized for phytoplasma detection and characterization.

The most common phytoplasma disease on bedding plants is aster yellows, which is caused by bacteria classified as '*Candidatus* Phytoplasma asteris'; they form a phylogenetically distinct group known as "AY phytoplasmas" (16SrI). The AY phytoplasmas are collectively associated with more than

Fig. 9. Aster yellows ('*Candidatus* Phytoplasma asteris') on a *Coreopsis* sp., showing virescence on flowers. (Courtesy M. L. Daughtrey—© APS)

Fig. 10. Aster yellows ('*Candidatus* Phytoplasma asteris') on annual vinca, showing witches'-brooms. (Courtesy A. R. Chase—© APS)

80 species of plants and have more than 30 insect vectors. In the United States, AY phytoplasmas are carried to their host plants primarily by the aster leafhopper (*Macrosteles quadrilineatus*) and transmitted during the feeding of this insect for its entire life span, once it has been infected. Migration of the aster leafhopper from the southern to the northern United States during each growing season allows widespread dissemination of the phytoplasma.

Aster yellows has severe effects on infected plants and can infect a variety of ornamentals, vegetables, field crops, and weeds. Bedding plants that are hosts include *Aconitum* sp. (monkshood), *Calendula officinalis, Callistephus chinensis* (China aster), *Catharanthus roseus* (annual vinca), *Cosmos bipinnatus, Cyclamen persicum,* and *Zinnia elegans.*

High temperatures inactivate the AY phytoplasma in insect vectors and host plants; thus, aster yellows is rare or absent in areas of the world with very hot climates. Leafhoppers exposed to a continuous temperature of 31°C for 10–12 days will be freed of the pathogen. A summer "hot spell" that lasts 2 weeks or more may thus curtail leafhoppers' ability to transmit the phytoplasma, and infected plants may show remission of symptoms under these conditions. Aster yellows is best managed by scouting for and removing infected plants, whether weeds or ornamentals. Management of the insect vector is not usually effective in preventing disease transmission.

In 2007, researchers in Sicily reported a phytoplasma in *Matthiola incana* (garden stock). Plants were stunted and rosetted, and flowers were very small and virescent. The causal organism was identified by PCR/RFLP techniques as belonging to another group of phytoplasmas: '*Candidatus* Phytoplasma aurantifolia.'

Selected References

Davino, S., Calari, A., Davino, M., Tessitori, M., Bertaccini, A., and Bellardi, M. G. 2007. Virescence of tenweeks stock associated to phytoplasma infection in Sicily. Bull. Insectol. 60:279-280.

Lee, I.-M., Gundersen-Rindal, D. E., Davis, R. E., Bottner, K. D., Marcone, C., and Seemüller, E. 2004. '*Candidatus* Phytoplasma asteris,' a novel phytoplasma taxon associated with aster yellows and related diseases. Int. J. Syst. Evol. Microbiol. 54:1037-1048.

McCoy, R. E., and Thomas, D. L. 1980. Periwinkle witches' broom disease in south Florida. Proc. Fla. State Hortic. Soc. 93:179-181.

Montano, H. G., Dally, E. L., Davis, R. E., Pimentel, J. P., and Brioso, P. S. T. 2001. First report of natural infection by "*Candidatus* Phytoplasma brasiliense" in *Catharanthus roseus.* Plant Dis. 85:1209.

(Prepared by A. R. Chase and M. L. Daughtrey)

Soft Rot Diseases Caused by *Pectobacterium* and *Dickeya* spp.

The genera *Pectobacterium* and *Dickeya* (formerly known as *Erwinia*) include soft rot bacteria that can cause important diseases of bedding plants. They are fast-growing, opportunistic bacteria and capable of causing serious losses within a few days.

Symptoms

Pectobacterium and *Dickeya* spp. produce enzymes that break down pectin, resulting in a soft, mushy tissue rot (Fig. 11). On an infected cutting, the base of the stem may be affected for several centimeters or more. During rooting in a warm, moist environment, rapid collapse of entire flats of cuttings may occur. When cuttings in flats are near completion of their growth cycle, lower leaves can become infected and act as entry points for bacteria, resulting in losses of entire plugs. Rarely, a soft rot may develop on seed-propagated bedding plants; cyclamen is a notable example (Fig. 12).

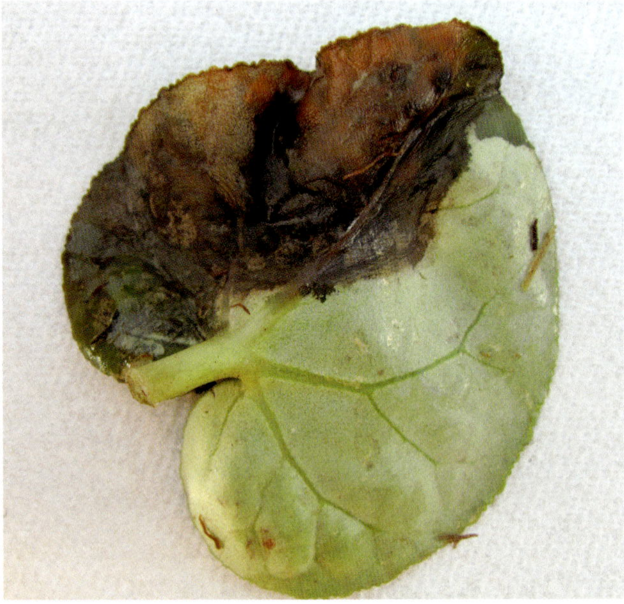

Fig. 11. Bacterial soft rot (*Pectobacterium* sp.) on a cyclamen leaf. (Courtesy A. R. Chase—© APS)

Fig. 12. Collapse of cyclamen plugs caused by bacterial soft rot (*Pectobacterium* or *Dickeya* sp.). (Courtesy A. R. Chase—© APS)

Causal Organisms

The species reported to cause soft rot of bedding plants (and many other ornamentals) include *Pectobacterium atrosepticum* (syn. *Erwinia carotovora* subsp. *atroseptica*) and *P. carotovorum* subsp. *carotovorum* (syn. *E. carotovora* subsp. *carotovora*), as well as *Dickeya chrysanthemi* pv. *chrysanthemi* (syn. *E. chrysanthemi* pv. *chrysanthemi*). *Pectobacterium* and *Dickeya* spp. are facultatively anaerobic, gram-negative rods that are motile by peritrichous flagella. *P. carotovorum* subsp. *carotovorum* and *P. atrosepticum* are closely related and can be distinguished chiefly by the inability of *P. atrosepticum* to grow at 36°C.

Host Range and Epidemiology

Many if not all bedding plants are likely susceptible under some conditions to one or more species of *Pectobacterium* or *Dickeya*. Bacterial soft rot has been observed on cyclamen, gazania, browallia, campanula, osteospermum, and stock (*Matthiola* spp.). These *Pectobacterium* and *Dickeya* spp. have broad host ranges, although each strain is often most virulent on the host from which it was isolated.

Little if any research has been conducted on the epidemiology of soft rot diseases of specific bedding plants; however, some findings are generally applicable. Soft rot bacteria may be associated with plants and plant debris, water, and soil and potting media. Surface and underground water have been shown to harbor *Pectobacterium* spp.; therefore, irrigation water should be considered both a potential source and a means of dissemination of these bacteria (especially under conditions in which water is recirculated). Insects are capable of disseminating the bacteria and can provide infection courts by feeding. Fungus gnat (*Bradysia* spp.) larvae, for example, have been seen in association with bacterial soft rot on cyclamen. One of the enigmas of bacterial soft rot is that *Pectobacterium* spp. may be present throughout the crop cycle without causing disease. Anaerobic conditions are sometimes a prerequisite to a disease outbreak; planting cyclamen corms too deeply has been associated with losses to *Pectobacterium* spp.

Movement of soft rot bacteria occurs over long distances through distribution of infected plant materials (vegetatively propagated bedding plants) (Fig. 13). In the greenhouse, water splashing, contaminated tools, and handling of infected plants are long-recognized means of efficient dissemination.

Once soft rot bacteria come in contact with the host, they may remain as epiphytes without causing disease. The soft rot *Pectobacterium* spp. are opportunistic pathogens and require a wounded or otherwise stressed host and favorable environmental conditions to infect and cause disease. High levels of nitrogen fertilization have been shown to increase resistance to soft rot bacteria in some plants; however, in poinsettia, high nitrogen levels significantly increased soft rot of cuttings in one trial. Therefore, the effectiveness of adhering to a particular nitrogen regime to reduce soft rot has not been sufficiently proven.

Fig. 13. Bacterial soft rot (*Pectobacterium* or *Dickeya* sp.) on gazania cuttings. (Courtesy A. R. Chase—© APS)

Although the soft rot *Pectobacterium* spp. are similar in many ways, temperature alone can affect their host range, symptom development, and virulence. *P. atrosepticum* has an optimum growth temperature of 27°C with a range of 3–35°C. *P. carotovorum* subsp. *carotovorum* has an optimum growth temperature of 28–30°C with a minimum of 6°C and a maximum of 37–42°C. *D. chrysanthemi* has the highest optimum growth temperature of the three major soft rot bacteria (34–37°C), and some strains can grow at temperatures higher than 45°C.

There is ample evidence that *P. carotovorum* can survive in the soil and rhizosphere; the population is greatly affected by host root exudates. Even soilless media can become contaminated and harbor *Pectobacterium* spp. for several months in the absence of a host.

Management

Only plants believed free of *Pectobacterium* spp. should be vegetatively propagated. Stock plants should not be too soft or have excessive or deficient levels of nitrogen. Cuttings should be removed with a sharp knife when plants are not suffering from a water deficit to minimize wounding and facilitate rapid healing. Cutting instruments should be disinfested regularly with a quaternary ammonium compound or 70% ethyl alcohol. (Sodium hypochlorite is corrosive to metal.) Cuttings should be collected in a clean, surface-disinfested container and transported to the propagation area as soon as possible to reduce stress from water loss. Bench surfaces should be cleaned of organic debris and then thoroughly disinfested with sodium hypochlorite or a greenhouse disinfestant containing a quaternary ammonium compound or peroxide material.

Care should be taken to limit growing medium moisture (less than 70–75% for commercial potting media) during propagation to limit development of bacterial soft rot. A higher moisture content will reduce rooting efficiency and increase soft rot, as has been shown for chrysanthemum. In addition, reducing or eliminating misting at night and carefully setting the timing and duration of misting can control development of soft rot on many cuttings.

The use of bactericides for prevention of soft rot has typically shown better results with quaternary ammonium compounds or streptomycin sulfate than with copper or peroxide products.

Selected References

Dickey, R. S. 1981. *Erwinia chrysanthemi:* Reaction of eight plant species to strains from several hosts and to strains of other *Erwinia* species. Phytopathology 71:23-29.

Haygood, R. A., Strider, D. L., and Echandi, E. 1982. Survival of *Erwinia chrysanthemi* in association with *Philodendron selloum,* other greenhouse ornamentals, and in potting media. Phytopathology 72:853-859.

Ma, B., Hibbing, M. E., Kim, H.-S., Reedy, R. M., Yedidia, I., Breuer, J., Breuer, J., Glasner, J. D., Perna, N. T., Kelman, A., and Charkowski, A. O. 2007. Host range and molecular phylogenies of the soft rot enterobacterial genera *Pectobacterium* and *Dickeya.* Phytopathology 97:1150-1163.

McCarter-Zorner, N. J., Franc, G. D., Harrison, M. D., Michaud, J. E., Quinn, C. E., Sells, I. A., and Graham, D. C. 1984. Soft rot *Erwinia* bacteria in surface and underground waters in southern Scotland and in Colorado, United States. J. Appl. Bacteriol. 57:95-105.

McGovern, R. J., Horst, R. K., and Dickey, R. S. 1985. Effect of plant nutrition on susceptibility of *Chrysanthemum morifolium* to *Erwinia chrysanthemi.* Plant Dis. 69:1086-1088.

Norman, D. J., Yuen, J. M. F., Resendiz, R., and Boswell, J. 2003. Characterization of *Erwinia* populations from nursery retention ponds and lakes infecting ornamental plants in Florida. Plant Dis. 87:193-196.

Perombelon, M. C. M., and Kelman, A. 1980. Ecology of the soft rot *Erwinias*. Annu. Rev. Phytopathol. 18:361-387.
Randhawa, P. S., and Semer, C. R., IV. 1986. Increased moisture content of propagation media enhances bacterial rot of chrysanthemum. Proc. Fla. State Hort. Soc. 99:251-253.

(Prepared by M. L. Daughtrey, R. L. Wick, and J. L. Peterson; revised by A. R. Chase and M. L. Daughtrey)

Southern Bacterial Wilt

In the United States, *Ralstonia solanacearum* causes a disease commonly known as "southern bacterial wilt" and is most problematic in tropical and subtropical environments. *R. solanacearum,* as a true soilborne pathogen, typically enters the host through the root system and ultimately causes a vascular wilt. Geranium (*Pelargonium* spp.) is the most common bedding plant that is a known host of this bacterium (Fig. 14). However, because soilless media are used for most bedding plant propagation and production, southern wilt is rarely a problem. The disease occurs primarily in landscape beds in frost-free areas that were previously contaminated by an infected plant.

Symptoms

Wilting and yellowing of the lower leaves, followed by necrosis, are symptoms common to many crops infected by *R. solanacearum* (Fig. 15). Southern bacterial wilt almost always results in plant death once symptoms begin to develop on the host, but the possibility of a latent infection has been demonstrated on several bedding plants. The vascular system is typically discolored, and when cut stems are suspended in water, bacterial streaming is usually abundant. Bacterial streaming helps to differentiate southern bacterial wilt from other wilts of geranium caused by *Verticillium* and *Fusarium* spp.

In florist's geranium (*Pelargonium* × *hortorum*), the lower leaves may wilt as soon as 2 weeks after infection. As the disease develops, the leaves turn chlorotic and then necrotic. The vascular system becomes discolored, and the stem rots from the inside out, ultimately turning dark brown to black. The roots also become necrotic as they die.

Causal Organism

R. solanacearum (syn. *Pseudomonas solanacearum*) is a species complex of gram-negative, rod-shaped, strictly aerobic, nonfluorescent bacteria in the *Ralstoniaceae. R. solanacearum* has been divided into four phylotypes that indicate original geographic origin and further divided into sequevars based on the DNA sequence of the endoglucanase gene. Some strains produce a diffusible, brown pigment in culture. The optimal growth temperature for most strains is 28–32°C.

A member of phylotype II, which originated in the Americas, is the most problematic of the *R. solanacearum* bacteria for the ornamentals industry, in part because it can thrive in more temperate areas and at higher altitudes in the tropics. This bacterium, regulated under the name "Race 3, biovar 2 (R3bv2)," is extremely destructive on potato, causing brown rot disease in Africa, Asia, and Latin America. Race 3, biovar 2 (which is more properly described as phylotype II and sequevar 1 or 2) was found in Europe in the mid-1990s and in the United States in 2002 on geraniums shipped into the country as unrooted cuttings. It is a quarantine pest in Europe and was listed as a Select Agent in the United States, making it subject to strict biosecurity regulation. R3bv2 can infect potato, tomato, eggplant, pepper, and geranium. The weed *Solanum dulcamara* is a host of R3bv2; additional solanaceous and nonsolanaceous weeds can harbor other strains of *R. solanacearum.*

Phylotype II, sequevar 7 of *R. solanacearum* (historically known as "Race 1") is endemic in the southern United States. It causes occasional garden losses on a variety of annual flowers, including ageratum, gerber daisy, impatiens, marigold, nasturtium, nicotiana, petunia, salvia, verbena, vinca (*Catharanthus roseus*), and zinnia.

Host Range and Epidemiology

The *R. solanacearum* species complex collectively has a wide host range that includes at least 270 different plants in 44 families. Bedding plant genera that are susceptible to one or more strains of *R. solanacearum* include *Browallia, Catharanthus, Chaenostoma* (bacopa), *Cyclamen, Dahlia, Gerbera, Impatiens, Osteospermum,* and *Pelargonium.* The pathogenicity of different sequevars to bedding plants has not been systematically investigated, but it is important to understand that sequevars of *R. solanacearum* other than the Select Agent (R3bv2, or phylotype II, sequevars 1 and 2) may cause disease in these genera. Differentiating between strains of *R. solanacearum* requires PCR testing and sequencing. In the United States, any sample that may possibly contain R3bv2 should be forwarded to the USDA–APHIS for testing.

Infection generally occurs through the root system in field-grown crops and gardens, but the pathogen is also waterborne, making the use of surface water sources and recirculating irrigation potentially dangerous for introducing and spreading bacteria. Additionally, the potential for contamination of vegetatively propagated crops, such as geranium, is a concern, because cut-

Fig. 14. Southern bacterial wilt (*Ralstonia solanacearum,* Race 3, biovar 2) on geranium. (Courtesy A. R. Chase—© APS)

Fig. 15. Geranium leaves starting to wilt because of systemic infection by *Ralstonia solanacearum,* which causes southern bacterial wilt. (Courtesy C. Allen—© APS)

tings are produced by specialist propagators in parts of the world in which various strains of the pathogen may be endemic.

High temperatures (30–35°C) and high levels of soil moisture are generally conducive to disease development. Seed transmission is a possible but unlikely means of infection. The bacterium has been shown to colonize foliage of pepper plants (*Capsicum* spp.) from infested seed. Splashing water can disperse epiphytic populations and bacteria exuded from infected plants. Several weed hosts have been reported to be symptomless carriers of *R. solanacearum,* and some bedding plants have been shown to be symptomlessly infected.

Management

Development of southern bacterial wilt and other soilborne diseases can be avoided by using soilless growing media and disease-free propagative materials. The use of soil fumigants is no longer possible in many ornamental settings worldwide and is largely ineffective against this pathogen. Culture-indexing procedures used in geranium clean stock production systems detect and eliminate *R. solanacearum.* The geranium industry has worked closely with the USDA–APHIS to establish testing procedures, which are used at offshore geranium greenhouses to guard against *R. solanacearum* being inadvertently introduced to the United States within symptomless cuttings.

Drenches of potassium salts of phosphorous acid (H_3PO_3) have been shown effective in protecting geraniums growing in soilless potting medium from infection by *R. solanacearum.* The phosphorous acid portion was effective both in vitro and in vivo. While other bactericides slowed disease progress, they were not as effective. Additionally, phosphorus pentoxide and phosphoric acid were not effective. The use of phosphorous acid materials prophylactically is not recommended for propagation of vegetative materials, because they could mask an infection in the mother block (the clean stock plants from which all cuttings are derived). Developing less susceptible cultivars of crop plants by traditional means will not likely be successful for a pathogen with a host range as wide as that of *R. solanacearum.* Host plant resistance studies have been published on geranium (zonal, regal, ivy, and scented), but most of the plants were very susceptible and died as a result of infection.

Selected References

Champoiseau, P. G., Jones, J. B., and Allen, C. 2009. *Ralstonia solanacearum* race 3 biovar 2 causes tropical losses and temperate anxieties. Online. Plant Health Progress. doi:10:1094/PHP-2009-0313-01-RV

Hayward, A. C. 1991. Biology and epidemiology of bacterial wilt caused by *Pseudomonas solanacearum.* Annu. Rev. Phytopathol. 29:65-87.

Ji, P., Allen, C., Sanchez-Perez, A., Yao, J., Elphinstone, J. G., Jones, J. B., and Momol, M. T. 2007. New diversity of *Ralstonia solanacearum* strains associated with vegetable and ornamental crops in Florida. Plant Dis. 91:195-203.

Jones, R. K. 1993. Southern bacterial wilt. Pages 242-245 in: Geraniums IV. J. W. White, ed. Ball Publishing, Chicago.

Kim, S. H., Olson, T. N., Schaad, N. W., and Moorman, G. W. 2003. *Ralstonia solanacearum* race 3, biovar 2, the causal agent of brown rot of potato, identified in geraniums in Pennsylvania, Delaware, and Connecticut. Plant Dis. 87:450.

Norman, D. J., Chen, J., Yuen, J. M. F., Mangravita-Novo, A., Byrne, D., and Walsh, L. 2006. Control of bacterial wilt of geranium with phosphorous acid. Plant Dis. 90:798-802.

Norman, D. J., Huang, Q., Yuen, J. M. F., Mangravita-Novo, A., and Byrne, D. 2009. Susceptibility of geranium cultivars to *Ralstonia solanacearum.* HortScience 44:1504-1508.

Strider, D. L. 1982. Susceptibility of geraniums to *Pseudomonas solanacearum* and *Xanthomonas campestris* pv. *pelargonii.* Plant Dis. 66:59-60.

Strider, D. L., Jones, R. K., and Haygood, R. A. 1981. Southern bacterial wilt of geranium caused by *Pseudomonas solanacearum.* Plant Dis. 65:52-53.

Swanson, J. K., Yao, J., Tans-Kersten, J., and Allen, C. 2005. Behavior of *Ralstonia solanacearum* race 3 biovar 2 during latent and active infection of geranium. Phytopathology 95:136-143.

Weibel, J., Tran, T. M., Bocsanczy, A. M., Daughtrey, M., Norman, D. J., Mejia, L., and Allen, C. 2016. A *Ralstonia solanacearum* strain from Guatemala infects diverse flower crops, including new asymptomatic hosts *Vinca* and *Sutera,* and causes symptoms in geranium, mandevilla vine, and new host African daisy (*Osteospermum ecklonis*). Plant Health Prog. 17:114-121.

Williamson, L., Nakaho, K., Hudelson, B., and Allen, C. 2002. *Ralstonia solanacearum* race 3, biovar 2 strains isolated from geranium are pathogenic on potato. Plant Dis. 86:987-991.

(Prepared by M. L. Daughtrey, R. L. Wick, and J. L. Peterson; revised by A. R. Chase and M. L. Daughtrey)

Diseases Caused by *Xanthomonas campestris* Pathovars

Xanthomonas spp. cause diseases of a wide range of plants. Following taxonomic revision at the end of the twentieth century, the familiar species *X. campestris* has been considered a pathogen solely of crucifers. The other xanthomonads affecting flower crops are found within *X. axonopodis* and *X. hortorum,* primarily. In some cases, certain subgroups of *Xanthomonas* spp. (designated pathovars) have been defined as having narrow host ranges. For example, *X. hortorum* pv. *pelargonii* is pathogenic only to *Geranium* and *Pelargonium* spp. A large number of bedding plants have been found infected with a *Xanthomonas*-like bacterium, but few have been researched sufficiently to include in this compendium.

Selected References

Rademaker, J. L. W., Louws, F. J., Schultz, M. H., Rossbach, U., Vauterin, L., Swings, J., and de Bruijn, F. J. 2005. A comprehensive species to strain taxonomic framework for *Xanthomonas.* Phytopathology 95:1098-1111.

Vauterin, L., Hoste, B., Kersters, K., and Swings, J. 1995. Reclassification of *Xanthomonas.* Int. J. Syst. Bacteriol. 45:472-489.

Vauterin, L., Rademaker, J., and Swings, J. 2000. Synopsis on the taxonomy of the genus *Xanthomonas.* Phytopathology 90:677-682.

Vauterin, L., Swings, J., and Kersters, K. 1991. Grouping of *Xanthomonas campestris* pathovars by SDS-PAGE of proteins. J. Gen. Microbiol. 137:1677-1687.

(Prepared by A. R. Chase and M. L. Daughtrey)

Bacterial Blight of *Matthiola* spp.

A bacterial blight of garden stock (*Matthiola incana*) has been reported in commercial seed and cut-flower plantings, as well as home gardens, in the coastal areas of California since 1933. Death or stunting of many plants and serious reductions in or even total losses of seed crops occurred in some seasons. Bacterial blight of *Matthiola* spp. has been reported primarily from the United States as both a greenhouse and a landscape problem.

Symptoms

Infected seedlings are stunted and may wilt and die when they are several inches tall. Stems may be rotted and show dark vascular discoloration internally; they sometimes display a yellowish bacterial ooze on the surface. On inoculated *M. incana,* areas of black stem necrosis developed into cankers; leaf wilt and death followed. On older plants, the lower leaves may show blighting, including chlorosis starting at the tips of leaves and darkening along the veins. Lengthwise cracks in stems have been described in stock grown as cut flowers.

Causal Organism

Xanthomonas campestris pv. *incanae* is a gram-negative, rod-shaped bacterium with cell dimensions of 0.4–1.0 × 0.7–2.0 µm; it forms colonies with yellow pigmentation in culture on yeast–dextrose–carbonate (YDC) agar. The species *X. campestris* is composed of bacteria associated with cruciferous plants, which include *Matthiola* spp. Further subdivision of *X. campestris* has occurred through the assignment of pathovar names based on host range and symptoms produced.

Strains of six *X. campestris* pathovars from crucifers were evaluated on a large range of species in the Brassicaceae. Three distinct diseases were observed: black rot of Brassicaceae, leaf spot of Brassicaceae and Solanaceae, and bacterial blight disease of garden stock. The pathogens associated with these three diseases were named *X. campestris* pv. *campestris, X. campestris* pv. *raphani,* and *X. campestris* pv. *incanae,* respectively.

Host Range and Epidemiology

X. campestris pv. *incanae* showed high specificity to *Matthiola* spp. in a host range trial comparing pathovars of *X. campestris.* Although a strain of *X. campestris* pv. *incanae* was shown to be pathogenic to wallflower (*Erysimum* × *cheiri* [syn. *Cheiranthus* × *cheiri*]) in this trial, the wallflower-infecting strain could not infect stock and vice versa (Fig. 16); thus, there may be multiple races of *X. campestris* pv. *incanae.*

X. campestris (including the pathovar that affects stock) is a seedborne bacterium. It can spread rapidly under favorable conditions, especially those that occur during plug production and in garden use in which overhead irrigation is employed. A PCR-based method for detection and identification of the pathovars of *X. campestris* that affect crucifers has been developed; a similar assay could provide sensitive, specific identification of *X. campestris* pv. *incanae* during production of stock seed.

Management

To reduce losses from bacterial blight, treating seed to remove superficial contamination would be ideal. Early tests showed that the disease could be partially controlled by immersing seed (preferably in small amounts in loose cheesecloth bags) in water at 53–55°C for 10 min, followed by prompt cooling in cold water. Commercial growers have reported that this treatment (with or without chlorine bleach) can produce a gelatinous mass of seeds that cannot be used unless they are rinsed quickly. The treated seed can be stored for several months.

Fig. 16. Leaf lesions on wallflower, symptomatic of bacterial blight caused by a pathovar of *Xanthomonas campestris.* (Courtesy A. R. Chase—© APS)

Field evidence indicates that *X. campestris* pv. *incanae* persists in the soil; thus, a 2- to 3-year rotation is recommended. Minimizing water splash and reducing the length of leaf wetness periods will slow disease spread during plug production and finishing, so that scouting efforts to detect and rogue out diseased plants can be more effective.

Efforts at chemical control with fosetyl-aluminum (FRAC 33), *Bacillus* spp. (*B. subtilis* and *B. amyloliquefaciens*), and other experimental products have not been very effective. The only active ingredients with consistently significant results are some of those containing copper (FRAC M1); different copper formulations can be expected to provide different levels of control. Copper and copper plus mancozeb (FRAC M3) materials will provide some management benefit in production, but a source of clean seed is essential.

Long-term solutions through breeding stock with related species should be possible. Accessions of *M. tricuspidata, M. longipetala,* and *M. aspera* were all highly resistant to *X. campestris* pv. *incanae* in a study in which an interspecific hybrid of *M. incana* with *M. tricuspidata* was similarly resistant. The results indicated that there may be one or more dominant genes for resistance in a wild relative of garden stock.

Selected References

Berg, T., Tesoriero, L., and Hailstones, D. L. 2005. PCR-based detection of *Xanthomonas campestris* pathovars in *Brassica* seed. Plant Pathol. 54:416-427.

Ecker, R., Zutra, D., Barzilay, A., Osherenko, E., and Rav-David, D. 1995. Sources of resistance to bacterial blight of stock (*Matthiola incana* R. Br.). Gen. Res. Crop Evol. 42:371-372.

Fargier, E., and Manceau, C. 2007. Pathogenicity assays restrict the species *Xanthomonas campestris* into three pathovars and reveal nine races within *X. campestris* pv. *campestris.* Plant Pathol. 56:805-818.

Minardi, P. U., Mazzucchi, U., and Parrini, C. 1988. Epidemics of bacterial blight of stock (*Matthiola incana* R. Br.) caused by *Xanthomonas campestris* pv. *incanae* in Tuscany. (Italian, English abstract.) Inf. Fitopatol. 38:43-46.

Vicente, J. G., Conway, J., Roberts, S. J., and Taylor, J. D. 2001. Identification and origin of *Xanthomonas campestris* pv. *campestris* races and related pathovars. Phytopathology 91:492-499.

(Prepared by A. R. Chase and M. L. Daughtrey)

Bacterial Blight of Geranium

Bacterial blight is the most destructive disease of florist's geranium (*Pelargonium* × *hortorum*). The disease was first noted in 1898 from Massachusetts (United States). In the 1950s and 1960s, annual losses of 10–15% were reported, as well as the first comprehensive descriptions of symptomology and control of the disease in the United States. The disease has been referred to as "bacterial stem rot," "bacterial wilt," and "bacterial leaf spot," but the more comprehensive term "bacterial blight" is preferable because of the wide spectrum of symptoms that may appear.

Despite the long-term recognition of bacterial blight and the widespread use of culture-indexing techniques, this disease remained a serious and frequent problem for geranium growers until sanitation measures were redoubled to prevent *Ralstonia solanacearum* race 3, biovar 2 from entering the geranium industry. Improving standards to prevent one bacterial disease also increased the likelihood that cuttings would be free from other bacterial diseases.

Symptoms

Symptoms of bacterial blight vary depending on the plant cultivar, environmental conditions, and strain of bacterium.

When the bacterium is disseminated by splashing water from rainfall or irrigation, small, water-soaked spots develop and are visible initially on the undersides of leaves. After a few days, these leaf spots also become apparent on the upper surfaces of leaves. The spots eventually become tan to dark brown, round, and 2–3 mm in diameter; adjacent spots may coalesce to form larger necrotic areas. The lesions are slightly sunken and have well-defined margins, which may develop chlorotic halos. In addition to round spots, large, wedge-shaped areas of chlorosis and necrosis often form following infection via the hydathodes at leaf edges (Figs. 17 and 18). These wedges are often mistaken for symptoms of the more common Botrytis blight (see the later section "Botrytis Blight").

Typically, the bacterium moves into the vascular system from the initial spots or wedges and may eventually cause the entire plant to wilt. In other cases, the bacterium does not become systemic, and only the infected leaf dies. When zonal geraniums (*P.* × *hortorum*) are systemically infected through the roots or brought into a greenhouse as asymptomatic infected cuttings, wilting of the lower leaves is the first symptom that develops.

The xylem may show dark discoloration in a systemically infected plant. Stem rot typically follows vascular infection, when the bacterium moves from the vascular system into the pith and out into the cortex. In an advanced infection, the affected stem sections may be brown or black. Systemically infected cuttings may or may not root, and those that do not root usually rot at the base.

The roots are not symptomatic with bacterial blight. One clue to the identification of bacterial blight is that the roots remain white and healthy looking, while the top of the plant shows chlorotic leaves and wilting.

Causal Organism

Bacterial blight of geranium is caused by *Xanthomonas hortorum* pv. *pelargonii* (syn. *X. campestris* pv. *pelargonii*). *X. hortorum* pv. *pelargonii* is a gram-negative, rod-shaped, obligately aerobic, yellow-pigmented bacterium in the *Xanthomonadaceae*. *X. hortorum* pv. *pelargonii* has the characteristics of *X. hortorum* but is restricted in host range to members of the Geraniaceae.

Cultural characteristics of *X. hortorum* pv. *pelargonii* on agar media help to differentiate it from other pathovars. *X. hortorum* pv. *pelargonii* is characteristically pale yellow, compared with the brighter yellow of most other xanthomonads. Occasionally, bright-yellow xanthomonads other than *X. hortorum* pv. *pelargonii* are cultured from geranium. Immunological methods are often used to identify blight infections caused by *X. hortorum* pv. *pelargonii*. Testing yellow bacterial isolates for pathogenicity by inoculating them into the stems of healthy cuttings is a good precautionary confirmation: If a soft, brown canker does not form within about 7 days, the bacterium is not likely *X. hortorum* pv. *pelargonii*.

Host Range and Epidemiology

The host range of *X. hortorum* pv. *pelargonii* is restricted to *Pelargonium* and *Geranium* spp. Hardy perennial geraniums produced by the nursery industry are one potential source of inoculum for greenhouse crops of florist's geranium. Cultivars of both florist's geranium (grown from cuttings or seeds) and ivy geranium (*P. peltatum*) are susceptible to bacterial blight. Symptoms are more difficult to recognize, however, on ivy geranium, because the leaves are firm and do not wilt in the manner of florist's geranium. Regal geranium (*P. domesticum*) is not susceptible, nor are *P. acerifolium*, *P. tomentosum*, and *P.* × *scarboroviae*. No significant resistance has been noted among the commercially propagated cultivars of florist's geranium and ivy geranium.

Leaf spot symptoms may occur within 7 days at 27°C and within 21 days at 16°C. Temperatures below 10°C and above 32°C may suppress symptom development, but the bacterium becomes active when ideal temperatures are reached. Many instances have been observed of infected stock plants that do not show symptoms of disease at the time cuttings are removed. Moisture and warmth supplied during rooting apparently trigger disease development.

The disease is highly contagious and spreads rapidly in an overhead-irrigated or subirrigated crop. *X. hortorum* pv. *pelargonii* is capable of infection through the root system, as well as through wounds and natural openings aboveground. The cutting knife is a common means of disease transmission from one stock plant to another. Additional important means of dissemination include splashing irrigation water, workers' hands removing infected leaves, and dripping water from infected ivy geraniums in hanging baskets to geraniums below them. The greenhouse whitefly (*Trialeurodes vaporariorum*) can transmit *X. hortorum* pv. *pelargonii* from diseased to healthy geraniums, but this does not appear to be a common or important method of spread. The foliage of plants other than geraniums can harbor populations of *X. hortorum* pv. *pelargonii*, but the importance of such plants as a source of this bacterium in greenhouse geranium production is not yet understood.

The bacterium does not survive in soil in the absence of infected host debris, but it can survive for at least 1 year in un-

Fig. 17. Small, round spots and wedge-shaped areas of chlorosis and necrosis are typical of infection by *Xanthomonas hortorum* pv. *pelargonii,* which causes bacterial blight of geranium. (Courtesy M. L. Daughtrey—© APS)

Fig. 18. Symptoms of bacterial blight (*Xanthomonas hortorum* pv. *pelargonii*) on infected geraniums in the landscape. (Courtesy A. R. Chase—© APS)

decomposed plant tissue. It can also survive on the foliage of nonhost plants and has been reported to survive epiphytically on geranium leaves for several months without causing symptoms. *X. hortorum* pv. *pelargonii* can also survive on wild and cultivated perennial *Geranium* spp. and overwinter as far north as Ithaca, New York. Overwintering in protected nursery crops of ornamental *Geranium* spp. and in mulched landscape plantings in cold temperate regions is particularly likely.

Management

Strict adherence to good sanitation practices and a carefully administered culture-indexing program are the primary propagator's sole tools for supplying clean cutting stock. Secondary geranium propagators must purchase stock from reputable sources and thus rely primarily on disease exclusion. In the event of a bacterial blight outbreak, diseased plants must be detected and rogued out promptly to minimize the potential crop loss for that season.

Hanging basket crops of ivy geraniums should not be placed directly above geranium crops, and hybrid geraniums grown from seed should be grown apart from vegetatively propagated cuttings. Geraniums from different propagators should also be kept separate. Losses are more likely and may be much more extensive in greenhouses in which geraniums of different types and from different sources are brought into close contact during production.

Titanium dioxide (TiO_2) has been studied for its potential to assist in management of bacterial blight of geranium. TiO_2 application at 25 and at 75 mM before inoculation with *X. hortorum* pv. *pelargonii* caused 53 and 67% reductions of lesion numbers, respectively, in one of two trials in Florida (United States).

No effective chemical control is known for bacterial blight of geranium, although copper products (FRAC M1) are commonly employed. Prompt removal of diseased plants and those in the vicinity of diseased plants is important for curbing an epidemic. Entire crops can be compromised if infection occurs in a crop with a recirculating irrigation system.

Selected References

Anderson, M. J., and Nameth, S. T. 1990. Development of a polyclonal antibody-based serodiagnostic assay for the detection of *Xanthomonas campestris* pv. *pelargonii* in geranium plants. Phytopathology 80:357-360.

Bugbee, W. M., and Anderson, N. A. 1963. Whitefly transmission of *Xanthomonas pelargonii* and histological examination of leaf spots of *Pelargonium hortorum*. Phytopathology 53:177-178.

Dimock, A. W. 1962. Obtaining pathogen-free stock by cultured cutting techniques. Phytopathology 52:1239-1241.

Dougherty, D. E., Powell, C. C., and Larsen, P. O. 1974. Epidemiology and control of bacterial leaf spot and stem rot of *Pelargonium hortorum*. Phytopathology 64:1081-1083.

Kennedy, B. W., Pfleger, F. L., and Denny, R. 1987. Bacterial leaf and stem rot of geranium in Minnesota. Plant Dis. 71:821-823.

Munnecke, D. E. 1954. Bacterial stem rot and leaf spot of *Pelargonium*. Phytopathology 44:627-632.

Munnecke, D. E. 1956. Development and production of pathogen-free geranium propagative material. Plant Dis. Rep. (Suppl.) 40:93-95.

Munnecke, D. E. 1956. Survival of *Xanthomonas pelargonii* in soil. Phytopathology 46:297-298.

Nameth, S. T., Daughtrey, M. L., Moorman, G. W., and Sulzinski, M. A. 1998. Bacterial bight of geranium: A history of diagnostic challenges. Plant Dis. 83:204-212.

Norman, D. J., and Chen, J. 2011. Effect of foliar application of titanium dioxide on bacterial blight of geranium and Xanthomonas leaf spot of poinsettia. HortScience 46:426-428.

Raju, B. C., and Olson, C. J. 1985. Indexing systems for producing clean stock for disease control in commercial floriculture. Plant Dis. 69:189-192.

Vauterin, L., Vantomme, R., Pot, B., Hoste, B., Swings, J., and Kersters, K. 1990. Taxonomic analysis of *Xanthomonas campestris* pv. *begoniae* and *X. campestris* pv. *pelargonii* by means of phytopathological, phenotypic, protein electrophoretic and DNA hybridization methods. Syst. Appl. Microbiol. 13:166-276.

(Prepared by M. L. Daughtrey, R. L. Wick, and J. L. Peterson; revised by A. R. Chase and M. L. Daughtrey)

Bacterial Leaf Spot of Ranunculus

In 1994, ranunculus grown commercially in Riverside and San Diego Counties in California (United States) showed symptoms of a new disease. A similar disease was reported from Italy in 2010. The affected leaves and stems showed irregular, necrotic spots, often with yellow halos, and some leaves became entirely chlorotic (Fig. 19). Water soaking, which is often a clue to a bacterial etiology, was only rarely associated with the leaf lesions. Leaves sometimes showed vein chlorosis along with black patches on the inner margins of leaflets. Symptoms progressed to the death of leaves and eventual plant collapse.

A *Xanthomonas campestris* was isolated from symptomatic tissues in both California and Italy. The pathogen was also recovered from both tubers and seeds, which indicates that the bacteria could be widely disseminated along with ranunculus propagules. When year-old tubers were examined in California, bacteria were recovered from 4 and 7% of cultivars Rose and Picotee, respectively. Seeds of 10 cultivars showed a range of 1.1–16.0% contamination with *X. campestris* when plated on Tween medium and incubated at 28°C for 7 days. Ranunculus plants showed symptoms from 3 to 22 days after inoculation with the bacterium; the speed of symptom development varied with environmental conditions and inoculation method. Humidity was important for disease development; spray-inoculated plants developed symptoms readily only under mist.

PCR and RFLP analysis indicated that the ranunculus strains from the two California counties were closely related and also similar to *X. campestris* pv. *campestris*. The bacterial strains isolated from leaves, seeds, and tubers of ranunculus failed to cause disease in inoculated cauliflower, however, which suggests that these ranunculus isolates represent a different pathovar from *X. campestris* pv. *campestris*. Further studies are needed to fully characterize this ranunculus-infecting xanthomonad. *X. campestris* was revised in 1995 to include only bacteria with hosts in the Brassicaceae, so both species and pathovar names may be needed for the ranunculus pathogen.

Fig. 19. Bacterial leaf spot (*Xanthomonas campestris*) of ranunculus. (Courtesy A. R. Chase—© APS)

Management of bacterial leaf spot on ranunculus will require roguing out symptomatic plants as they are detected. Keeping the greenhouse environment below 85% relative humidity and avoiding extended periods of leaf wetness will be beneficial.

Selected References

Azad, H. R., Vilchez, M., Paulus, A. O., and Cooksey, D. A. 1996. A new ranunculus disease caused by *Xanthomonas campestris*. Plant Dis. 80:126-130.

McGuire, R. G., Jones, J. B., and Sasser, M. 1986. Tween media for semiselective isolation of *Xanthomonas campestris* pv. *vesicatoria* from soil and plant material. Plant Dis. 70:887-891.

Vauterin, L., Hoste, B., Kersters, K., and Swings, J. 1995. Reclassification of Xanthomonas. Int. J. Syst. Bacteriol. 45:472-489.

(Prepared by A. R. Chase and M. L. Daughtrey)

Bacterial Leaf Spot and Flower Spot of Zinnia

A bacterial leaf spot and flower spot on *Zinnia* spp. was first reported in Italy in 1929 but then not reported again from Europe until 2008. Within the United States, the disease was formally reported in 1973 from Ohio and North Carolina, in 1977 from California, and in 1985 from Louisiana; it is frequently seen in landscape plantings in the United States. The disease has been observed in countries around the world, including Australia, Brazil, China, India, Hungary, Malawi, Rhodesia, Sierra Leone, and, most recently, Korea.

Symptoms

Xanthomonas leaf spot on zinnia begins as infection of the cotyledons, which develop angular or rounded lesions. The lesions are usually water soaked and dark brown to black until they dry and turn tan; yellow halos are sometimes evident (Figs. 20 and 21). Bacterial lesions rarely coalesce to any significant degree. Infected plants in plug trays often escape notice until the plugs are ready for sale. Infected cotyledons are most evident just a few weeks after seeding; they may later drop from the plant, and the presence of the pathogen will sometimes go unnoticed until the crop is flowering.

Angular leaf spots are prevalent on plants both in production and in landscape use, and they are often confused with the symptoms of Alternaria leaf spot. Mixed infections are common in both settings.

In the landscape, under typical summer conditions and if rainfall is plentiful, flowers and stems also become infected with the bacterium, developing similar lesions to those found on leaves.

Causal Organism

The cause of bacterial leaf spot and flower spot is a yellow-pigmented *Xanthomonas campestris* (syns. *Xanthomonas campestris* pv. *zinniae; X. nigromaculans* f. sp. *zinniae*). (*Note:* Should be considered *Xanthomonas campestris sensu lato* until further studied.) Using repetitive extragenic palindromic-polymerase chain reaction (rep-PCR) fingerprinting and sequence analysis of the 16S–23S rDNA spacer region, a 2003 study determined that 22 strains of *X. campestris* pv. *zinniae* collected from seed and plants were closely related to each other but clearly distinct from other *Xanthomonas* spp., including *X. campestris* pv. *campestris, X. axonopodis* pv. *vesicatoria, X. vesicatoria,* and *X. hortorum* pv. *vitians*. Fingerprinting done with rep-PCR was shown to have the potential for reliable, practical, routine detection of *X. campestris* pv. *zinniae.* Two primer sets—Xgyr1BF/Xgyr1BR and A1/B1, for the *gyrB* gene and the 16S–23S rRNA internal transcribed spacer (ITS) region, respectively—are used to identify the pathogen.

Host Range and Epidemiology

The pathogen is known to be seedborne in zinnia; thus, there is a mechanism in its long-distance spread. The host range of this *X. campestris* causing leaf spot of zinnia was recently found to extend beyond that host. A host range study reported in 2003 indicated that isolates of *X. campestris* pv. *zinniae* caused leaf spots on tomato but not on cabbage, lettuce, pepper, and radish. Also, xanthomonads pathogenic to cabbage, lettuce, pepper, and radish (*X. campestris* pv. *campestris, X. axonopodis* pv. *vesicatoria, X. vesicatoria,* and *X. hortorum* pv. *vitians*) did not cause leaf spots on zinnia.

Management

There are wide ranges in susceptibility of *Zinnia* spp. and cultivars to bacterial leaf spot. Since the late 1970s, quite a few studies have been conducted that identify possible sources of resistance in breeding lines. *Z. angustifolia* (syn. *Z. linearis*) was identified as resistant. Hybrids of *Z. marylandica, Z. violacea,* and *Z. angustifolia* were screened for resistance

Fig. 20. Bacterial leaf spot (*Xanthomonas campestris* pv. *zinniae*) of zinnia. (Courtesy A. R. Chase—© APS)

Fig. 21. Vein-bounded leaf lesions on zinnia, caused by infection by *Xanthomonas campestris* pv. *zinniae,* which causes bacterial leaf spot. (Courtesy A. R. Chase—© APS)

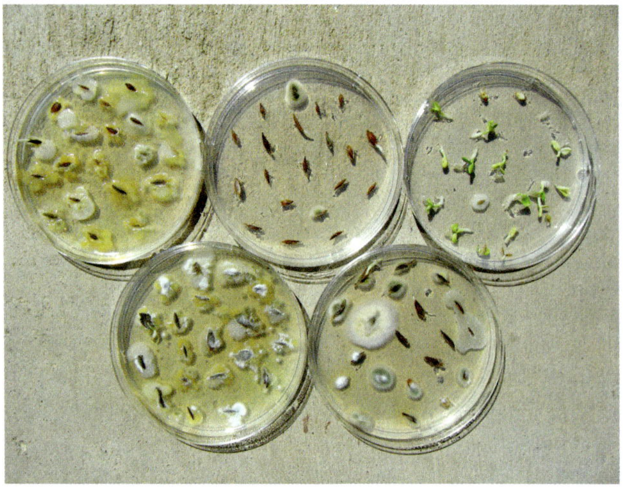

Fig. 22. Zinnia seeds showing infestation with a yellow-pigmented xanthomonad. (Courtesy A. R. Chase—© APS)

to bacterial leaf and flower spot. Crosses derived from crossing *Z. marylandica* with *Z. angustifolia* were highly resistant to the disease. Fifty-seven cultivars of zinnia were studied for 17 weeks to determine their resistance to Alternaria leaf spot (*Alternaria zinniae*), powdery mildew (*Golovinomyces cichoracearum* [syns. *Erysiphe cichoracearum; E. cichoracearum* var. *cichoracearum*]), and bacterial leaf and flower spot under landscape conditions in the southern United States. At week 4, all the cultivars were acceptable in appearance, but by week 10, only 11 of 57 cultivars were acceptable. At week 17, all the cultivars were unacceptable. These results suggest that using cultivars of commercially available zinnias will not be an effective method of managing losses in the landscape until a concerted effort is made by plant breeders to develop disease-resistant plants.

Care should be taken to separate tomato and zinnia crops in the greenhouse. A seedborne bacterial leaf spot problem in zinnia could potentially be transferred to tomato transplants in the same facility by handling or by splashing water during irrigation.

Eradication of the pathogen on seed using sodium hypochlorite was first reported in 1979 (Fig. 22). A 30-min soak in 10,500 ppm of sodium hypochlorite (dilute bleach) was shown to be very effective in cleaning contaminated seed. More recent research showed that a 20-min soak in hot water (52°C) was more effective than a bleach soak for a period less than 2 h. A soak in 2% peroxyacetate for 1 min was not effective. The use of standard bactericides during propagation and production can be helpful, but results are inconsistent and never 100% effective. The use of pathogen-free seed is the most successful method of controlling losses on zinnia.

Selected References

Boyle, T. H., and Wick, R. L. 1996. Responses of *Zinnia angustifolia* × *Z. violacea* backcross hybrids to three pathogens. HortScience 31:851-854.

Gombert, L., Windham, M., and Hamilton, S. 2001. Evaluation of disease resistance among 57 cultivars of *Zinnia*. HortTechnology 11:71-74.

Jones, J. J., and Strider, D. L. 1979. Susceptibility of zinnia cultivars to bacterial leaf spot caused by *Xanthomonas nigromaculans* f. sp. *zinniae*. Plant Dis. Rep. 63:449-453.

Sahin, F., Kotan, R., Abbasi, P. A., and Miller, S. A. 2003. Phenotypic and genotypic characterization of *Xanthomonas campestris* pv. *zinniae* strains. Eur. J. Plant Pathol. 109:165-172.

Strider, D. L. 1979. Detection of *Xanthomonas nigromaculans* f. sp. *zinniae* in zinnia seed. Plant Dis. Rep. 63:869-873.

Strider, D. L. 1979. Eradication of *Xanthomonas nigromaculans* f. sp. *zinniae* in zinnia seed with sodium hypochlorite. Plant Dis. Rep. 63:873-876.

(Prepared by A. R. Chase and M. L. Daughtrey)

Disease Caused by *Xylella fastidiosa*

Periwinkle wilt, caused by *Xylella fastidiosa,* was described from Florida (United States) in 1978 following the appearance of symptoms on several greenhouse-grown plants of *Catharanthus roseus* (annual vinca or Madagascar periwinkle). Foliage of infected plants showed blotchy yellowing, marginal chlorosis and wilting of older leaves, and sometimes vein clearing, followed by dropping of the lower leaves. A pleomorphic, rod-shaped bacterium with a rippled cell wall was found in the xylem, and the sharpshooter *Oncometopia nigricans* was found capable of transmitting this xylem-limited bacterium to healthy periwinkle after 5–7 days of acquisition feeding. A later study demonstrated transmission by another sharpshooter, *Homalodisca coagulata*. This strain of *X. fastidiosa* causing periwinkle wilt is closely related but not identical to the causal agents of phony disease of peach and plum leaf scorch.

In 1998, plants of Madagascar periwinkle in a garden in Brazil were observed to have small leaves, short internodes, and dieback. Sap expressed from a plant stem with pliers was examined with a light microscope at 400× magnification, revealing quantities of rod-shaped bacteria. After culturing for 10 days on buffered cysteine–yeast extract and periwinkle wilt agar media, colonies typical of *X. fastidiosa* were observed, and indirect immunofluorescence tests with an antibody specific to *X. fastidiosa* confirmed this identification. However, symptoms for this first detection of *X. fastidiosa* infection of Madagascar periwinkle in Brazil were different from those described in the earlier U.S. report.

C. roseus has been used as an experimental host for the strain of *X. fastidiosa* that causes citrus variegated chlorosis (CVC). *C. roseus* plants mechanically inoculated with this strain showed stunted growth and deformed leaves after 2 months; symptoms eventually progressed to severe leaf chlorosis. The bacterium has been shown to reproduce within periwinkle.

Problems with *X. fastidiosa* have not been noted on annual bedding plants in commercial greenhouse production, where leafhoppers (including sharpshooters) are rarely a concern. The pathogen is not detected by standard culturing or examination of tissue with a light microscope; thus, PCR and ELISA techniques should be utilized if a disease caused by a *Xylella* sp. is suspected.

Selected References

Davis, M. J., Raju, B. C., Brlansky, R. H., Lee, R. F., Timmer, L. W., Norris, R. C., and McCoy, R. E. 1983. Periwinkle wilt bacterium: Axenic culture, pathogenicity, and relationships to other gram-negative, xylem-inhabiting bacteria. Phytopathology 73:1510-1515.

McCoy, R. E., Thomas, D. L., Tsai, J. H., and French, W. J. 1978. Periwinkle wilt, a new disease associated with xylem delimited rickettsialike bacteria transmitted by a sharpshooter. Plant Dis. Rep. 62:1022-1026.

Monteiro, P. B., Renaudin, J., Jagoueix-Eveillard, S., Ayres, A. J., Garneir, M., and Bové, J. M. 2001. *Catharanthus roseus,* an experimental host plant for the citrus strain of *Xylella fastidiosa*. Plant Dis. 85:246-251.

Ueno, B., Funada, C. K., Yorinori, M. A., and Leite, R. P., Jr. 1998. First report of *Xylella fastidiosa* on *Catharanthus roseus* in Brazil. Plant Dis. 82:712.

(Prepared by A. R. Chase and M. L. Daughtrey)

Diseases Caused by Fungi

The most common biotic diseases on bedding plants are caused by fungi: heterotrophic organisms that usually have highly branched, tubular bodies and produce spores for dissemination and reproduction. Fungi create a variety of enzymes that break down substrates, including plant cells, thus allowing nutrients to be absorbed.

Plant diseases caused by fungi are typified by a wide range of symptoms, including leaf and flower spots, leaf and blossom blights, stem cankers, galls, root and crown rots, and vascular wilts. Many fungal pathogens can be observed on infected plant tissues, particularly when the fungi are sporulating. This is especially true of rust and powdery mildew fungi, which are particularly conspicuous.

Fungal pathogens are often introduced to a greenhouse or a landscape via infected plant materials, such as cuttings, plugs, seeds, and plants. In some cases, infested soil particles are blown or tracked into the greenhouse. Sometimes, spores are blown directly into the greenhouse from outside; more rarely, insects vector fungi. Resting structures such as oospores, chlamydospores, and sclerotia allow a fungus to remain dormant in the soil and in plant debris under benches. Pathogen contamination on the greenhouse floor may be brought up to the crop level—again, via insects or air currents or even on the end of a dropped hose—leading to the start of another cycle of disease.

Fungi known to infect bedding plants are listed in Appendix II. Readers will note that in both the appendix and the text in this section, alternative names of fungi are identified primarily as "synonyms," rather than as specific life cycle stages (e.g., "anamorph," "teleomorph"). This practice reflects the policy for naming fungi issued in 2011 by the International Botanical Congress: namely, that the use of multiple names for the same fungus be abandoned. (Readers interested in learning more about this policy are referred to the 2011 article by Hawksworth [see Selected References].) Going forward, accommodating this new policy may mean that the most familiar names of some fungi will change. Changes to the names of fungi, as well as other updates to taxonomy and nomenclature, will be made in the Common Names of Plant Diseases list for bedding plants, which is available on the website of The American Phytopathological Society (APS). The source used by APS to confirm current fungal taxonomy is the U.S. National Fungus Collections Fungal Database. Readers are advised to use this source, as well (see Selected References).

The following sections address most of the important fungal diseases of bedding plants. Many of these diseases have not been studied in depth, but we have chosen to remark on their importance here.

Table 3 identifies fungicides by FRAC group and rates their performance against diseases in research trials.

Selected References

Farr, D. F., and Rossman, A. Y. Fungal Databases, U.S. National Fungus Collections. U.S. Department of Agriculture, Agricultural Research Service. https://int.ars-grin.gov/fungaldatabases/

Hawksworth, D. L. 2011. A new dawn for the naming of fungi: Impacts of decisions made in Melbourne in July 2011 on the future publication and regulation of fungal names. IMA Fungus 2:155-162.

(Prepared by A. R. Chase and M. L. Daughtrey)

Alternaria Leaf Spot

Alternaria leaf spot is one of the most common diseases of seed-propagated bedding plants. Significant research has been conducted on some hosts, while little is known about problems on other plants, regardless of how common the diseases may be.

Alternaria leaf spot has been reported on *Zinnia* spp. throughout the world. It was originally reported from Denmark, England, and the United States. Alternaria leaf spot of zinnia is arguably one of the most recognized and most common leaf spot diseases found on bedding plants.

One of the Alternaria leaf spot diseases on which significant research has been performed is that on florist's geranium (*Pelargonium* × *hortorum*) and ivy geranium (*P. peltatum*). It has been reported from Europe and the U.S. states of California and Florida. This disease has also been observed on regal geranium (*P. domesticum*) and rose geranium (*P. graveolens*) in the United States. Disease development is fostered by production conditions that are stressful to the host. This disease is very common in landscapes on the U.S. West Coast in settings with overhead irrigation.

An Alternaria leaf spot has also commonly been reported on marigolds (*Tagetes* spp.) throughout the United States and Mexico. Similarly, Alternaria leaf spot has been reported on China pink (*Dianthus chinensis*) and other *Dianthus* spp. throughout the world. The pathogen on *Dianthus* spp. (*Alternaria dianthicola*) was reported as early as 1950 from California. This disease is not common on dianthus because cultivar resistance makes its appearance relatively rare, but additional species of *Alternaria* reported from dianthus may be encountered. A number of other common Alternaria leaf spots have not been fully described, including those affecting impatiens (Fig. 23) and annual vinca (*Catharanthus roseus*) (Fig. 24).

Symptoms

On zinnia, spots on leaves, stems, and petals are initially tiny (a few millimeters in diameter), circular, dark brown to purple, and water soaked. At maturity, round to angular spots are dark black and have purple or red margins. Spots are first observed

Fig. 23. Alternaria leaf spot (*Alternaria* sp.) on garden impatiens leaves. (Courtesy M. L. Daughtrey—© APS)

TABLE 3. Fungicides Listed by FRAC Group and Their Performance on Pathogens/Diseases in Research Trials[a,b]

(FRAC Group) Active Ingredient[c]	*Alternaria*	Bacteria	*Botrytis*	*Cercospora*	*Colletotrichum*	*Corynespora*	Downy Mildews	*Fusarium*	*Myrothecium*
(1) Thiophanate-methyl	N/P		P/S	S/E				S/G	S/VG
(2) Iprodione	VG/E		E	S	N/S			VG	VG
(3) Propiconazole	F/VG			G/VG		N/G		G	G
(3) Myclobutanil	G/E			VG/E	N/VG				S/G
(3) Triadimefon	VG		G				P/E		
(3) Triflumizole	G/E		N/VG	VG	N/VG			G	VG
(3) Tebuconazole									
(3) Metconazole				VG/E	VG				
(3) Triticonazole	VG/E		G		S/G			G	VG
(4) Mefenoxam							E		
(5) Piperalin									
(11) Trifloxystrobin	G/E		VG	S/E			G/VG	S/G	S/G
(11) Fluoxastrobin	E		S	VG/E	VG/E			P	VG/E
(11) Pyraclostrobin								E	
(11) Fenamidone							VG/E		
(11) Azoxystrobin	VG/E		F/G	VG/E	N/VG	VG/E	VG/E	G/VG	VG/E
(12) Fludioxonil	E		VG/E	S/E	N/G		N	G/E	VG/E
(14) Pentachloro-nitrobenzene (PCNB)			N						
(14) Etridiazole							E		
(17) Fenhexamid	F		VG/E				N	F	
(19) Polyoxin D									
(21) Cyazofamid							E		
(33) Fosetyl aluminum	N	F	S				E	N	
(33) Mono- and di-potassium salts of phosphorous acid	N	P					G/E	N/VG	
(33) Potassium phosphite		G					VG/E		
(40) Mandipropamid							VG/E		
(40) Dimethomorph							E		
(43) Fluopicolide							VG/E		
(44) *Trichoderma harzianum* T22	N						P/F	N/VG	N/VG
(M1) Copper octanoate	G	G	F		G	VG	P/VG		
(M1) Copper hydroxide		G	F	S			P/VG		
(M1) Copper pentahydrate	F/E	VG/E	P/G	S/E	VG/E		VG	N/VG	N/VG
(M3) Mancozeb	G/VG		G/E	VG/E	S/E		G/VG		
(M5) Chlorothalonil	VG/E		VG/E	VG/E	S/E		S	VG/E	VG/E
(NC) *Streptomyces lydicus* W108	N	N/S	N/G				N/S	N/G	
(NC) *Bacillus subtilis* QST 713	N/P	G	S	VG	G	N	S/G	P	
(NC) Dimethyldidecyl-ammonium chloride (DDAC)	VG	VG	S	VG/E	N		S	G	
(NC) Potassium bicarbonate		N	S/VG			N	N/E		
(NC) Extract of giant knotweed	P	S	P	S/G			N/G		
(NC) Clarified hydrophobic extract of neem oil			F/G				P/F		
Premixes									
(1 and 2) Thiophanate-methyl and iprodione			S/E	VG	G				
(1 and M3) Thiophanate-methyl and mancozeb	E								
(1 and M5) Thiophanate-methyl and chlorothalonil	VG		P/G		E		N	N	N
(3 and M5) Propiconazole and chlorothalonil				E	VG	VG			
(4 and 12) Mefenoxam and fludioxonil								P	
(40 and 45) Dimethomorph and ametoctradin							VG/E		
(44 and NC) *Trichoderma harzianum* T22 and *T. viride* G41									
(7 and 11) Boscalid and pyraclostrobin	VG/E		VG/E	VG	VG/E		VG/E	S/G	S/E
(9 and 12) Cyprodinil and fludioxonil	VG/E		VG	G	S/E	S		N	VG/E
(M1 and M2) Copper hydroxide and mancozeb	G	G	F				N/VG		N

[a] Courtesy A. R. Chase and M. L. Daughtrey—© APS.
[b] Performance codes: N = None; S = Some; P = Poor; F = Fair; G = Good; VG = Very good; E = Excellent.
[c] NC = Not classified.

(FRAC Group) Active Ingredient[c]	*Phyllosticta*	*Phytophthora*	Powdery Mildews	*Pythium*	*Rhizoctonia*	Rusts	*Sclerotinia*	*Sclerotium*	*Thielaviopsis*
(1) Thiophanate-methyl	N	N	P/VG	N	G/E	VG/E			VG/E
(2) Iprodione					G/E		S/E		
(3) Propiconazole	S/G		VG/E			G/E			
(3) Myclobutanil	VG/E		VG/E		VG	VG/E			N
(3) Triadimefon			G/E		F	G/E			
(3) Triflumizole	N		VG/E	N	F/VG	G/E		G	S/G
(3) Tebuconazole									
(3) Metconazole			E		VG/E				
(3) Triticonazole			E		VG/E	E		S/E	N
(4) Mefenoxam		G/E		VG/E					
(5) Piperalin			VG/E						
(11) Trifloxystrobin	N	N/VG	VG/E	P/G	G	F/E	VG/E		N
(11) Fluoxastrobin			N	VG	VG/E				N
(11) Pyraclostrobin		G/E		G/E	E				
(11) Fenamidone		VG/E		VG					
(11) Azoxystrobin	N/G	VG	VG/E	S/VG	VG/E	E	VG	G/E	N
(12) Fludioxonil	S/VG	N	N	N	E				N/VG
(14) Pentachloro-nitrobenzene (PCNB)					G/E			G	N
(14) Etridiazole		G/E		VG/E					
(17) Fenhexamid		F	F/VG	N	F	F/G	P/E		N
(19) Polyoxin D									VG/E
(21) Cyazofamid		VG/E		VG/E					
(33) Fosetyl aluminum		VG/E	P/S	N/E					
(33) Mono- and di-potassium salts of phosphorous acid		VG/E	S/VG	N/E					
(33) Potassium phosphite		N/E	VG/E	S/E					
(40) Mandipropamid		VG/E							
(40) Dimethomorph		VG/E		N					
(43) Fluopicolide		VG/E		N/S					
(44) *Trichoderma harzianum* T22		P/G	F	P/VG	N/E	F/VG			N
(M1) Copper octanoate		P/G	P/VG	S	P/S	P/G	N		
(M1) Copper hydroxide				F/VG	S	N		N	
(M1) Copper pentahydrate		P/VG	G/E	F/VG	P/G	P/VG	N		N
(M3) Mancozeb		F	S/VG		P/F	VG/E			
(M5) Chlorothalonil	G	N/VG		N/E	E	S	E		
(NC) *Streptomyces lydicus* W108		P/S	N/S	N/VG	N	N/S			N
(NC) *Bacillus subtilis* QST 713		S	G/VG	S	N				
(NC) Dimethyldidecyl-ammonium chloride (DDAC)		N	S/VG	N	N				N
(NC) Potassium bicarbonate			VG/E			S/VG			
(NC) Extract of giant knotweed		N	S/VG					N	
(NC) Clarified hydrophobic extract of neem oil			VG/E	VG/E	F	VG/E			
Premixes									
(1 and 2) Thiophanate-methyl and iprodione					E				
(1 and M3) Thiophanate-methyl and mancozeb			P/E		G (foliar)				
(1 and M5) Thiophanate-methyl and chlorothalonil	VG/E	VG		E		VG	G		
(3 and M5) Propiconazole and chlorothalonil			E			G			
(4 and 12) Mefenoxam and fludioxonil		VG		VG/E	VG/E				
(40 and 45) Dimethomorph and ametoctradin		VG/E							
(44 and NC) *Trichoderma harzianum* T22 and *T. viride* G41		P/G		P/VG	S				
(7 and 11) Boscalid and pyraclostrobin	E	N/VG	VG/E	N/G	VG/E	G/E	VG/E	VG/E	N
(9 and 12) Cyprodinil and fludioxonil			G		E	N	E		
(M1 and M2) Copper hydroxide and mancozeb			F/VG		S	G			

Fig. 24. Alternaria leaf spot (*Alternaria zinniae*) on annual vinca leaves. (Courtesy A. R. Chase—© APS)

Fig. 25. Alternaria flower spot (*Alternaria zinniae*) on zinnia petals. (Courtesy A. R. Chase—© APS)

Fig. 26. Alternaria leaf spot (*Alternaria zinniae*) on zinnia cotyledons. (Courtesy A. R. Chase—© APS)

on the oldest leaves and flowers in a landscape bed (Fig. 25); mixed infections with bacteria (*Xanthomonas campestris* pv. *zinniae*) on the same plant and even in the same spot are common. In propagation, spots can be seen on the cotyledons (Fig. 26). Severely infected leaves turn chlorotic and then wilt and die. On mature plants, the centers of spots on leaves and petals are often light tan to white and elliptical in shape.

On geranium, symptoms include small (1–2 mm), blisterlike, water-soaked spots that appear on the undersides of older leaves. Spot centers may become sunken and brown; sometimes, each spot has a diffuse, yellow halo. Eventually, spots develop on the upper leaf surfaces, and they may enlarge to irregularly shaped areas 6–10 mm wide. Zonate patterns in lesions and merging of individual spots can also occur. Dark spores are produced on the necrotic spots, particularly on fallen leaves.

On marigold, leaf, stem, and petal symptoms begin as tiny (a few millimeters in diameter), circular, brown to purple, water-soaked spots. Mature spots are black and have purple or red margins. They can be seen on the cotyledons on some infected plants during plug production. On more mature plants, spots are initially found near the bases on the oldest leaves.

On *Dianthus* spp., leaf spots are initially dark, water soaked, and a few millimeters in diameter. They mature to a reddish or olivaceous-brown color and enlarge and sometimes coalesce to kill portions of the leaves. Alternaria leaf spot is easily confused with fairy ring leaf spot (caused by *Cladosporium echinulatum* [syns. *Didymellina dianthi; Heterosporium echinulatum; Mycosphaerella dianthi*]), which is more common in *Dianthus* spp. (see the later section "Fairy Ring Leaf Spot").

Causal Organisms

Alternaria alternata (syn. *A. tenuis*) is a widely distributed saprophyte that occasionally causes disease on a wide range of plants. Colonies are effuse and range in color from gray to black or olivaceous-black. *A. alternata* has simple or branched, brown conidiophores (up to 50 × 3–6 µm) that are grouped or single. The muriform, light-golden-brown conidia (20–63 × 9–18 µm) are smooth or rough walled and have very short, pale beaks. They are produced in long chains that may be branched.

A. nobilis (syn. *A. dianthi*), *A. dianthicola, A. saponariae,* and *A. ellipsoidea* have all been reported on *Dianthus* spp. Spores of *A. nobilis* measure 69–90 (to 119) × 17–21 (to 25) µm; each has a distinctive beak cell that is 17–53 µm long. Conidia form short chains on host tissue.

A. tagetica causes the disease on marigold, and *A. zinniae* causes the disease on cosmos, sunflower, marigold, and zinnia, among other hosts. An average *A. zinniae* conidium is 20–24 × 170–210 µm including the beak, which is twice as long as the body.

Host Range and Epidemiology

A. alternata has a number of bedding plant hosts, including *Dahlia* hybrids, gerber daisy (*Gerbera jamesonii*), annual vinca, florist's geranium, and balsam impatiens (*Impatiens balsamina*). The pathogen has not been formally identified on garden impatiens (*I. walleriana*). The wide host range of *A. alternata* suggests that this species may be responsible for some of the bedding plant diseases that have been attributed to an "*Alternaria* sp." over the years. A wet environment favors infection by this and other *Alternaria* spp.

A. saponariae is known to affect *Dianthus* spp. and some close relatives; it is commonly found on carnation (*D. caryophyllus*) and China pink. *A. nobilis* affects China pink, maiden pink (*D. deltoides*), and Sweet William (*D. barbatus*); its optimal temperature for growth is 25–30°C. *A. dianthicola,* which is found on Sweet William and China pink, has been proven seedborne on carnation, and seed transmission is likely for other Alternaria diseases of bedding plants, as well.

A. tagetica is known to affect only marigold, including both African (*Tagetes erecta*) and French (*T. patula*) types, as well as *Tagetes* hybrids. The disease requires a minimum of 4 h of moisture for development; a period of foliage wetness longer than 8 h can result in an epidemic. *A. tagetica* was proven to be seedborne in 1983, but attempts to infect other members of the Asteraceae (*Calendula officinalis, Callistephus chinensis, Chrysanthemum maximum, Cosmos bipinnatus, Gazania linearis* [syn. *G. longiscapa*], and *Zinnia elegans*) were not successful.

The most familiar Alternaria leaf spot pathogen is likely *A. zinniae*. Although most commonly reported from zinnia worldwide, it has also been reported from species of *Helianthus, Tagetes, Impatiens,* and *Callistephus*, among other flowers. The optimal temperature for growth of *A. zinniae* in the laboratory is about 27°C.

Management

The most critical step for managing Alternaria leaf spot is to use pathogen-free propagation materials, whether cuttings or seeds. *A. zinniae* has been proven to be seedborne on zinnia. Extensive work at North Carolina State University has shown the effectiveness of treating seeds with bleach in controlling this disease (as well as another disease caused by a *Xanthomonas* sp.).

Sanitation steps include removing infected leaves to reduce inoculum and avoiding temperature stress and extended periods of leaf wetness during production. Wide temperature fluctuations during shipping and keeping plants in shipping boxes for extended periods should also be avoided.

Researchers have explored the development of Alternaria leaf spot, along with powdery mildew (*Golovinomyces cichoracearum* [syn. *Erysiphe cichoracearum*]) and bacterial leaf spot (*Xanthomonas campestris* pv. *zinniae*), on zinnia cultivars. None of the 57 cultivars tested showed season-long resistance to any of these diseases. Fungicide or bactericide efficacy was also studied with the same three diseases, and none of the products tested safely and effectively controlled all the diseases. Tests with Alternaria leaf spot alone showed the best efficacy with azoxystrobin (FRAC 11). In some trials, cupric hydroxide was severely phytotoxic.

Chlorothalonil (FRAC M3) and iprodione (FRAC 2) are registered in the United States for control of *A. alternata* on some bedding crops and have been shown effective. Mancozeb (FRAC M3) has been shown effective against *Alternaria* spp. but should not be used on French marigold to avoid phytotoxicity. (African marigold is not sensitive to mancozeb.) Many other products, including strobilurins (FRAC 11) (especially azoxystrobin) and those containing fludioxonil (FRAC 12), are equally effective on bedding plants for Alternaria leaf spot and have less-pronounced residues. In landscape plantings of marigold, fungicide applications may be necessary on a 2-week interval. Azoxystrobin proved very effective when applied at the highest labeled rates on a 2- or 3-week interval.

On some crops, including annual vinca and impatiens, work has been conducted on cultivar responses to Alternaria leaf spot. However, information on the most current cultivars is sadly lacking.

Selected References

Baker, K. F., and Davis, L. H. 1950. Some diseases of ornamental plants in California caused by species of *Alternaria* and *Stemphylium*. Plant Dis. Rep. 34:403-413.

Chase, A. R. 1994. Resistance of vinca cultivars to Alternaria leaf spot, 1993. Biol. Cult. Tests 9:166.

Cotty, P. J. 1986. *Alternaria tagetica* on marigold in New Jersey. Plant Dis. 70:1159.

Cotty, P. J., Misaghi, I. J., and Hine, R. B. 1983. Production of zinniol by *Alternaria tagetica* and its phytotoxic effect on *Tagetes erecta*. Phytopathology 73:1326-1328.

Dimock, A. W., and Osborn, J. H. 1943. An *Alternaria* disease of zinnia. Phytopathology 33:372-381.

Gombert, L., Windham, M., and Hamilton, S. 2001. Evaluation of disease resistance among 57 cultivars of zinnia. HortTechnology 11:71-74.

Hagan, A. K., and Akridge, J. R. 2004. Chemical control of Alternaria leaf spot, powdery mildew and bacterial leaf spot on zinnia. Proc. South. Nursery Assoc. Res. Conf. 51:187-190.

Hagan, A. K., Akridge, J. R., and Rivas-Davila, M. E. 2004. Reaction of marigold selections to Alternaria leaf spot. Proc. South. Nursery Assoc. Res. Conf. 49:260-263.

Hagan, A. K., Akridge, J. R., and Rivas-Davila, M. E. 2010. Application rate and treatment interval influence efficacy of Heritage fungicide for the control of Alternaria leaf spot on marigold. J. Environ. Hort. 28:81-84.

Hagan, A. K., Olive, J. W., and Stephenson, J. 2005. Alternaria leaf spot on impatiens and its control. Proc. South. Nursery Assoc. Res. Conf. 50:260-263, 265-267.

Holcomb, G. E., Owings, A., and Witcher, A. 2004. Reaction of rainbow pink cultivars to Alternaria leaf spot, 2003. Biol. Cult. Tests 19:O003.

Holcomb, G. E., Owings, A., and Witcher, A. 2005. Reaction of vinca (Madagascar periwinkle) cultivars to Alternaria leaf spot, 2004. Biol. Cult. Tests 20:O0004.

Hotchkiss, E. S., and Baxter, L. W., Jr. 1983. Pathogenicity of *Alternaria tagetica* on *Tagetes*. Plant Dis. 67:1288-1290.

(Prepared by A. R. Chase and M. L. Daughtrey)

Anthracnose Caused by *Colletotrichum gloeosporioides*

Anthracnose on bedding plants is not as common as some other foliar diseases, and it has been studied extensively on very few bedding plant crops. Dusty miller, begonia, vinca, dahlia, gerber daisy, geranium, primrose, and verbena are occasionally infected with anthracnose fungi (Fig. 27). More commonly affected is cyclamen, for which the disease can severely disfigure both leaves and petals. Severe outbreaks of anthracnose have been occasionally seen on *Cyclamen persicum* in the U.S. states of Florida, Indiana, Massachusetts, Missouri, New Jersey, New York, North Carolina, Ohio, Pennsylvania, Texas, and Virginia. On some crops, anthracnose is seen only in production.

Symptoms

Anthracnose on cyclamen causes small, round, brown leaf spots (Fig. 28), which resemble one of the symptoms caused by *Impatiens necrotic spot virus* (INSV) (see the later section "Tomato Spotted Wilt, Impatiens Necrotic Spot, and Other Tospovirus Diseases"). The leaf spots caused by anthracnose

Fig. 27. Anthracnose (*Colletotrichum gloeosporioides*) on gerber daisy leaves. (Courtesy A. R. Chase—© APS)

are round, brown, and sunken; they may be numerous and coalesce to form large patches of necrosis. Leaves may be entirely blighted, and young pedicels and unexpanded leaves beneath the canopy may be killed. Flowers may also develop brown lesions, which commonly coalesce.

Causal Organisms

Colletotrichum gloeosporioides (syn. *Glomerella cingulata*) has black, ostiolate, beaked, obpyriform to subglobose perithecia, which are at least partially immersed in plant tissue; there are hairs around the ostioles. Ascospores are hyaline, unicellular, ellipsoid to fusiform, and less than 20 µm long. They are biseriate within the unitunicate and elongate to cylindrical asci. Both periphyses and paraphyses are present within the perithecium.

C. gloeosporioides is the anamorph of the species; it produces spores in acervular conidiomata that sometimes have setae. It is very similar to *Cryptocline cyclaminis* (syn. *Gloeosporium cyclaminis*), which causes another anthracnose disease on cyclamen (see the next section, "Anthracnose of Cyclamen Caused by *Cryptocline cyclaminis*").

Since the late 1990s, another species of *Colletotrichum*, *C. theobromicola* (syn. *C. fragariae*), has been identified causing anthracnose on cyclamen in both Florida and North Carolina. Isolates from strawberry, silver date palm (*Phoenix sylvestris*), and cyclamen were cross-pathogenic in one Florida study.

Host Range and Epidemiology

C. gloeosporioides causes anthracnose symptoms on a wide range of ornamentals and may also be a saprophyte on injured tissue. It is a cosmopolitan fungal species that can be introduced into the greenhouse on numerous crops. Spore inoculum may also conceivably originate from infected native or landscape plants and be introduced through open vents by wind-driven rain or insects. The fungus is spread by splashing water, and wet leaf surfaces are conducive to development of infection. Symptoms ordinarily become apparent during the spring or summer, when greenhouse conditions are warm and plants are frequently irrigated.

The host range of *C. theobromicola* has expanded and includes cyclamen.

Fig. 28. Anthracnose (*Colletotrichum gloeosporioides*) on cyclamen leaves. (Courtesy A. R. Chase—© APS)

Management

Overhead irrigation should be used as infrequently as possible during the production cycle, and irrigating should be done early in the day to allow foliage to dry before nightfall. Subirrigation will minimize spread. Good air circulation should be maintained. Plants should be scouted for the appearance of leaf spot symptoms, particularly while they are small and closely spaced or at any time they have become crowded.

Fungicides used for management of cyclamen anthracnose in Europe include chlorothalonil (FRAC M3) and thiophanate-methyl (FRAC 1). In the United States, chlorothalonil (alone and in combination with thiophanate-methyl), myclobutanil (FRAC 3), pyraclostrobin (FRAC 11) (alone and in combination with boscalid [FRAC 7]), and copper pentahydrate (FRAC M1) have been effective.

Control of anthracnose on dusty miller has been reported from Florida trials. Optimal control was reported with chlorothalonil or azoxystrobin (FRAC 11), and poor control was reported with thiophanate-methyl or copper pentahydrate.

Selected References

Liu, B., Munster, M., Johnson, C., and Louws, F. J. 2011. First report of anthracnose caused by *Colletotrichum fragariae* on cyclamen in North Carolina. Plant Dis. 95:1480.

MacKenzie, S. J., Mertely, J. C., Seijo, T. E., and Peres, N. A. 2008. *Colletotrichum fragariae* is a pathogen on hosts other than strawberry. Plant Dis. 92:1432-1438.

McGovern, R. J., and Seijo, T. E. 2000. Evaluation of fungicides for control of Colletotrichum blight in dusty miller, 1999. Fung. Nemat. Tests 55:541.

McMillan, R. T., Jr., and Graves, W. R. 1996. Periwinkle twig blight caused by *Colletotrichum dematium* on *Catharanthus roseus*. Proc. Fla. State Hort. Soc. 109:19-20.

Norman, D. J. 1997. First report of *Colletotrichum gloeosporioides* on *Cyclamen persicum* in Florida. Plant Dis. 81:227.

(Prepared by M. L. Daughtrey, R. L. Wick, and J. L. Peterson; revised by A. R. Chase and M. L. Daughtrey)

Anthracnose of Cyclamen Caused by *Cryptocline cyclaminis*

In Europe, anthracnose caused by *Cryptocline cyclaminis* is the most serious disease of cyclamen. It is much less common in the United States, but this anthracnose can be confused with that caused by *Colletotrichum gloeosporioides* (syn. *Glomerella cingulata*) (see the previous section).

Symptoms

Anthracnose caused by *C. cyclaminis* produces round, brown spots on leaves and stunting, deformity, and necrosis of immature petioles and pedicels at the center of the plant. The vascular system may also show discoloration (Fig. 29).

Symptoms on immature leaves are easily confused with those shown on plants injured by feeding of a tarsonemid mite, the cyclamen mite (*Phytonemus pallidus*). Sometimes, the presence of waxy, pale orange–pink masses of spores helps to distinguish this disease from mite injury.

Causal Organism

Hyphae of *C. cyclaminis* (syn. *Gloeosporium cyclaminis*) are hyaline to pale brown and 3–4 µm wide. The acervuli (100–150 µm in diameter) are at first covered by the cuticle and then erupt as small, white to pale-orange pustules. Acervuli are often clustered and are sometimes confluent. The pseudoparenchymatous stroma, which is 35–60 µm thick and partially im-

Fig. 29. Vascular discoloration in cyclamen corms, symptomatic of anthracnose (*Cryptocline cyclaminis*). (Courtesy A. R. Chase—© APS)

mersed in the host tissue, is made up of isodiametric or slightly elongate, hyaline to pale-brown cells. Conidia form successively from phialides on straight to flexuous, smooth-walled conidiophores (17–28 × 3–5 µm), each of which has a small, flared collarette at the apex. Conidia are unicellular (12–16 × 4–6 µm) and hyaline, oblong to ellipsoidal, guttulate, and truncated at the base.

Host Range and Epidemiology

Cyclamen persicum is the only reported host of this pathogen. The members of the genus *Cryptocline* are facultative parasites and cause leaf spots or stem lesions only occasionally.

Disease development is favored by high levels of humidity and leaf wetness. Infection severity increases as the temperature increases from 14 to 26°C.

The sticky spores are disseminated by splashing water and possibly by insects and by workers' hands. Seed transmission may occur, as well.

Management

Minimizing leaf wetness duration and high temperatures—particularly during the initial stages of crop production, when plants are closely spaced—are both important for control. Thorough and routine scouting and removal of symptomatic plants are recommended to minimize losses, which can escape detection until late in crop production.

In Europe, *C. cyclaminis* has shown some resistance to benzimidazole fungicides (FRAC 1).

Selected References

Krebs, E.-K. 1986. Brennfleckenkrankheit ist *Cryptocline cyclaminis*. (In German.) Gb + Gw 86:1868-1869.

Krebs, E.-K. 1987. Brennfleckenkrankheit ist sicher bekämpfbar. (In German.) Gb + Gw 87:1731-1733.

Rampanini, G. 1991. La coltivazione del ciclamino: Non tutto, ma quasi. Clamer Informa. Suppl. 9. (In Italian.) Pentagono Editrice, Milan, Italy.

(Prepared by M. L. Daughtrey, R. L. Wick, and J. L. Peterson; revised by A. R. Chase and M. L. Daughtrey)

Botrytis Blight

Botrytis cinerea has a worldwide distribution and is ubiquitous in greenhouses. Botrytis blight (sometimes called "gray mold") is one of the most common diseases of greenhouse crops and is also very common on certain annuals in landscape plantings.

Symptoms

Infection by *B. cinerea* causes a range of symptoms, including spots and blight on leaf and petal tissues, as well as crown rot, stem cankers, cutting rot, and damping-off. Wounded or senescent tissue is especially susceptible to invasion (Fig. 30), but healthy tissue may also become colonized.

Lesions caused by *B. cinerea* are often identified in gardens by the characteristic gray, fuzzy sporulation (Fig. 31). However, spores develop only under humid conditions. Leaf lesions often develop a zonate pattern. Flower petals may have tiny flecks of discoloration or become completely blighted (Fig. 32). Stems may die back starting at pruning wounds or develop tan to brown cankers from the bases of petioles of blighted leaves (Fig. 33).

One of the most commonly affected bedding plants is primrose (*Primula* spp.). This crop can show symptoms anytime during propagation and production and into use in the landscape or in hanging baskets. Infected petals and flowers collapse onto the leaves, where the pathogen rapidly infects and destroys the leaf blades and finally invades the stems and crowns.

Another bedding plant commonly affected by Botrytis blight is wax begonia (*Begonia semperflorens*). On this host, the disease is most commonly manifested as large leaf spots and stem rot. Botrytis blight is almost always present when conditions favor *B. cinerea*. Infection can result in a rapid collapse of the plant.

Fig. 30. Botrytis blight (*Botrytis cinerea*) on geranium flowers. (Courtesy M. L. Daughtrey—© APS)

Fig. 31. Sporulation of *Botrytis cinerea,* the pathogen that causes Botrytis blight, on infected calibrachoa. (Courtesy A. R. Chase—© APS)

Crops of petunia (*Petunia hybrida*) are also commonly affected with Botrytis blight—anytime from their initiation as plugs through greenhouse production and finally in the landscape, where flower blight is very common. Petiole rot, leaf spot, and crown rot are quite likely during production in the greenhouse or shade house.

Geranium (*Pelargonium* spp.) is the bedding plant most commonly affected by Botrytis blight. Losses are particularly high in vegetatively propagated florist's geranium (*P.* × *hortorum*). Stock plants are grown closely together and have multiple branches and a dense canopy, which are created with growth regulators to maximize cutting productivity. The lower leaves of these stock plants senesce, resulting in a massive buildup of inoculum. Wounding of stock plants when cuttings are harvested can result in stem blight. In both production and landscape, flower blight is very common.

Causal Organism

B. cinerea (syn. *Botryotinia fuckeliana*) is a hyphomycete with straight, brown, septate conidiophores, which are generally alternately branched. Hyaline conidia are borne in botryose clusters on conidiophores and appear grayish-brown in mass. Conidia (8–14 × 6–9 µm) are ellipsoid to ovoid. *Botrytis* spp. are pleomorphic: The conidial anamorph is in the form-genus *Botrytis*, the microconidial anamorph is in *Myrioconium*, and the sclerotial anamorph is in *Sclerotium*. The teleomorph, *B. fuckeliana*, is only rarely observed in the form of apothecia on sclerotia. Colonies of *B. cinerea* on potato dextrose agar are at first off white and become gray to brown with the development of spores.

Fig. 32. Botrytis blight (*Botrytis cinerea*) on primrose petals. (Courtesy A. R. Chase—© APS)

Fig. 33. Stem canker on zinnia caused by Botrytis blight (*Botrytis cinerea*). (Courtesy A. R. Chase—© APS)

Sclerotia are black, hard, tightly appressed to the substrate, irregular in size and shape, and 1–15 mm long or confluent. Each sclerotium has a dark, pseudoparenchymatous rind of nearly isometric cells that are approximately 5–10 µm in diameter; they enclose a medulla of tightly knit, hyaline, thick-walled hyphae.

The optimal growth temperature for this fungus in culture is 24–28°C, but some growth occurs from 0 to 35°C.

Host Range and Epidemiology

B. cinerea has a very broad host range, including all the bedding plants covered in this compendium to some degree. Infection may occur directly or through natural openings or wounds by means of conidial germ tubes or by hyphal growth from colonized dead plant parts or other organic debris that contacts healthy tissue.

Conidia of *B. cinerea* are released by a hygroscopic mechanism in association with a rapid change in relative humidity and require air currents or splashing water for dispersal within the greenhouse. In production areas for geranium stock plants, peak concentrations of conidia have been observed with harvesting cuttings, spraying pesticides, cleaning plants, and even drip-tube watering. Releases of 504–1,297 conidia per cubic meter per hour have been recorded during cutting harvests of florist's geranium.

The conidia germinate in a water film that contains solutes, and a higher concentration of nutrients increases the likelihood of an aggressive infection. In one study, even after *B. cinerea* conidia were maintained under dry conditions for as long as 14 months, they retained their ability to infect flowers once a film of moisture was available. Germination and lesion development on gerber daisy (*Gerbera jamesonii*) have been observed at 4–25°C but not at 30°C.

When crops of flowering plants are grown in hanging baskets over a crop susceptible to Botrytis blight, fallen petals may serve as an energy source for the fungus and facilitate infection of healthy tissue on the crop below.

Management

Nutritional management of Botrytis blight on ornamentals is not usually practical. However, avoiding an excessive rate of nitrogen application has been shown to reduce disease incidence and severity on a number of crops. In addition, applications of calcium chloride have been shown to reduce susceptibility of some flowers (and fruit) to Botrytis infection, especially postharvest. Results with altering levels of potassium have been variable.

Cultivars of a single species may vary remarkably in their susceptibility to Botrytis blight, but major gene resistance has not been identified for any plant species. Making practical observations of the least susceptible cultivars of a particularly sensitive genus may help growers manage the disease; unfortunately, no general information is available about which cultivars are less prone to Botrytis blight.

Potential injury from fungicides (especially copper [FRAC M1]), unsightly fungicide residues, and the development of fungicide resistance make environmental management a critical first line of defense against Botrytis blight. Critical attention to managing leaf wetness duration and relative humidity (particularly the uses of ventilation and heat at sunset) will help drive moisture-laden air out of the greenhouse. Any means of improving air movement around plants—including adequate spacing, open-mesh benching, and horizontal airflow systems—should be used. Fungicide sprays are usually not needed in a greenhouse that has excellent environmental management.

Chemical control has been complicated by the appearance of widespread thiophanate-methyl (FRAC 1) resistance and a less pervasive iprodione (FRAC 2) resistance in populations of *B. cinerea* in the greenhouse flower industry in Europe and North America. In 2013, resistance to fenhexamid (FRAC 17) was reported but did not appear to spread from the northeastern United States. Fungicide-insensitive strains are exchanged among greenhouses along with the exchange of propagative materials.

Other fungicides registered in the United States for control of *Botrytis* spp. in greenhouses are effective against *B. cinerea,* including chlorothalonil (FRAC M3) (prohibited for use on open flowers because of possible damage), a combination of fludioxonil (FRAC 12) and cyprodinil (FRAC 9), and a combination of pyraclostrobin (FRAC 11) and boscalid (FRAC 7). No instances of resistance to chlorothalonil or the combination products have been reported in the ornamentals industry. The use of tank mixtures or rotations among FRAC groups is critical for resistance management. Product labels provide information about approved uses and tank mixtures on bedding plants.

Selected References

Hausbeck, M. K., and Pennypacker, S. P. 1991. Influence of grower activity and disease incidence on concentrations of airborne conidia of *Botrytis cinerea* among geranium stock plants. Plant Dis. 75:798-803.

Jarvis, W. R. 1980. Epidemiology. Pages 219-250 in: The Biology of Botrytis. J. R. Coley-Smith, K. Verhoeff, and W. R. Jarvis, eds. Academic Press, London.

Jarvis, W. R. 1992. Managing Diseases in Greenhouse Crops. American Phytopathological Society, St. Paul, MN.

Moorman, G. W., and Lease, R. J. 1992. Benzimidazole- and dicarboximide-resistant *Botrytis cinerea* from Pennsylvania greenhouses. Plant Dis. 76:477-480.

Moorman, G. W., and Lease, R. J. 1992. Residual efficacy of fungicides used in the management of *Botrytis cinerea* on greenhouse-grown geraniums. Plant Dis. 76:374-376.

Pappas, A. C. 1982. Inadequate control of grey mould on cyclamen by dicarboximide fungicides in Greece. Z. Pflanzenkrankh. Pflanzenschutz 89:52-58.

Tompkins, C. M., and Hansen, H. N. 1948. Cyclamen petal spot caused by *Botrytis cinerea,* and its control. Phytopathology 38:114-117.

Trolinger, J. C., and Strider, D. L. 1984. Botrytis blight of *Exacum affine* and its control. Phytopathology 74:1181-1188.

Trolinger, J., and Strider, D. L. 1985. Botrytis diseases. Pages 17-101 in: Diseases of Floral Crops. D. L. Strider, ed. Praeger, New York.

Vali, R. J., and Moorman, G. W. 1992. Influence of selected fungicide regimes on frequency of dicarboximide-resistant and dicarboximide-sensitive strains of *Botrytis cinerea*. Plant Dis. 76:919-924.

(Prepared by M. L. Daughtrey, R. L. Wick, and J. L. Peterson; revised by A. R. Chase and M. L. Daughtrey)

Cercospora Leaf Spot of Pansy

One of the most common diseases of bedding plants is Cercospora leaf spot on pansy (*Viola × wittrockiana*). It can be found in production and landscape beds and is sometimes found in mixed infections with anthracnose fungi.

Symptoms

Infected leaves develop pale-green to yellow spots that are 1–3 mm in diameter and sometimes have sunken centers. The spots become dark gray or purplish, and their edges have a feathery appearance (Figs. 34 and 35), distinguishing them from the round, smooth-edged lesions produced by Mycocentrospora leaf spot on pansy (see the later section "Mycocentrospora Leaf Spot of Pansy"). If left untreated, Cercospora leaf spots will continue to expand and coalesce, perhaps covering the entire leaf. Leaves often turn yellow and collapse.

Cercospora leaf spot occurs most frequently in landscape plantings during the early to middle spring. Outbreaks may also develop at any time under greenhouse or nursery production of pansies. Spores are produced when conditions are wet to very humid, and they are easily moved to nearby healthy foliage by wind currents and splashing water droplets. New leaf spots appear about 1 week after the occurrence of spore dispersal and infection.

Causal Organism

Cercospora violae is a hyphomycete within a very large genus that has been reorganized according to molecular analysis of the internal transcribed spacer (ITS) region and additional gene loci; it is phylogenetically distinct. *C. violae* forms dense clumps of 2–16 pale-brown to brown conidiophores that have 1–10 septa and are up to 175 μm long; they develop on necrotic lesions that form on *Viola* hosts. The conidiogenous cells are primarily terminal and show sympodial proliferation. The cylindrical, obclavate, or acicular conidia are borne singly; have 0–18 septa; are hyaline and obtuse at the apex; and measure 35–195 × 2.5–5.0 μm.

Host Range and Epidemiology

Cercospora leaf spot caused by *C. violae* occurs on nearly all pansy and viola (*V. cornuta*) cultivars. However, in trials in the

Fig. 34. Cercospora leaf spot (*Cercospora violae*) on pansy leaves; some spots have coalesced. (Courtesy A. R. Chase—© APS)

Fig. 35. Black, feathery lines around the edges of spots, associated with Cercospora leaf spot (*Cercospora violae*) on pansy. (Courtesy M. L. Daughtrey—© APS)

southeastern United States, some cultivars were more heavily damaged than others. The following cultivars were affected the worst by Cercospora leaf spot: Bingo Light Rose, Crown Scarlet, Crown White, Crown Yellow Splash, Crystal Bowl Scarlet, Delta Pure Rose, Delta Pure Yellow, Delta Red with Blotch, Majestic Giant Blue Shades, and Penny Blue. Cercospora leaf spot diseases on impatiens in Argentina and on bells-of-Ireland (*Moluccella laevis*) in California (United States) have been attributed to *C. apii,* a compound species. Generally, however, Cercospora leaf spot is not common on bedding plants other than viola and pansy.

Some Cercospora diseases are known to be seedborne, increasing the likelihood of their introduction to a greenhouse or a garden. Spread of the pathogen also occurs from handling plants and through splashing water from irrigation or rainfall.

Management

In greenhouse disease management trials, strobilurins (FRAC 11), certain sterol inhibitors (FRAC 3), and copper (FRAC M1) have provided very good to excellent prevention of Cercospora leaf spot on pansy. Landscape trials with fungicides have indicated that azoxystrobin (FRAC 11) is effective when used on a weekly interval. (Growers should note that product labels require rotating azoxystrobin with other fungicides that have different modes of action.) Excellent control has also been observed with chlorothalonil (FRAC M3) or thiophanate-methyl (FRAC 1) used on a 14-day interval.

Selected References

Dal Bello, G. M., and Wolcan, S. M. 2002. An outbreak of *Cercospora violae* on pansy in Argentina. Austral. Plant Pathol. 31:423-424.

Groenewald, J. Z., Nakashima, C., Nishikawa, J., Shin, H.-D., Park, J.-H., Jama, A. N., Froenewald, M., Braun, U., and Crous, P. W. 2013. Species concepts in *Cercospora:* Spotting the weeds among the roses. Stud. Mycol. 75:115-170.

Hagan, A. K., and Akridge, J. R. 1998. Reaction of cultivars of pansy to powdery mildew and Cercospora leaf spot, 1997. Biol. Cult. Tests 13:62.

Hagan, A. K., and Rivas-Davila, M. E. 2003. Reaction of pansy and viola selections to Cercospora leaf spot, 2002. Biol. Cult. Tests 18:O0003.

Holcomb, G. E., and Cox, P. 2003. Reaction of pansy, panola and viola cultivars to Cercospora leaf spot, 2002. Biol. Cult. Tests 18:O014.

Wolcan, S. M., Dal Bello, G. M., and Sisterna, M. 2006. Cercospora leaf spot on *Impatiens* spp. in Argentina. Plant Pathol. 55:581.

(Prepared by A. R. Chase and M. L. Daughtrey)

Choanephora Wet Rot

Choanephora wet rot has been reported to cause flower blight on petunia in the United States, Japan, and Nepal; on dahlia in Korea and Nepal; and on zinnia in Nepal. It has also been seen on marigold, nasturtium, and annual vinca in the U.S. Southeast and occasionally on the West Coast.

Symptoms

Plants develop symptoms within 48 h after infection; water-soaked spots appear on all aboveground portions of the plant, and complete collapse rapidly follows. Fungal sporulation is often conspicuous on rotted tissue (Fig. 36). Petunias may wilt and die without showing fungal sporulation, but strands of the fungus may be present. Flower blight may occur without symptoms developing on the rest of the plant.

Causal Organisms

The most common causal organism is *Choanephora cucurbitarum.* Both sporangiola (conidia) and sporangia are produced, and they may be found together on the host or in culture.

Fig. 36. Flower blight on Rose-of-Sharon, caused by Choanephora wet rot (*Choanephora cucurbitarum*). (Courtesy A. R. Chase—© APS)

Sporangiola are ellipsoid, pale brown to reddish-brown with distinct longitudinal striations, and about 15–20 × 9–14 µm. Sporangiospores are broadly ellipsoid, pale to dark reddish-brown, and indistinctly striate with hyaline appendages at both ends. They measure 16–34 × 7–12 µm. Brown zygospores—which are smooth, spherical, and 55–90 µm in diameter—may be observed in culture.

The closely related species *C. infundibulifera* has also been reported from the United States and Nepal.

Host Range and Epidemiology

C. cucurbitarum has a wide host range that includes trees, shrubs, and vegetables, as well as annuals such as begonia, coleus, dahlia, sunflower, nasturtium, marigold, vinca, and zinnia. The host range of *C. infundibulifera* includes petunia, dahlia, and zinnia.

Choanephora wet rot occurs during hot, humid summer weather. Resistance in cultivars of marigold and annual vinca to *C. cucurbitarum* has been reported from Louisiana. The marigold cultivars least severely affected with flower blight in a 1996 trial were Aspen Flame, Aspen Orange, Durango Red, Aspen Spry, Bonanza Bolero, and Aspen Red. In a 2004 study, the cultivars with the lowest levels of infection on annual vinca were Titan Polka Dot, Victory Deep Pink, Victory Lavender, and Victory Pure White.

Management

Preventing losses from Choanephora wet rot through the use of resistant cultivars is impractical, because cultivars change so rapidly and the level of resistance is very low. Fungicide applications using a wide range of products have failed to control this disease under commercial conditions.

Increasing plant spacing and using fans to promote air circulation in the greenhouse may help control the disease, especially during hot, wet conditions. Stress caused by extreme heat, as well as drought and injury from chemical phytotoxicity, should be avoided, because these conditions are conducive to an outbreak of disease.

Selected References

Holcomb, G. E. 2003. First report of petunia blight caused by *Choanephora cucurbitarum* in the United States. Plant Dis. 87:751.

Holcomb, G. E., and Buras, H. 1996. Reaction of marigold cultivars to Choanephora flower blight, 1995. Biol. Cult. Tests 12:65.

Holcomb, G. E., Owings, A., and Witcher, A. 2005. Reaction of marigold cultivars to Choanephora flower blight, 2004. Biol. Cult. Tests 20:O0006.

Holcomb, G. E., Owings, A., and Witcher, A. 2005. Reaction of vinca (Madagascar periwinkle) cultivars to Choanephora flower spot/blight, 2004. Biol. Cult. Tests 20:O0003.
Park, J. H., Choi, I. Y., Park, M. J., Han, K. S., and Shin, H. D. 2016. First report of Choanephora flower blight on *Dahlia pinnata* caused by *Choanephora cucurbitarum* in Korea. Plant Dis. 100:534.

(Prepared by A. R. Chase and M. L. Daughtrey)

Corynespora Leaf Spot

Only a few bedding plants are known to be hosts of *Corynespora cassiicola,* including coleus, salvia, and verbena (Fig. 37). Corynespora leaf spot causes significant losses when it occurs in both production and the landscape.

Symptoms

Corynespora leaf spot is present in propagation, production, and the landscape. It is sometimes found on unrooted cuttings of salvia, beginning as tiny, sunken lesions that may escape notice during propagation. Angular spots rapidly enlarge into black, irregularly shaped lesions that may form anywhere on leaves or stems of infected salvia plants.

Causal Organism

Isolates of *C. cassiicola* (syn. *Helminthosporium cassiicola*) from azalea were described in 1966. Conidiophores were brown to dark brown, single, erect, and unbranched. Singly borne conidia were 3–13 pseudoseptate and measured 72–206 × 13–20 µm. Each conidium had a distinct hilum with a diameter of 3.0–6.3 µm. Colonies on V8 juice agar were velvety and dark grayish-brown and generally produced conidia in 7–14 days. On potato dextrose agar, optimum growth occurred at 22–26°C.

Host Range and Epidemiology

A variety of plants are hosts for *C. cassiicola,* including foliage plants (*Aeschynanthus pulcher, Aphelandra squarrosa, Ficus benjamina,* and *Saintpaulia ionantha*), woody ornamentals (*Rhododendron obtusum, Hydrangea macrophylla*), and bedding plants (*Begonia* sp., *Catharanthus roseus* [annual vinca], *Impatiens balsamina, Plectranthus scutellarioides* [coleus], *Salvia splendens* (red salvia), *Verbena hybrida,* and *Zinnia elegans*), as well as vegetable and field crops. Most isolates show no host specialization.

Variation in the temperature preferences of *C. cassiicola* isolates has been observed; although most isolates are found in tropical and subtropical regions, some are more adapted to temperate climates. Some isolates are very host specific, while others are pathogens of multiple plant species. Isolates may be saprophytic on the tissue of one plant and pathogenic to one or more other species. There is a lot of genetic diversity within *C. cassiicola;* it has been divided into six clonal lineages.

Fig. 37. Corynespora leaf spot (*Corynespora cassiicola*) on a *Salvia* sp. (Courtesy A. R. Chase—© APS)

As with most foliar diseases, high levels of moisture and humidity are conducive to development of Corynespora leaf spot. Wounding is necessary for infection on *Aphelandra* but is not usually required on other hosts.

Conidia of the pathogen are airborne and also spread by splashing water and the movement of infested seed. *C. cassiicola* may remain viable in plant debris in the greenhouse for 2 years.

Management

In a series of tests from the early 1990s, five cultivars of red salvia were evaluated for resistance to *C. cassiicola.* The cultivar Red Hot Sally was significantly more susceptible than the Empire colors (Light Salmon, Lilac, and White), which were moderately susceptible, and the cultivar Fuego, which was only slightly susceptible.

Fungicide trials on other hosts have shown that fungicides effective on *Alternaria* spp. and other dematiaceous hyphomycetes (FRAC 2, M3, and M5) are very effective on *C. cassiicola,* as well.

Selected References

Chase, A. R. 1994. Resistance of *Salvia* cultivars to Corynespora stem rot and leaf spot, 1993. Biol. Cult. Tests 9:164.
Dixon, L. J., Schlub, R. L., Pernezny, K., and Datnoff, L. E. 2009. Host specialization and phylogenetic diversity of *Corynespora cassiicola.* Phytopathology 99:1015-1027.
Sobers, E. K. 1966. A leaf spot disease of azalea and hydrangea caused by *Corynespora cassiicola.* Phytopathology 56:455-457.

(Prepared by A. R. Chase and M. L. Daughtrey)

Fairy Ring Leaf Spot

Fairy ring leaf spot of *Dianthus barbatus* (Sweet William) was first seen in Scotland in 1936. The disease has since been widely reported worldwide on *D. caryophyllus* (carnation) and on *D. chinensis* (China pink).

Symptoms

Spots are small (1–4 mm) and round to elliptical in shape. They appear bleached, because they have white to tan centers; each spot has a black to purplish margin (Fig. 38). Lesions can

Fig. 38. Fairy ring leaf spot (*Cladosporium echinulatum*) on *Dianthus* sp. leaves. (Courtesy A. R. Chase—© APS)

form anywhere on leaves, including the tips, and lesions on the lower leaf surfaces can have a greasy appearance.

Fairy ring leaf spot is sometimes called "bird's eye leaf spot" because of the pronounced ring that may surround the center of the lesion. Olive-green, powdery spores produced on brown conidiophores may be evident at the centers of lesions.

Causal Organism

Fairy ring leaf spot is caused by *Cladosporium echinulatum* (syns. *Didymellina dianthi; Heterosporium echinulatum; Mycosphaerella dianthi*). Conidia found on lesions are large, echinulate, yellow-brown, straight or somewhat curved, and constricted at the septa; they have 1–4 septa and measure 15–65 × 10–17 µm.

The black, irregularly spherical perithecia are 100–270 µm in diameter and have stout beaks that are 10–30 µm long. Sometimes, relatively few perithecia actually contain asci. Under moist conditions, tufts of conidiophores develop on the beaks of some of the sterile perithecia and bear conidia with the morphology of *C. echinulatum*. Fertile perithecia contain 8–18 fascicled, thin-walled, hyaline, very irregularly shaped asci. Paraphyses are absent.

The apex of the ascus is as much as 10 µm thick. Most asci enclose 8 torpedo-shaped, thin-walled, colorless ascospores measuring 22–31 × 7– 9 µm; they are septate, with an upper cell measuring 1–6 µm (average 3.3 µm) and a longer, narrower basal cell with a slight constriction at the septum.

Host Range and Epidemiology

In addition to Sweet William, carnation, and China pink, fairy ring leaf spot can affect other members of the Caryophyllaceae, including *Gypsophila, Cerastium,* and *Saponaria* spp., as well as a few plants outside the family, such as salvia and iris. In infection experiments in which Sweet William was highly susceptible, China pink and carnation were very resistant, indicating that specialized races of the fungus may be present on different hosts. There is a wide range of susceptibility in bedding plant cultivars, which is why outbreaks are often seen to occur on single cultivars.

A long period of leaf wetness is needed for spores to germinate: 15–20 h at 18°C. The time between inoculation of greenhouse carnation and the appearance of symptoms was found to be 6–9 days. Low temperatures favor disease development, and a series of bright days has been observed to halt an outbreak.

Management

In Colombia, a report on management of fairy ring leaf spot on miniature carnation showed that weekly applications of two protectant fungicides, dichlofluanid (FRAC M6) and propineb (FRAC M3), were more effective than weekly applications of two systemic products, penconazole and triforine (FRAC 3). In 2004, a curative study on China pink found that the highest reduction in disease severity occurred with weekly sprays of azoxystrobin (FRAC 11) and chlorothalonil (FRAC M3). Other effective products were copper (FRAC M1), fludioxonil (FRAC 12), and iprodione (FRAC 2). Studies performed in Chile on carnation showed that sodium bicarbonate at 5 g/L and a strain of *Trichoderma virens* were more effective than mancozeb (FRAC M3), copper pentahydrate (FRAC M1), and mineral oil alone and in combination. A study in Japan concluded that tetraconazole (FRAC 3) and fluazinam (FRAC 29) were the most effective of the 14 fungicides tested. Mancozeb and tebuconazole (FRAC 3) were effective when sprayed at a 5-day interval.

Preventive measures are, as usual, more effective than curative applications. Treatments of *Dianthus* spp. for the control of rust and Alternaria leaf spot will help to manage fairy ring leaf spot, as well.

Selected References

Bensch, K., et al. 2010. Species and ecological diversity within the *Cladosporium cladosporioides* complex (*Davidiellaceae, Capnodiales*). Stud. Mycol. 67:1-94.

Burt, C. C. 1936. A leaf-spot disease of Sweet William caused by *Heterosporium echinulatum*. Trans. Brit. Mycol. Soc. 20:207-215.

Novoa, C., Casallas, J., Briceno, G., de Orozco, A., M., and de Granada, E. G. 1992. Some aspects on the biology of *Heterosporium echinulatum* (Berk) Cooke. Acta Hortic. 307:233-240.

Penaranda, L. A., and Torres, H. J. 1992. Chemical control of the fairy ring leaf spot of miniature carnation caused by *Heterosporium echinulatum*. Acta Hortic. 307:247-250.

Pierce, L., Sciaroni, R. H., and McCain, A. H. 1982. Leaf spot of sweet William [*Dianthus barbatus, Heterosporium echinulatum*]. Univ. Calif. Coop. Ext. 57:2-3.

Sandoval, C., Terreros, V., and Schiappacasse, F. 2009. Control of *Cladosporium echinulatum* in carnation using bicarbonates and *Trichoderma*. Cien. Inv. Agrar. 36(3):487-498.

(Prepared by A. R. Chase and M. L. Daughtrey)

Fusarium Root, Crown, and Stem Rots

Fusarium spp. cause root, crown, and stem rots, and some are also capable of causing systemic vascular wilt disease (see the later section "Vascular Wilts Caused by *Fusarium oxysporum*"). These fungi are widely distributed in soils and are commonly isolated from the roots and stems of many plants. *Fusarium* spp. are commonly encountered as plant pathogens, but many strains are strictly saprophytic, making diagnosis uncertain.

Symptoms

Lesions caused by *Fusarium* spp. are found on the roots and stems below the surface of the rooting medium; they are dark black and soft. As disease develops, the resulting root and crown rot may discolor, stunt the growth of, and eventually kill the plant (Fig. 39).

Crown rot of gerber daisy caused by *F. oxysporum* has been described in Holland, Poland, and the United States. Stems of infected plants darken and turn black and decay. Wilting precedes plant death. Fusarium diseases are frequently encountered on other flowering potted plants, including chry-

Fig. 39. Fusarium root rot (*Fusarium* sp.) on annual vinca. (Courtesy A. R. Chase—© APS)

santhemum. Roots become soft and exhibit a brown to black discoloration. Stems infected by *F. solani* may develop cream to pale-orange sporodochia bearing conidia and also perithecia bearing ascospores. Many cutting-propagated bedding plants suffer occasionally from diseases caused by a *Fusarium* sp. (Fig. 40).

Causal Organisms

F. solani is a complex of species that are facultative parasites. This group includes the most common *Fusarium* spp. that cause root, crown, and lower-stem rots and damping-off on plants in many families. However, other *Fusarium* spp. may cause similar diseases in bedding plants.

Members of the *F. solani* complex produce abundant cylindrical to falcate microconidia that are aseptate to 1 septate, along with larger, similarly shaped, septate macroconidia (40–100 × 5.0–7.5 µm) and chlamydospores. In culture, *F. solani* produces a whitish colony, and some blue to bluish-brown discoloration often forms in the agar.

F. oxysporum has abundant microconidia (5–12 × 2.2–3.5 µm) but sometimes few macroconidia; at times, the latter are formed in sporodochia and have 3–7 septa. When spores with 3 septa were measured, they were 27–46 × 3.0–4.5 (to 5.0) µm. The mycelium in culture is usually white to peach with some tinges of purple.

F. avenaceum also causes damping-off on various hosts and a crown rot on lisianthus (most often on the taller, cut-flower cultivars) that is sometimes referred to as a "wilt." The pathogen, which is seedborne on many hosts, produces large masses of orange, long, slender macroconidia on lisianthus stem lesions that extend up to 35 cm from the soil surface. Macroconidia have 3–7 septa at maturity, and those with 4–6 septa measure 44–70 × 4–5 µm. Microconidia and mycelial chlamydospores are absent. Colonies on agar are various colors, including peach, red, olive, and pomegranate purple but not blue and violet.

Host Range and Epidemiology

The host ranges are extensive for the *Fusarium* spp. that commonly cause root and stem rot on bedding plants, so once a pathogen has been introduced, it will affect a series of crops if sanitation efforts are not successful. The *Fusarium* spp. that cause vascular wilt are much more host specific.

In the absence of a living host, *Fusarium* spp. persist saprophytically as chlamydospores, resistant hyphae, or spores. Chlamydospores usually form in close association with infected plant tissue. Macroconidia washed into the soil can transform into chlamydospores. Chlamydospores are stimulated to germinate by exudates from plant roots. Both chlamydospores and spores germinate over a wide range of soil pH values and at an optimum temperature range of 25–28°C. *Fusarium* spp. survive longer in dry soils than in wet soils.

Spores are disseminated by insects and movement of diseased plants in the trade, as well as splashing and recirculating irrigation water and splashing rain.

Fig. 40. Wilting and chlorosis of leaves are early symptoms of Fusarium root rot (*Fusarium* sp.) on dianthus. (Courtesy A. R. Chase—© APS)

Management

Management of Fusarium diseases is best accomplished by preventing the fungi from contaminating the plant production area. *Fusarium* spp. usually enter the greenhouse via contaminated media, soiled containers, and plant materials. It is also probable that insects, such as fungus gnats and shore flies, introduce *Fusarium* spp. into growing media. Sanitation practices should be initiated at propagation time and continued through the crop cycle. Soiled hands and tools should not come in contact with soilless media.

Plant pathogens such as *Fusarium* spp. develop rapidly when introduced into soilless media and rooting cubes because of the lack of competing microorganisms. Pasteurized field soil is also subject to rapid colonization by soilborne plant pathogens, including *Fusarium* spp.

In general, cleanliness of the plant-growing area is recommended. Diseased plants and plant debris should be removed as they appear on the bench, space between plants should be increased to improve aeration, and excessive splashing during watering should be avoided. General sanitation is especially critical with *Fusarium* spp., because the production of spores that can be spread by water makes *Fusarium* spp. more easily spread than fungi that do not routinely make spores, such as *Rhizoctonia* spp.

Fungicides are rarely helpful against Fusarium diseases.

Selected References

Coleman, J. J. 2016. The *Fusarium solani* species complex: Ubiquitous pathogens of agricultural importance. Mol. Plant Pathol. 17:146-158.

Geiser, D. M., et al. 2013. One name, one fungus: Defining the genus *Fusarium* in a scientifically robust way that preserves longstanding use. Phytopathology 103:400-408.

Manning, W. J., Vardaro, P. M., and Cox, E. A. 1973. Root and stem rot of geranium cuttings caused by *Rhizoctonia* and *Fusarium*. Plant Dis. Rep. 57:177-179.

Nelson, P. E., Toussoun, T. A., and Cook, R. J., eds. 1981. *Fusarium:* Diseases, Biology, and Taxonomy. Pennsylvania State University Press, University Park.

Nelson, P. E., Toussoun, T. A., and Marasas, W. F. O. 1983. *Fusarium* Species: An Illustrated Manual for Identification. Pennsylvania State University Press, University Park.

(Prepared by M. L. Daughtrey, R. L. Wick, and J. L. Peterson; revised by A. R. Chase and M. L. Daughtrey)

Fusarium Crown and Stem Rot of Lisianthus

Fusarium avenaceum is globally distributed; it occurs primarily in temperate climates but is found occasionally in humid tropical climates. *F. avenaceum* causes crown rot on a variety of plants, including cauliflower and fennel; it was found to cause crop losses on lisianthus (*Eustoma grandiflorum*) in the 1990s (Fig. 41). Fusarium crown and stem rot of lisianthus is a destructive disease, especially in California and Florida (United States). It has been reported from Italy, as well.

Symptoms

F. avenaceum causes a crown and stem rot on lisianthus that is sometimes referred to as a "wilt." Infected plants initially develop a duller-green coloration than normal and appear to be water stressed; the lower leaves and stems turn light tan (Fig. 42). Stem cankers extend 10–15 cm (even 35 cm) from the soil surface, where they originate; they are sometimes decorated with white mycelium and often bear masses of orange-colored *Fusarium* spores. Infected plants ultimately die.

Causal Organism

F. avenaceum (syn. *Gibberella avenacea*) produces no microconidia, and no chlamydospores form in the mycelium. However, *F. avenaceum* forms abundant, slender macroconidia that are slightly curved; each has a notched basal cell and an elongated apical cell that may be bent over. Macroconidia have 3–7 septa and sometimes form within sporodochia; those with 4–6 septa measure 44–70 × 4–5 µm.

The teleomorph, *G. avenacea,* has variably colored perithecia, which may be black, gray-green, or hyaline; it has not been described on lisianthus. Colony morphology is variable; peach pigmentation is common and blue is absent.

Host Range and Epidemiology

F. avenaceum is commonly isolated from soil and from a wide range of plants, including carnation, chrysanthemum, and lisianthus, as well as some cereals. Species from more than 150 genera are hosts.

Pathogenicity tests of *F. avenaceum* isolates obtained from several hosts other than lisianthus were able to cause disease on lisianthus, suggesting that *F. avenaceum* may be pathogenic on lisianthus regardless of its host plant origin. These findings have management implications and suggest that any of the many hosts that support *F. avenaceum* may serve as sources of inoculum for lisianthus crops.

Fig. 41. Fusarium stem rot (*Fusarium avenaceum*) on lisianthus. (Courtesy A. R. Chase—© APS)

Fig. 42. Light-tan lower leaves and stems of lisianthus, characteristic of Fusarium stem rot (*Fusarium avenaceum*). (Courtesy A. R. Chase—© APS)

Lisianthus cultivar sensitivity to *F. avenaceum* was reported in 2000. Forty-six cultivars were grouped according to flower color (blue/purple, pink, and white). Symptoms appeared in almost half of the cultivars within 55 days of inoculation. The most disease-resistant cultivars were Ventura Deep Blue, Hallelujah Purple, Bridal Pink, and Heidi Pure White. Interestingly, these cultivars were developed by four different genetics companies. The researchers concluded that all the cultivars were susceptible to *F. avenaceum.*

F. avenaceum is known to be seedborne on a number of hosts. Because sporulation occurs on stem cankers, this pathogen is not limited to transmission by movement of contaminated soil or seed; spores can be moved aerially by wind and may also be splashed to new locations by rainfall and irrigation.

Vectoring of the fungus is another option. Adult shore flies, fungus gnats, and moth flies—all common greenhouse insects—have been shown to efficiently vector *F. avenaceum* to lisianthus. All three insects may be surface contaminated with spores; shore flies also ingest spores and excrete viable macrospores in their frass. Symptoms began to appear within 25 days following a spore drench application to the growing medium of healthy lisianthus, while symptoms appeared beginning 40 days after contaminated insects were introduced.

Management

Fungicide trials on *F. avenaceum* showed that the best results occurred with fludioxonil (FRAC 12) applied as a drench once prior to inoculation; strobilurins (FRAC 11) were not effective. In a separate trial, products were applied as drenches once before inoculation and once 3 weeks after inoculation; optimal control was achieved with triflumizole (FRAC 3) and fludioxonil. Thiophanate-methyl (FRAC 1), myclobutanil (FRAC 3), and azoxystrobin (FRAC 11) provided significant but less effective control. Further trials on *F. avenaceum* showed a very high level of control with azoxystrobin and trifloxystrobin (FRAC 11), while fludioxonil proved the best treatment (i.e., 100% effective in preventing crown rot from developing). Growers should always check fungicide labels for approved uses, including application method, rate, and interval.

Selected References

Booth, C., and Waterston, J. M. 1964. *Fusarium avenaceum.* CMI Descriptions of Pathogenic Fungi and Bacteria, No. 25. CABI, Kew, Surrey, UK.

El-Hamalawi, Z. A., and Stanghellini, M. E. 2005. Disease development on lisianthus following aerial transmission of *Fusarium avenaceum* by adult shore flies, fungus gnats, and moth flies. Plant Dis. 89:619-623.

Harbaugh, B. K., and McGovern, R. J. 2000. Susceptibility of forty-six *Lisianthus* cultivars to Fusarium crown and stem rot. HortTechnology 10(4):816-819.

Koike, S. T., Gordon, T. R., and Lindow, S. E. 1996. Crown rot of *Eustoma* caused by *Fusarium avenaceum* in California. Plant Dis. 80:1429.

McGovern, R. J., Seijo, T. E., and Harbaugh, B. K. 2002. Evaluation of fungicides for reduction of Fusarium crown and stem rot in lisianthus, 2001. Fung. Nemat. Tests 57:OT16.

McGovern, R. J., Seijo, T. E., Myers, D. S., and Harbaugh, B. K. 2002. Evaluation of fungicides and biocontrols for reduction of Fusarium wilt in lisianthus, 2001. Fung. Nemat. Tests 57:OT17.

McGovern, R. J., Seijo, T. E., Myers, D. S., and Harbaugh, B. K. 2002. Evaluation of fungicides for reduction of Fusarium wilt in lisianthus, 2001. Fung. Nemat. Tests 57:OT18.

Pecchia, S. 1999. Lisianthus crown and stem rot caused by *Fusarium avenaceum* in central Italy. Plant Dis. 83:304.

(Prepared by A. R. Chase and M. L. Daughtrey)

Vascular Wilts Caused by *Fusarium oxysporum*

Host-specialized strains (*formae speciales*) of *Fusarium oxysporum* cause vascular wilt diseases on *Cyclamen persicum*, *Dahlia* hybrids, lisianthus (*Eustoma grandiflorum*), gerber daisy (*Gerbera jamesonii*), and *Ranunculus asiaticus*. Other strains of *F. oxysporum* that are not host specific will infect the roots and stems of these and many other bedding plant hosts.

Symptoms

Plants infected with wilt-inducing *F. oxysporum* typically show leaf chlorosis and other symptoms of water deficiency. Leaf scorch is common and more severe on older foliage. Highly susceptible cultivars are severely stunted or killed, while others may show moderate symptoms. Symptoms often appear throughout a crop after a period of high temperatures.

Causal Organism

F. oxysporum, which is considered a species complex, usually forms macroconidia, microconidia, and chlamydospores in culture. Macroconidia, which are slightly sickle shaped, are not always numerous; they have 3–5 septa and are initially produced on branched, lateral phialides and later often within sporodochia. The basal cell of each macroconidium is foot shaped, and the apical cell is attenuated; a macroconidium with 3 septa measures 27–46 × 3.0–4.5 (to 5.0) µm. Microconidia, which are oval or kidney shaped, are produced in false heads and are usually one celled; a single microconidium measures 5–12 × 2.2–3.5 µm. Monophialides bearing microconidia are shorter than those of *F. solani*. *F. oxysporum* also forms chlamydospores (intercalary or terminal), either single or paired.

Colonies of *F. oxysporum* on potato dextrose agar are typically white with fibrous margins, but appearance varies. A purplish tint forms in most isolates, and it is most visible from the underside of the colony. Colonies of *F. subglutinans* have a similar appearance, but the chlamydospores and the monophialidic conidiophores in *F. oxysporum* help to distinguish the species. Sporodochia are cream, tan, or orange. Mutation in culture may lead to forms with less sporulation and a greater amount of aerial mycelium or to forms with a wet appearance and yellow discoloration caused by production of macroconidia in pionnotes.

The use of PCR testing to supplement morphological observations can be extremely helpful for identifying *Fusarium* spp.

Host Range and Epidemiology

Formae speciales of *F. oxysporum* have limited host ranges—often, restricted to one plant species. For example, *F. oxysporum* f. sp. *eustomae* is known to be pathogenic only to lisianthus.

Fusarium vascular wilt generally starts with invasion of the root in the zone of elongation. The fungus then moves into xylem vessels and releases conidia, which travel ahead of the hyphae within the plant. The fungus eventually grows out through the cortex to the plant surface, where sporulation occurs. Spores on the plant surface can be moved by handling, insects, and splashing irrigation water, resulting in the spread of inoculum. A low incidence of disease in the early phases of production can thus become widespread in the mature crop.

High greenhouse temperatures provide optimal conditions for development of Fusarium wilt. Fusarium wilt of gerber daisy is most often seen during the midsummer, when temperatures are 23–25°C. Dramatic summer symptom development is also characteristic of Fusarium wilt of cyclamen.

Fusarium wilt is influenced by nutritional factors. A high level of ammonium nitrogen is conducive to disease, particularly when the level of potassium is low. Using nitrate nitrogen exclusively can suppress disease but also decreases plant height and flowering. Insufficient calcium favors development of Fusarium wilt; thus, other competitive cations (e.g., sodium) may increase the incidence of wilt, because they have the effect of reducing calcium uptake.

Fusarium wilt is also affected by trace elements. Symptoms increase with boron deficiency. It is possible to control Fusarium wilt in some cases by adjusting the soil pH to 6.5–7.5 with limestone; doing so reduces the availability of certain trace elements. *F. oxysporum* is affected by the deficiency of micronutrients in the soil, although the host plant can extract nutrients from the rhizosphere. If an excess amount of iron, manganese, zinc, or ammonium nitrogen is added to the soil, the benefit of adding limestone will be counteracted.

Management

Using nitrate nitrogen fertilizer and raising the soil pH with lime have been effective for management of *F. oxysporum* on *Chrysanthemum morifolium* and China aster (*Callistephus chinensis*). Although drenches with benzimidazole fungicides (FRAC 1) provide some disease suppression, fungicides are generally not sufficiently effective against Fusarium wilt to warrant their use.

Selected References

Bertoldo, C., Gilardi, G., Spadaro, D., Gullino, M. L., and Garibaldi, A. 2015. Genetic diversity and virulence of Italian strains of *Fusarium oxysporum* isolated from *Eustoma grandiflorum*. Eur. J. Plant Path. 141:83-97.

Booth, C. 1970. *Fusarium oxysporum*. CMI Descriptions of Pathogenic Fungi and Bacteria, No. 211. CABI, Kew, Surrey, UK.

Engelhard, A. W. 1975. Aster Fusarium wilt: Complete symptom control with an integrated fungicide-NO_3-pH control system. (Abstr.) Proc. Am. Phytopathol. Soc. 2:62.

Engelhard, A. W., and Woltz, S. S. 1973. Fusarium wilt of chrysanthemum: Complete control of symptoms with an integrated fungicide-lime-nitrate regime. Phytopathology 63:1256-1259.

Fantino, M. G., Pasini, C., and Contarini, M. R. 1985. Lotta contro la fusariosi vascolare del ranunculo. (In Italian.) Colt. Prot. 8/9:63-66.

Garibaldi, A., Gullino, M. L., and Rapetti, S. 1983. Tracheofusariosi: Una nuova malattia del ranunculo. (In Italian.) Colt. Prot. 12:23-24.

Garibaldi, A., Minuto, A., Shiniti Uchimura, M., and Gullino, M. L. 2008. Fusarium wilt of Gerbera caused by a *Fusarium* sp. in Brazil. Plant Dis. 92:655.

Gordon, W. L. 1965. Pathogenic strains of *Fusarium oxysporum*. Can. J. Bot. 43:1309-1318.

Gullino, M. L., Katan, J., and Garibaldi, A., eds. 2012. Fusarium Wilts of Greenhouse Vegetable and Ornamental Crops. American Phytopathological Society, St. Paul, MN.

(Prepared by M. L. Daughtrey, R. L. Wick, and J. L. Peterson; revised by A. R. Chase and M. L. Daughtrey)

Fusarium Wilt of Cyclamen

Fusarium wilt is an important and widespread vascular wilt disease of *Cyclamen persicum*. The disease was initially reported from Germany in 1930 and then from the United States in 1949 and the Netherlands in 1977. It is now common throughout the United States and Europe.

Symptoms

Symptoms may appear in cyclamen anytime during production. When the roots of young seedlings have rotted, the leaves may yellow and wilt before there is any indication of vascular discoloration in the corms. However, growers are often distressed to observe chlorosis and wilting for the first time during the final stages of crop production. When temperature conditions favor disease expression, an infected cyclamen plant develops yellow patches and wilts, leaf by leaf, progressing until the entire plant has been destroyed (Fig. 43). This progression of symptoms is characteristic of Fusarium wilt of cyclamen and helps to differentiate it from cyclamen stunt, caused by *Ramularia cyclaminicola*.

Cutting across the corms of cyclamen with Fusarium wilt discloses purple, reddish-brown, or nearly black discoloration of the xylem (Fig. 44). The corm remains firm unless it has simultaneously been invaded by soft rot bacteria. Root rot may also occur on mature plants. The integrity of the corm combined with the vascular discoloration clearly distinguish Fusarium wilt from bacterial soft rot, caused by *Pectobacterium* and *Dickeya* spp.

A "frosty" coating of sporulation may be seen on the bases of infected petioles and pedicels.

Causal Organism

Fusarium wilt of cyclamen is caused by *Fusarium oxysporum* f. sp. *cyclaminis*. *F. oxysporum*, which is considered a species complex, usually forms macroconidia, microconidia, and chlamydospores in culture. Macroconidia, which are slightly sickle shaped, are not always numerous; they have 3–5 septa and are initially produced on branched, lateral phialides and later often within sporodochia. The basal cell of each macroconidium is foot shaped, and the apical cell is attenuated; a macroconidium with 3 septa measures 27–46 × 3.0–4.5 (to 5.0) µm. Microconidia, which are oval or kidney shaped, are produced in false heads and are usually one celled; a single microconidium measures 5–12 × 2.2–3.5 µm. Monophialides bearing microconidia are shorter than those of *F. solani*. *F. oxysporum* also forms chlamydospores (intercalary or terminal), either single or paired.

Fig. 43. Advanced wilt and collapse of a cyclamen plant, caused by Fusarium wilt (*Fusarium oxysporum* f. sp. *cyclaminis*). (Courtesy A. R. Chase—© APS)

Fig. 44. Cross section of a cyclamen corm, showing vascular discoloration caused by Fusarium wilt (*Fusarium oxysporum* f. sp. *cyclaminis*). (Courtesy A. R. Chase—© APS)

Colonies of *F. oxysporum* on potato dextrose agar are typically white with fibrous margins, but appearance varies. A purplish tint forms in most isolates, and it is most visible from the underside of the colony. Colonies of *F. subglutinans* have a similar appearance, but the chlamydospores and the monophialidic conidiophores in *F. oxysporum* help to distinguish the species. Sporodochia are cream, tan, or orange. Mutation in culture may lead to forms with less sporulation and a greater amount of aerial mycelium or to forms with a wet appearance and yellow discoloration caused by production of macroconidia in pionnotes.

The use of PCR testing to supplement morphological observations can be extremely helpful for identifying *Fusarium* spp.

Host Range and Epidemiology

F. oxysporum f. sp. *cyclaminis* is host specific and thus affects only cyclamen. Seed transmission has not been conclusively demonstrated, but the pathogen has been isolated from organic debris in seed packets. The fungus may be introduced to the greenhouse with seed or in infected plug seedlings used for transplanting. Plants with latent infections can easily be moved from the propagator to the finished-plant producer and finally into the landscape.

F. oxysporum f. sp. *cyclaminis* may grow over a wide temperature range of 6–35°C, but 28°C is optimal. Crop losses occur primarily during warm times of the year. The fungus also grows well at 18–20°C, which is ideal for cyclamen production. Disease development is influenced by inoculum level as well as temperature. One study found that at 23°C, 25% of the cyclamen grown in *Fusarium*-contaminated soil developed symptoms 2 weeks after transplanting, and 100% of cyclamen were symptomatic after 4 weeks. At 17°C, symptom development was delayed until 6 weeks after transplanting, and at 15°C, infection was not apparent until after 7 weeks. Symptoms are delayed by a low inoculum level, but in a trial using inoculum of only 10 spores per 1 ml, 60% of inoculated plants showed symptoms at 17 weeks.

In experiments involving ebb-and-flood irrigation, spread of the pathogen from diseased to healthy cyclamen was not observed. Only adding very high levels of inoculum to the recirculating nutrient solution resulted in disease development in studies of cyclamen. One investigation determined that infection occurred only when the irrigation solution held a concentration of 10^5 *Fusarium* spores per 100 ml; roguing out diseased plants when they were first detected kept the spore concentration below this threshold.

Management

No cyclamen cultivars are resistant to Fusarium wilt, and chemical controls are not very effective. Trials performed in California since the early 2000s have consistently shown optimal control with preventive treatments of strobilurins (FRAC 11), sterol inhibitors (FRAC 3), and the phenylpyrrole fludioxonil (FRAC 12). Thiophanate-methyl (FRAC 1) has provided less control, and very little reproducible control has been achieved with phosphonates (FRAC 33).

Integration of chemical fungicides with other types of products has also been extensively tested. Using biocontrol agents after applying fungicides in various combinations was found to result in significant decreases in disease development over time. Fludioxonil paired with various biologicals (i.e., fungal, actinomycete, and bacterial agents) yielded the best results. Applying acibenzolar-S-methyl (FRAC P) (not currently labeled on ornamentals in the United States) on cyclamen wilt alone or in combination with a variety of fungicides has also been tested. Acibenzolar-S-methyl significantly reduced the disease when used alone, although disease symptoms were not eliminated at any rate tested. Moreover, treatments with this chemical in other trials have caused significant stunting of cyclamen.

Treatments with bioantagonists and biologically active growing media have shown some promise for disease suppression. Using composted hardwood and pine bark media has been proven beneficial for Fusarium wilt control in cyclamen. This effect may be attributable partly to chemical effects as the bark decomposes and partly to the microflora and their biocontrol properties.

Using dolomitic limestone to increase the pH of the growing medium above 6 inhibits Fusarium disease. Disease is suppressed by nitrate nitrogen in the fertilizer, while sources of ammonium nitrogen enhance disease development.

Sanitation is considered the key method for managing Fusarium wilt of cyclamen. Crops should be monitored closely, and symptomatic plants should be promptly removed from the greenhouse bench. Disease spread via splashing irrigation water, recirculating irrigation solution, and insect activity can all be reduced by regular monitoring and roguing of diseased plants. It is imperative to keep the inoculum concentration very low within the nutrient solution in an ebb-and-flood system, and doing so requires plant removal. Contaminated soil must be steam pasteurized or fumigated. Control of fungus gnats and other greenhouse insects may be critical to halt movement of the pathogen from a resident saprophytic soil population.

Selected References

Elmer, W., and McGovern, R. J. 2004. Efficacy of integrating biologicals with fungicides for the suppression of Fusarium wilt of cyclamen. Crop Prot. 23:909-914.

Elmer, W., and McGovern, R. J. 2006. Effects of acibenzolar-S-methyl in the management of Fusarium wilt of cyclamen. Crop Prot. 25:671-676.

Garibaldi, A. 1988. Research on substrates suppressive to *Fusarium oxysporum* and *Rhizoctonia solani*. Acta Hortic. 221:271-277.

Grouet, D. 1985. Vascular Fusarium disease of cyclamen. Phytoma 372:49-51.

Gullino, M. L., Minuto, A., Gilardi, G., and Garibaldi, A. 2002. Efficacy of azoxystrobin and other strobilurins against Fusarium wilts of carnation, cyclamen, and Paris daisy. Crop Prot. 21:57-61.

Rattink, H. 1986. Some aspects of the etiology and epidemiology of Fusarium wilt on cyclamen. Med. Fac. Landbouwwet. Rijksuniv. Gent. 51:617-624.

Rattink, H. 1990. Epidemiology of Fusarium wilt in cyclamen in an ebb and flow system. Neth. J. Plant Pathol. 96:171-177.

Tompkins, C. M., and Snyder, W. C. 1972. Cyclamen wilt in California and its control. Plant Dis. Rep. 56:493-497.

(Prepared by M. L. Daughtrey, R. L. Wick, and J. L. Peterson; revised by A. R. Chase and M. L. Daughtrey)

Itersonilia Petal Blight of China Aster

Itersonilia perplexans has been detected in Australia, Austria, Canada, Great Britain, Greece, Japan, New Zealand, the Netherlands, Portugal, the United States, and Uruguay. It causes flower blight in anemone, China aster, dahlia, florist's chrysanthemum, and globe artichoke, and in sunflower, it causes various other symptoms, including seedling blight, leaf spots and necrosis, and root cankers.

Symptoms

On China aster, petal spots start as pinpoint-sized, reddish to brown specks; they can enlarge to encompass the entire petal, commonly making the petal tan and streaked. In chrysanthemum, only the flowers are affected, but in China aster, damping-off is likely, as well.

Causal Organism

The basidiomycete *I. perplexans* has both yeast and hyphal forms in culture. Hyphal isolates are hyaline and form clamp connections or pseudoclamps, as well as clustered or solitary subglobose, ovoid or pear-shaped, inflated sporogenous cells; they measure 11–14 × 12–16 µm and may germinate directly to produce hyphae. Alternatively, sporogenous cells may form germ tubes and produce broadly lunate ballistospores, which are ovoid to pyriform and bilaterally symmetrical; they measure 10–16 × 6.0–10.5 µm. Ballistospores germinate with germ tubes, secondary ballistospores, or appressoria. Chlamydospores measuring 13–20 × 10.0–13.5 µm are also produced by isolates from some hosts. In certain media, hyphal strains may form yeast colonies; they are flat and light cream to yellow-brown and have a slightly unpleasant odor.

Host Range and Epidemiology

Outbreaks of Itersonilia petal blight occurred at a cut-flower production site in southwestern Florida (United States) in the late 1990s, resulting in extensive postharvest losses in China aster and sunflower. The causal fungus is also known to cause flower blight in dahlia and chrysanthemum, and it has weed and vegetable crop hosts. Isolates have variable host ranges. For example, the Florida *I. perplexans* isolates from China aster and sunflower were pathogenic to one another's hosts and to chrysanthemum and gerber daisy, causing significant flower or seedling blight, but neither was pathogenic to edible burdock. An isolate that caused a high incidence of seedling blight in edible burdock caused flower blight in chrysanthemum and gerber daisy at low and high incidences, respectively, but did not produce petal blight in China aster or seedling blight in sunflower. Isolates from hosts in the Apiaceae have been shown not to infect plants in the Asteraceae.

I. perplexans may persist as a saprophyte, but weed hosts may also be sources for disease outbreaks by providing inoculum reservoirs. The pathogenicities of two *I. perplexans* isolates recovered from the weed *Emilia fosbergii* were tested on detached flowers of China aster along with aster and sunflower isolates. All four isolates produced severe petal blight in China aster.

Sporulation and infection of *I. perplexans* are favored by rainfall, high relative humidity (>70%), and cool temperatures (10–15°C); disease occurs over the range of 1–21°C. Ballistospores are forcibly ejected and also windborne. In addition, low temperatures have been observed not to halt growth or sporulation; infected inflorescences of chrysanthemum were found after 31 days of frost.

Management

Foliage should be kept as dry as possible, and plants should be adequately spaced to allow rapid leaf drying. Removing infected crop debris is also recommended.

An evaluation of fungicides on China aster flowers compared the effects of potassium bicarbonate (FRAC NC) and azoxystrobin (FRAC 11) with the effects of myclobutanil (FRAC 3) and propiconazole (both FRAC 3). All the fungicides significantly reduced the progress of petal blight, although azoxystrobin was less effective than potassium bicarbonate, myclobutanil, and propiconazole in both trials.

Selected References

Boekhout, T. 1991. Systematics of *Itersonilia:* A comparative phenetic study. Mycol. Res. 95:135-146.
Dosdall, L. T. 1956. Petal blight of chrysanthemums incited by *Itersonilia perplexans*. Phytopathology 46:231-232.
Gandy, D. G. 1966. *Itersonilia perplexans* on chrysanthemums: Alternate hosts and ways of overwintering. Trans. Brit. Mycol. Soc. 49:499-507.
Horst, R. K., and Nelson, P. E., eds. 1997. Compendium of Chrysanthemum Diseases. American Phytopathological Society, St. Paul, MN.
McGovern, R. J., and Seijo, T. E. 1999. Petal blight of *Callistephus chinensis* caused by *Itersonilia perplexans*. Plant Dis. 83:397.
McGovern, R. J., Horita, H., Stiles, C. M., and Seijo, T. E. 2006. Host range of *Itersonilia perplexans* and management of Itersonilia petal spot of China aster. Online. Plant Health Progress. doi:10.1094/PHP-2006-1018-02-RS
McRitchie, J. J., Kimbrough, J. W., and Engelhard, A. W. 1973. Itersonilia petal blight of chrysanthemum in Florida. Plant Dis. Rep. 57:181-182.
Seijo, T. E., McGovern, R. J., and Marenco de Blandino, A. 2000. Petal blight of sunflower caused by *Itersonilia perplexans*. Plant Dis. 84:1153.

(Prepared by A. R. Chase and M. L. Daughtrey)

Mycocentrospora Leaf Spot of Pansy

The first description of Mycocentrospora leaf spot of pansy (formerly known as "Centrospora leaf spot") came from Switzerland in 1921. The disease was first reported in the United States in 1928 from the San Francisco Bay area of California, and by 1939–1940, it had reached epidemic proportions in the northern part of the state. The disease continues to occur sporadically in bedding plant production and garden plantings. The pathogen, *Mycocentrospora acerina,* has been found in North America and Europe, as well as Australia and New Zealand.

Symptoms

Spots are small and round to elliptical; they have blue-black margins and tan to nearly white centers (Figs. 45 and 46). Spots can appear anywhere on leaves, petioles, and even flowers; on lower leaves, they can have a greasy appearance. The spots rapidly enlarge to reach up to 5 mm wide. Sometimes, they have concentric rings, giving them a "target" appearance that is similar to the characteristic symptom of anthracnose, which is caused by several *Colletotrichum* species. Chlorotic halos may develop around the spots on older leaves.

With a susceptible pansy cultivar and optimal conditions for disease development, plants may be killed within 1 week.

Causal Organism

M. acerina (syns. *Centrospora acerina; Cercospora acerina*) is the pathogen that causes Mycocentrospora leaf spot of pansy. Colonies on agar appear hyaline initially and become bluish-green, brown, or red. Hyaline, septate conidiophores are as big as 65 × 4–7 µm. Conidia are hyaline or pale olivaceous. A single conidium is 150–200 (min. 60; max. 290) × 8–15 µm; it tapers sharply to 1–2 µm at the apex and 4–5 µm at the base. Each conidium has 8–11 (min. 4; max. 24) septa and a truncated base, to which is sometimes attached a downward-directed, tapered appendage that is septate and 30–150 × 2–3 µm.

The minimum and maximum temperatures for growth in culture are 0 and 31°C, respectively; the optimal temperature for growth is 17°C. Conidia are short lived, but chlamydospores allow inoculum to persist in the soil.

Host Range and Epidemiology

M. acerina has a very wide host range, including carrot, celery, parsley, and parsnip. Viola and pansy are the only significant natural hosts among bedding plant crops, but a number of ornamentals have been shown to be hosts under experimental greenhouse conditions.

The most severe disease occurs when temperatures are 15–20°C and rainfall or overhead irrigation is abundant. Conidia rapidly lose viability when exposed to dry air. Rain or irrigation stimulates spore production and splashes spores from plant to plant. Application of high-nitrogen fertilizers was found to increase disease severity in caraway.

Mycocentrospora leaf spot can become severe on pansy in as little as 12 h following inoculation under ideal conditions. Although studies have not been conducted of disease spread on pansy (or any other bedding plant), it likely mirrors spread of the same pathogen on carrot and dill. In those crops, the pathogen is seedborne and can also be soilborne.

Fig. 45. Early symptoms of Mycocentrospora leaf spot (*Mycocentrospora acerina*) on pansy leaves. (Courtesy A. R. Chase—© APS)

Fig. 46. Advanced symptoms of Mycocentrospora leaf spot (*Mycocentrospora acerina*) on pansy leaves. (Courtesy A. R. Chase—© APS)

The relatively infrequent appearance of the pathogen in the U.S. pansy industry indicates that seed infestation is at least uncommon. However, the pathogen has been found in pansy seed from Alaska, California, Georgia, North Carolina, and Washington.

Management

Increased plant spacing is recommended to reduce disease severity. Reducing the length of time that foliage remains wet will slow disease progression. Using reliably clean seed sources will make it very unlikely for the disease to be encountered in greenhouse production.

Preventive treatments with fludioxonil (FRAC 12), chlorothalonil (FRAC M5), and azoxystrobin (FRAC 11) have been shown effective at reducing Mycocentrospora leaf spot. Iprodione (FRAC 2) provided better control than these chemicals, but the best control was seen with trifloxystrobin and pyraclostrobin (both FRAC 11).

Selected References

Neergaard, P., and Newhall, A. G. 1951. Notes on the physiology of *Centrospora acerina* (Hartig) Newhall. Phytopathology 41:1021-1033.

O'Neil, T. M. 1985. *Mycocentrospora acerina* causing leaf spot and dieback of anemone seedlings. Plant Pathol. 34:632-635.

Sutton, B. C., and Gibson, I. A. S. 1977. *Mycocentrospora acerina.* CMI Descriptions of Pathogenic Fungi and Bacteria, No. 537. CABI, Kew, Surrey, UK.

Tompkins, C. M., and Hansen, H. N. 1950. Pansy leafspot, caused by *Centrospora acerina,* host range, and control. Hilgardia 19(12):383-397.

(Prepared by A. R. Chase and M. L. Daughtrey)

Myrothecium Diseases

Diseases caused by *Paramyrothecium roridum* (previously *Myrothecium roridum*) have been described on a wide variety of agricultural plants, including ornamental, vegetable, and agronomic crops. A petiole rot disease was reported many years ago on pansy, but it has been fairly uncommon since the 1940s. Myrothecium leaf spot also occurs on some bedding plants propagated from cuttings, including salvia and New Guinea impatiens. On these crops, the leaf spots are closely associated with wounding and typically found on the cuttings upon opening the shipping container.

Symptoms

Lesions generally develop at the edges and tips of leaves and at broken leaf veins on New Guinea impatiens (Fig. 47). Spots are dark brown to black and initially appear water soaked (Fig. 48). Examination of the lower leaf surface generally reveals signs of the pathogen: black sporodochia encircled by a white fringe of mycelium. Sporodochia form in concentric rings within the necrotic area on the lower leaf surface.

In contrast, on pansy, a petiole or stem rot develops, and on snapdragon, a stem rot is also described. Crown rot on pansy starts with development of light-brown, water-soaked stem tissue, which is followed by plant collapse at the soil line. Sudden wilting of the whole plant occurs if the stem becomes girdled.

Similar symptoms may result from infection by *Phytophthora nicotianae* (see the later section "Diseases Caused by *Phytophthora nicotianae*").

Causal Organisms

P. roridum (syn. *Myrothecium roridum*) was first described in 1790. Sporodochia are sessile, discoid, circular to irregularly shaped, and 0.1–2.0 mm in diameter. They are initially dark green and become black with white rims (Fig. 49); they are without setae. Conidiophores are erect, branched once or twice, and septate; each branch terminates in phialides. Phialides are slenderly clavate, straight, and hyaline; they are usually arranged in whorls of 3–7. Conidia are cylindrical or slightly tapered at the ends and guttulate; they are at first hyaline and

Fig. 47. Development of a Myrothecium leaf spot (*Paramyrothecium roridum*) at a wound on a New Guinea impatiens leaf. (Courtesy A. R. Chase—© APS)

Fig. 48. Leaf spots on New Guinea impatiens leaves caused by infection with *Paramyrothecium roridum.* (Courtesy A. R. Chase—© APS)

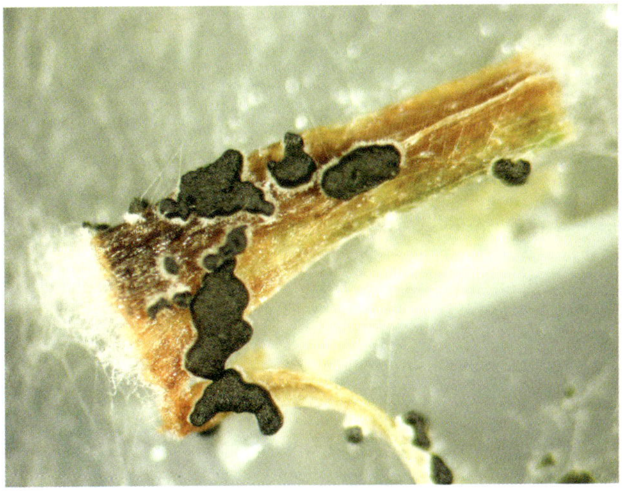

Fig. 49. *Paramyrothecium roridum* sporodochia on a pansy petiole. (Courtesy M. L. Daughtrey—© APS)

become pale green. A single conidium measures 5.0–11.5 × 1.5–2.5 µm. The conidial mass is initially dark green; it becomes black and viscid, finally drying to a dull black.

P. roridum grows as a white colony (pink on the reverse side) and sporulates well on many standard culture media, including V8 juice agar and potato dextrose agar. The first evidence of sporulation usually occurs within 10 days of initiating a culture, although some isolates lose their ability to produce abundant conidia in vitro. Sporodochia form in concentric rings in culture, much as they do on infected plant tissue.

The closely related species *M. verrucaria* is the cause of leaf spot on zinnia in India. It has relatively flat sporodochia and ovate spores.

Host Range and Epidemiology

The host range of *P. roridum* is very broad and includes plants from many families. Pansy and New Guinea impatiens are the most frequently affected bedding plant hosts. No evidence of host specificity has been found, although a wide range in virulence of isolates and host susceptibility occurs naturally. For resistant and slightly susceptible plants, wounding of tissue is necessary for infection; highly susceptible plants become infected easily without obvious wounding.

Avoiding wounding is one of the key means of limiting crown rot on pansy. A report from the southeastern United States in the mid-1990s cautioned against using mechanical transplanters, which may wound the crowns of the tiny plug seedlings, providing *P. roridum* with easy points of entry. Growers in northern U.S. greenhouses are most likely to see Myrothecium disease soon after the accidental wounding that occurs when rooted cuttings or plugs have been boxed and shipped to them.

The role of temperature in the development of Myrothecium leaf spot has been studied on the dumb cane (*Dieffenbachia maculata*) cultivar Perfection. Temperatures along a bench in a greenhouse with evaporative cooling varied as much as 4°C from the pad to the fan end of a greenhouse. The number of lesions caused by *P. roridum* was found to be significantly higher near the cooler pad than the warmer fan end under Florida (United States) growing conditions. Disease severity has been found to vary as much as 200% over the year as temperatures vary, with the highest levels of severity occurring during the spring and fall. The optimal temperature range for lesion development is 21–27°C; temperatures of 30°C and higher inhibit lesion formation.

Host nutrition has also been found to influence disease severity on the cultivar Perfection. When plants were fertilized with a slow-release formulation of 19-6-12 for 2 months prior to inoculation, increased fertilization resulted in a linear increase in the number of lesions formed, regardless of time of year. Studies of both temperature and fertilizer effects on disease expression indicated that of the two factors, temperature plays a more significant role in development of Myrothecium leaf spot in the greenhouse.

Management

Measures to minimize the severity of Myrothecium leaf spot include avoiding temperatures between 21 and 27°C when possible, minimizing wounding of plants, and fertilizing plants at recommended levels. Growers should take care to protect plants before any opportunity for physical damage to occur.

The most effective fungicides for prevention of Myrothecium leaf spot or petiole rot contain chlorothalonil (FRAC M5), fludioxonil (FRAC 12), or strobilurins (FRAC 11). Fungicides should not be applied more than once a week, and rotation among chemicals with different modes of action is critical for resistance management.

Selected References

Chase, A. R. 1983. Influence of host plant and isolate source on severity of Myrothecium leaf spot of foliage plants. Plant Dis. 67:668-671.

Chase, A. R., and Poole, R. T. 1984. Development of Myrothecium leaf spot of *Dieffenbachia maculata* 'Perfection' at various temperatures. Plant Dis. 68:488-490.

Chase, A. R., and Poole, R. T. 1985. Host nutrition and severity of Myrothecium leaf spot of *Dieffenbachia maculata* 'Perfection.' Sci. Hortic. 25:85-92.

Hausbeck, M. K., Woodworth, J. A., and Harlan, B. R. 2004. Evaluation of fungicides and a biopesticide for the control of Myrothecium leaf spot on New Guinea impatiens, 2003. Fung. Nemat. Tests 59:OT018.

Mangandi, J. A., Seijo, T. E., and Peres, N. A. 2007. First report of *Myrothecium roridum* causing Myrothecium leaf spot on *Salvia* spp. in the United States. Plant Dis. 91:772.

Mullen, J. M., Hagan, A. K., and Barnes, L. W. 1995. Crown rot of pansy caused by *Myrothecium roridum*. Plant Dis. 79:1250.

Preston, N. C. 1947. The parasitism of *Myrothecium roridum* Tode. I. Trans. Brit. Mycol. Soc. 30:242-251.

Wilhelm, S., Gunesch, W., and Baker, K. F. 1945. Myrothecium crown and stem canker of greenhouse snapdragons in Colorado. Plant Dis. Rep. 29:700-702.

(Prepared by A. R. Chase and M. L. Daughtrey)

Phomopsis Stem and Leaf Blight of Lisianthus

A *Phomopsis* sp. has been reported from Florida (United States) as causing a leaf blight on lisianthus (*Eustoma grandiflorum*). Significant outbreaks occurred in 1997 and 1998 at two potted-plant production facilities. Mature plants were affected, showing stem necrosis that rapidly developed into leaf blight and collapse of affected plant parts. Dark pycnidia were observed on the infected areas. The pycnidia contained biguttulate, 7.2 × 2.2 µm spores (alpha conidia) but no beta conidia. Symptoms occurred 11 days after wounded tissue was inoculated and 4 days later on nonwounded tissue.

Means of managing this disease include avoiding wounding of plants and minimizing leaf wetness.

Selected References

Alfieri, S. A., et al. 1994. Diseases and Disorders of Plants in Florida. Bull. No. 14. Division of Plant Industry, Gainesville, FL.

McGovern, R. J., Seijo, T. E., and Harbaugh, B. K. 2000. Outbreaks of stem and leaf blight of *Eustoma grandiflorum* caused by a *Phomopsis* sp. in Florida. Plant Dis. 84:491.

(Prepared by A. R. Chase and M. L. Daughtrey)

Phyllosticta Leaf Spot of *Salvia* spp.

Phyllosticta leaf spot has been evaluated on a number of *Salvia* spp. in Louisiana (United States). The pathogen has not been determined, and no official description of the disease is available.

Results of trials in 1996 and 1999 indicated resistance in some species of *Salvia*, including *S. uliginosa* (bog sage), *S. guaranitica* 'Argentina Skies,' *S. involucrata* 'Bethellii,' *S. lycioides* (canyon sage), and *S. greggii* (autumn sage) 'White Autumn' and 'Purple Autumn.' Very low levels of disease were found in *S. guaranitica* 'Purple Majesty,' *S. microphylla* 'La Trinidad Pink' and 'San Carlos Festival,' *S. darcyi* (syn. *S. orbesbia*) (Galena red sage), *S. greggii* 'Maraschino,' *S. madrensis* (forsythia sage), *S.* × *superba* 'Blue Queen,' and *S. regal* (royal sage). In contrast, several cultivars of *S. farinacea* ('Rhea,'

'Argent White,' 'Victoria Blue,' and 'Strata') and *S. coccinea* ('Lady in Red,' 'Snow Nymph,' and 'Cherry Blossom') were very susceptible.

Selected References

Holcomb, G. E., and Buras, H. 1997. Reaction of perennial *Salvia* species and cultivars to Phyllosticta leaf spot, 1996. Biol. Cult. Tests 12:69.

Holcomb, G. E., and Cox, P. 2000. Reaction of perennial *Salvia* species and cultivars to Phyllosticta leaf spot, 1999. Biol. Cult. Tests 15:75.

(Prepared by A. R. Chase and M. L. Daughtrey)

Powdery Mildew Diseases

The powdery mildew fungi are obligate parasites, which means they need living plant hosts to complete their life cycles. These fungi do not usually kill their hosts, but the scattered, white colonies they create on plant surfaces quickly render infected ornamentals unsightly and unmarketable. Collectively, the many powdery mildew fungi infect a wide range of plants, including many bedding plants (Table 4). Powdery mildews are especially important on *Begonia* spp., *Dahlia* spp., gerber daisy (*Gerbera jamesonii*), *Petunia hybrida,* blue salvia (*Salvia farinacea*), *Verbena* spp., pansy and viola (*Viola* spp.), and *Zinnia elegans.* The bedding plants covered in this compendium are hosts to the powdery mildew genera *Erysiphe, Golovinomyces, Leveillula,* and *Podosphaera* (Table 5).

Symptoms

Conspicuous white, powdery colonies on leaves, stems, petals, and other aboveground parts are the signs of infection that make most powdery mildew diseases easy for growers and gardeners to identify. However, genera of powdery mildew fungi cannot for the most part be distinguished on the basis of symptoms. All the powdery mildew fungi except *Leveillula* spp. produce a superficial vegetative growth of hyphae on the plant, and this growth tends to have a consistent appearance.

Because hyaline conidia are typically borne singly or in chains on upright conidiophores arising from the mycelium, a sporulating colony will often have a white, sugary appearance. In some cases, the fungus may cause chlorosis or necrosis on an area on the upper leaf surface but little if any discernible mycelium; in these instances, the diagnosis of a powdery mildew disease may require the use of a microscope. Occasionally, when the environment is not very favorable to disease development, the upper leaf surface may show patches of chlorosis or necrosis while recognizable powdery mildew colonies are visible on the lower leaf surface. *L. taurica* infection is internal to leaves; thus, infection by this unique powdery mildew fungus is also first perceived as irregular patches of chlorosis on upper leaf surfaces, delaying disease identification.

On *Coreopsis* spp., blue salvia, and zinnia, the disease usually begins as typical powdery, white spots on both leaf surfaces (Figs. 50 and 51); spots can also appear on the petioles, stems, and flowers. Severely affected leaves and flowers become dry and die. On zinnia, especially, the production of fungal mycelia and conidia may reach such a volume as to make the plant look flocked.

On begonia, small, greasy-looking spots sometimes appear on the lower leaf surface. Spots more typically occur in dense, white patches on the upper leaf surface, or the mycelium may coat portions of the leaves, stems, and flowers. Infected tissue may eventually turn brown and die.

On dahlia, powdery mildew generally develops over the entire leaf surface, forming a cobwebby mat of hyphae (Fig. 52). Conidia soon follow, giving the mycelium a more cottony, white appearance. Severely affected leaves become distorted and may fall prematurely.

Both bedding and cut-flower varieties of gerber daisy are very susceptible to several different powdery mildews. Leaves are often completely covered with white mycelium and conidia, giving leaf surfaces an overall gray, powdered appearance. Severely infected leaves turn yellow and die. It is common to find the undersurfaces of flower petals and calyxes covered with white, powdery growth (Fig. 53).

TABLE 4. Powdery Mildew Fungi That Affect Bedding Plants[a]

Host	Fungus/Fungi
Begonia spp.	*Erysiphe begoniicola, E. polygoni, Golovinomyces cichoracearum, G. orontii*
Bellis perennis (English daisy)	*Golovinomyces cichoracearum*
Campanula spp. (campanula, bellflower)	*Golovinomyces cichoracearum*
Catharanthus roseus (annual vinca)	*Erysiphe aquilegiae, Leveillula taurica*
Cleome houtteana (spider flower)	*Erysiphe cruciferarum, Leveillula taurica*
Coreopsis spp. (tickseed)	*Golovinomyces cichoracearum*
Cyclamen persicum	*Golovinomyces orontii, Pseudoidium cyclaminis*
Dahlia spp.	*Erysiphe orontii, E. polygoni, Golovinomyces cichoracearum, Neoerysiphe cumminsiana, Podosphaera fuliginea*
Gerbera jamesonii (gerber daisy)	*Fibroidium heliotropii-indici, Golovinomyces cichoracearum, Leveillula taurica, Podosphaera fuliginea, P. fusca*
Petunia hybrida	*Euoidium longipes, Golovinomyces orontii, Oidium neolycopersici, Podosphaera xanthii*
Plectranthus scutellarioides (coleus)	*Golovinomyces biocellatus*
Primula spp. (primrose)	*Erysiphe polygoni*
Ranunculus spp.	*Erysiphe aquilegiae, E. aquilegiae* var. *ranunculi, E. pisi* var. *pisi, E. polygoni, Golovinomyces cichoracearum, Podosphaera macularis*
Salvia farinacea (blue salvia)	*Golovinomyces biocellatus, G. cichoracearum, Neoerysiphe galeopsidis*
Salvia splendens (red salvia)	*Golovinomyces biocellatus*
Torenia fournieri (torenia, wishbone flower)	*Podosphaera xanthii*
Verbena spp.	*Euoidium longipes, Golovinomyces cichoracearum, G. monardae, G. orontii, Leveillula taurica, Podosphaera fusca, P. xanthii*
Viola × *wittrockiana* (pansy)	*Podosphaera fuliginea, P. macularis*
Zinnia elegans	*Golovinomyces asterum* var. *asterum, G. biocellatus, G. cichoracearum, G. spadiceus, Podosphaera aphanis* var. *aphanis, P. fuliginea, P. fusca*

[a] Courtesy A. R. Chase and M. L. Daughtrey—© APS.

Causal Organisms

The powdery mildew fungi are ascomycetes in the Erysiphaceae. They have been classified into six tribes and 17 genera based on characteristics of both the sexual and asexual spore structures. The sexual stage is not usually seen in the greenhouse but may appear on bedding plants at the end of the growing season outdoors. For some powdery mildew fungi, the sexual stage has not been described; only single-celled conidia borne on upright conidiophores disseminate this type of fungus.

The term "*Oidium* sp." was long used as a catch-all for powdery mildew fungi that had not been further identified because of the absence of the sexual stage, which was for many years essential for classification. The subtle characteristics of the

TABLE 5. Important Genera of Powdery Mildew Fungi That Affect Bedding Plants[a]

Genus (Tribe)	Description
Erysiphe (Erysipheae)	Includes powdery mildews that were previously classified under the names *Microsphaera* and *Uncinula*, as well as some that were known as *Erysiphe*. Conidia are borne singly (*Pseudoidium* type), rather than in chains, although they may accumulate in "false chains" in a still atmosphere. A common powdery mildew on *Begonia* spp. in Africa, Asia, Europe, and New Zealand that was previously called *Microsphaera begoniae* is now named *E. begoniicola*. This powdery mildew is also probably present in North America. A second species, *E. polygoni*, has been reported widely from *Dahlia variabilis* in the United States and also occurs on a number of other plants, including vegetables.
Golovinomyces (Golovinomyceteae)	*G. cichoracearum* (syn. *Erysiphe cichoracearum*) is a powdery mildew with more than 2,500 hosts, including many ornamentals in the Asteraceae. There are two spores per ascus in the chasmothecium; conidia are borne in chains (*Euoidium* type). This is the powdery mildew reported most often on *Zinnia elegans*, but other species of powdery mildew occur on zinnia, as well. *G. orontii* (syn. *Oidium begoniae*) is reported on *Begonia tuberhybrida* and other begonias, as well as *Petunia hybrida* and *Verbena hybrida;* each of these plants can also host other powdery mildews. *G. spadiceus* has also been reported from *Dahlia pinnata* worldwide.
Leveillula (Phyllactinieae)	Unique among the powdery mildews found on bedding plants in that it has internal mycelium; the signs of the fungus first appear on the lower surface of the leaf. The conidia are of the *Ovulariopsis* type and borne singly; conidiophores emerge from stomates. Hosts of *L. taurica* reported from around the world include *Calendula officinalis*, *Celosia cristata*, *Catharanthus roseus*, *Cleome* spp., *Cosmos sulphureus*, *Dianthus* spp., *Eustoma grandiflorum*, *Gerbera jamesonii*, *Impatiens balsamina*, *Phlox drummondii*, *Salvia* spp., *Tropaeolum majus*, and *Verbena* spp., as well as tomato and many herbaceous perennials.
Podosphaera (Cystotheceae)	Includes powdery mildews that were previously classified under the name *Sphaerotheca*. Has chasmothecia with single asci. Conidia are produced in chains and contain fibrosin bodies; fibrosin bodies can be seen in conidia mounted in water but are best observed when fresh conidia are soaked in 2–3% potassium hydroxide prior to observation. A representative sample of conidia must be examined, because not all conidia will contain fibrosin bodies. *P. xanthii* is one of the most important members of this genus to bedding plant growers; it affects *Verbena* spp. as well as other flower and vegetable hosts.

[a] Courtesy A. R. Chase, M. L. Daughtrey, and N. Shishkoff.

Fig. 50. Powdery mildew on coreopsis, caused by an *Oidium* sp. (Courtesy A. R. Chase—© APS)

Fig. 51. Powdery mildew on blue salvia, caused by an *Oidium* sp. (Courtesy A. R. Chase—© APS)

asexual sporulation of these fungi have since been recognized for their taxonomic value, and powdery mildew fungi in their conidial forms are classified as *Euoidium, Oidium, Ovulariopsis,* and *Pseudoidium* spp.

The sexual spores of the powdery mildew fungi are produced in chasmothecia (syn. cleistothecia). In the past, the characteristics used to identify these fungi were the morphology of chasmothecial appendages, number of asci within the chasmothecium, number and shape of conidia, type of conidial germination, germ tube characteristics, and host plant species. Now (as of 2018), identification of the powdery mildews is partly DNA based, using the sequence of the internal transcribed spacer (ITS) region, as well as the morphology of the sexual and asexual structures.

Host Range and Epidemiology

Although some powdery mildew fungi have broad host ranges, others are host specific and may be confined to a particular plant species, genus, or family. For example, *Golovinomyces cichoracearum* has a very wide host range and affects many members of the Asteraceae. *Podosphaera xanthii* is well known on cucurbits such as pumpkin and squash but also readily attacks *Verbena hybrida,* petunia, *Bidens* spp., and *Calendula* spp. Worldwide, a number of *Erysiphe, Golovinomyces,* and *Leveillula* spp. are reported on *Salvia* spp., and these bedding plants in mass landscape plantings are likely to develop powdery mildew in some areas of the United States.

Most powdery mildew fungi form a network of hyphae over the plant surface, from which they penetrate epidermal cells and derive nutrients via fungal structures called "haustoria." (*Leveillula* spp. are unique in having internal growth of hyphae followed by sporulation through stomates.) In the greenhouse, the asexually produced conidia initiate most plant infections and are responsible for secondary spread. Conidia are easily disseminated by air currents and splashing water.

Generally, the movement of powdery mildew from one crop to another is not the major difficulty in managing this disease in the greenhouse. Rather, the greatest challenge is to curtail disease development on a highly susceptible host. The humidity and crowding that are typical of a greenhouse are conducive to powdery mildew development whenever the temperature requirements for a particular pathogen are met. Depending on the genus of the powdery mildew fungus, conidia can germinate during periods of high or low relative humidity but rarely germinate in free water. Extended periods of dry conditions and free moisture are both detrimental to conidia. Most powdery mildews on flower crops that have been studied under greenhouse conditions develop best at a level of relative humidity greater than 95% and at an optimal temperature near 20°C. Development of an epidemic may be quite rapid when both temperature and humidity are favorable for formation of abundant conidia and successful infection. Given these environmental preferences, powdery mildew outbreaks outdoors in the northern United States are generally noticed during the late summer to fall, rather than the rainy spring.

Powdery mildew fungi usually survive in the greenhouse on crop or weed hosts and move from greenhouse to greenhouse as plant materials are exchanged. The role of chasmothecia in the survival of these fungi on bedding plants grown in the landscape is not well understood. In the landscape, introduction of powdery mildew fungi from weed hosts may be a significant concern. For example, native populations of balsam impatiens (*Impatiens balsamina*) have been shown infected with *Podosphaera fuliginea* (syn. *Sphaerotheca fuliginea*) in India.

Fig. 52. Powdery mildew on dahlia, caused by an *Oidium* sp. (Courtesy A. R. Chase—© APS)

Fig. 53. Powdery mildew on the calyx of a gerber daisy flower, caused by an *Oidium* sp. (Courtesy A. R. Chase—© APS)

Management

One of the most important steps that growers and landscapers can take to minimize powdery mildew is to use resistant cultivars. Many studies have reported cultivar effects on the severity of powdery mildew on plants such as begonia, gerber daisy, pansy, verbena, and zinnia. Plants within a series may vary in disease susceptibility. Gerber daisy cultivar Festival Semi Double Orange was less susceptible than others in the Festival series in two trials. Researchers noted that in one trial, Light Eye Pink and Semi Double Orange flowers were not infected, whereas Dark Eye Cherry flowers showed heavy infection under the same conditions. Several begonia cultivars—*Begonia semperflorens–cultorum* hybrid 'Charm,' *B.* × *cheimantha* 'Lady Mac,' and *B. serratipetala* 'Serratipetala'—are resistant to powdery mildew. The responses of viola and pansy cultivars to *Podosphaera* (syn. *Sphaerotheca*) spp. are shown in Tables 6 and 7.

Because it is sometimes difficult to avoid the environmental conditions conducive to powdery mildew development, the use of fungicides has traditionally been the preferred control measure. Resistance to initially effective FRAC 1 products, including benomyl and thiophanate methyl, occurred rapidly after their introduction in the 1970s, but products in this FRAC group may still be useful in some situations in rotation with other chemicals. Fungicides with the greatest effect against the powdery mildews include strobilurins (FRAC 11) and demethylation inhibitors (FRAC 3), as well as contact materials such as potassium bicarbonate and horticultural oils. Oils should be used with care; application in the greenhouse may cause phytotoxicity, because high humidity impedes prompt drying.

It is also possible to reduce the incidence and severity of powdery mildew with a number of biological control products and spray adjuvants. Biological active ingredients with efficacy against powdery mildew fungi include *Bacillus amyloliquefaciens, B. subtilis,* and *Streptomyces lydicus* and the biopesticide derived from *Reynoutria sachalinensis.*

Selected References

Bélanger, R. R., Bushnell, W. R., Dik, A. J., and Carver, T. L. W. 2002. The Powdery Mildews: A Comprehensive Treatise. American Phytopathological Society, St. Paul, MN.

Braun, U. 1987. A Monograph of the Erysiphales (Powdery Mildews). J. Cramer, Berlin.

Braun, U., and Cook, R. T. A. 2012. Taxonomic Manual of the Erysiphales (Powdery Mildews). CBS-KNAW Fungal Biodiversity Centre, Utrecht, the Netherlands.

Daughtrey, M., Hodge, K. T., and Shishkoff, N. 2017. The powdery mildews. Pages 191-204 in: Plant Pathology Concepts and Labora-

TABLE 6. Severity of Powdery Mildew (*Podosphaera fuliginea*) on Pansy Cultivars[a]

Very Low Severity	Low to Moderate Severity	Moderate Severity	Very High Severity
Bingo Deep Purple	Bingo Blue/Blotch	Bingo Light Rose	Beaconsfield
Crown Yellow Splash	Bingo White/Blotch	Crown Blue	Delta Violet Face
Delta Pink Shades	Clear Sky Primrose	Crown Purple	Majestic Giant Purple
Delta Pure Rose	Crown Cream	Crown Scarlet	Universal Plus
Delta Pure Yellow	Crown Orange	Crown White	
Majestic Giant White/Blotch	Crown Rose	Crown Yellow	
Maxim Orange	Delta Pure Primrose	Delta Blue Blotch	
	Delta White/Rose Wing	Delta Red Blotch	
	Imperial Frosty Rose	Delta True Blue	
	Majestic Giant Scarlet and Bronze	Golden Crown	
	Majestic Giant Yellow/Blotch	Majestic Giant Blue/Blotch	
	Maxim Blue and Yellow		
	Maxim Sherbet		
	Skyline Blue		

[a] Courtesy A. R. Chase and M. L. Daughtrey—© APS.

TABLE 7. Severity of Powdery Mildew (*Podosphaera* sp.) on Viola Cultivars[a]

Very Low Severity	Low to Moderate Severity	Moderate Severity	Very High Severity
Grimes Special Mix	Sorbet Blueberry Cream	Penny Azure Wing	Alpine Summer
Sorbet Blackberry Cream	Sorbet Coconut	Penny Violet Beacon	Alpine Sun
Velour Blue Blotch	Sorbet Lemon Chiffon	Penny Violet Flare	Alpine Wing
Velour Purple	Sorbet Plum Velvet	Sorbet Sunny Royale	Penny Primrose
Velour White	Velour Blue	Sorbet Yellow Frost	Sorbet Lavender Ice
Velour Yellow		Sorbet YTT	
Violin Red and Yellow			
Violin Royal Picotee			

[a] Courtesy A. R. Chase and M. L. Daughtrey—© APS.

tory Exercises, 3rd ed. B. H. Ownley and R. N. Trigiano, eds. Taylor and Francis, Milton Park, Oxfordshire, UK.

Denton, G. J., Beal, E., and Denton, J. O. 2014. First report of powdery mildew on *Solenostemon*. New Dis. Rep. 30:18. doi:10.5197/j.2044-0588.2014.030.018

Farr, D. F., and Rossman, A. Y. Fungal Databases, U.S. National Fungus Collections, ARS, USDA. http://nt.ars-grin.gov/fungaldatabases

Garibaldi, A., Minuto, A., and Gullino, M. L. 2008. First report of powdery mildew caused by *Golovinomyces cichoracearum* on English Daisy (*Bellis perennis*) in Italy. Plant Dis. 92:484.

Glawe, D. A., Grove, G. G., and Nelson, M. 2006. First report of powdery mildew of *Coreopsis* species caused by *Golovinomyces cichoracearum* in the Pacific Northwest. Online. Plant Health Progress. doi:10.1094/PHP-2006-0405-01-BR

Hagan, A. K., and Akridge, J. R. 1998. Reaction of cultivars of pansy to powdery mildew and Cercospora leaf spot, 1997. Biol. Cult. Tests 13:62.

Holcomb, G. E. 1999. First report of powdery mildew caused by an *Oidium* sp. on *Torenia fournieri*. Plant Dis. 83:878.

Holcomb, G. E., Buras, H., and Cox, P. 1998. Reaction of pansy and viola cultivars to leaf spot and powdery mildew, 1997. Biol. Cult. Tests 13:63.

Hou, H. H., and Lee, C. S. 1979. Powdery mildew (*Sphaerotheca fuliginea*) of *Dahlia pinnata* in Taiwan. Plant Prot. Bull. 21:441-443.

Liberato, J. R., and Cunnington, J. H. 2006. First record of *Erysiphe aquilegiae* on a host outside the Ranunculaceae. Australas. Plant Pathol. 35:291-292.

Kloos, W. E., George, C. G., and Sorge, L. K. 2005. Inheritance of powdery mildew resistance and leaf macrohair density in *Gerbera hybrida*. HortScience 40:1246-1251.

Moyer, C., and Peres, N. A. 2008. Evaluation of biofungicides for control of powdery mildew of gerbera daisy. Proc. Fla. State Hort. Soc. 121:389-394.

Reis, A., Boiteux, L. S., and Paz-Lima, M. L. 2007. Powdery mildew of ornamental species caused by *Oidiopsis haplophylli* in Brazil. Summa Phytopathol. 33(4):405-408. doi:10.1590/S0100-54052007000400015

Sconyers, L. E., and Hausbeck, M. K. 2005. Evaluation of African daisy cultivars for resistance to powdery mildew, 2004. Biol. Cult. Tests 20:O016.

Sconyers, L. E., and Hausbeck, M. K. 2005. Evaluation of biological and biorational products in managing powdery mildew of African daisy, 2004. Fung. Nemat. Tests 60:OT001.

(Prepared by A. R. Chase and M. L. Daughtrey)

Begonia Powdery Mildew

Although *Begonia* spp. may be affected by powdery mildew, the disease is rare on the seed-propagated wax begonia (*B. semperflorens*), which is the type used most often as a bedding plant. In maritime areas with temperate climates, powdery mildew can cause severe losses in plantings of wax begonia (Fig. 54). This disease is much more common on hiemalis or Rieger begonia (*B.* × *hiemalis*) and other types used less frequently as bedding plants.

Studies of the pathogen, *Erysiphe begoniicola* (syns. *Microsphaera begoniae; Oidium begoniae* var. *macrosporum*), have demonstrated that conidia germinate on glass slides and the leaf surfaces of hiemalis begonia at temperatures between 4 and 32°C. The fastest germination occurred at 23–25°C, and temperatures greater than 28°C caused marked reductions in growth and sporulation of the pathogen. A decrease in relative humidity decreased conidial germination only slightly. The presence of free water killed submerged conidia within 10–30 min. Sporulation was responsive to diurnal cycles.

Tests have also been conducted on the effects of split night temperatures (17 versus 10°C) on powdery mildew of begonia. Extensive periods of lower temperatures were shown to eventually decrease powdery mildew severity on begonia when the time at 10°C was at least 12 h.

Fig. 54. Powdery mildew on wax begonia, caused by *Erysiphe begoniicola.* (Courtesy A. R. Chase—© APS)

Fig. 55. Powdery mildew on petunia, caused by *Euoidium longipes.* (Courtesy M. L. Daughtrey—© APS)

Selected References

Quinn, J. A., and Powell, C. C., Jr. 1981. Identification and host range of powdery mildew of begonia. Plant Dis. 65:68-70.

Quinn, J. A., and Powell, C. C., Jr. 1982. Effects of temperature, light, and relative humidity on powdery mildew of begonia. Phytopathology 72:480-484.

Sammons, B., Rissler, J. F., and Shanks, J. B. 1982. Development of gray mold of poinsettia and powdery mildew of begonia and rose under split night temperatures. Plant Dis. 66:776-777.

(Prepared by A. R. Chase and M. L. Daughtrey)

Petunia Powdery Mildew

In Europe and the United States, most of the powdery mildews on petunia are attributed to *Podosphaera xanthii.* In Europe, petunia is also infected by *Golovinomyces orontii,* and other powdery mildew pathogens are detected occasionally. Petunia leaves infected by powdery mildew may be first noticed because of associated chlorosis; close inspection is often needed to review the hyphae of the fungus on the leaf surface at an early stage of disease development.

The first North American report of powdery mildew on petunia caused by *Euoidium longipes* (syn. *Oidium longipes*) was made in 2008 (Fig. 55). A vegetatively propagated petunia cultivar in a commercial greenhouse in New Jersey (United States) showed extensive growth of a powdery mildew, which may have originated offshore where the cuttings were produced. Although the susceptibilities of a range of different cultivars were not tested, isolates of the powdery mildew were obtained from both the Wave and the Surfinia series of petunia. Field surveys in Europe also found this powdery mildew on petunia in Hungary and Austria. Phylogenetic analysis of the internal transcribed spacer (ITS) sequences showed that the closest known relative of *E. longipes* is *E. lycopersici.*

Cross-inoculation with *E. longipes* from petunia resulted in heavy infection of tobacco cultivar Xanthi, whereas eggplant and tomato cultivars were moderately susceptible. Thus, the potential for this petunia powdery mildew to spread to other solanaceous crops has been demonstrated. Petunias were also successfully inoculated with multiple different powdery mildews, showing that the same leaves can be infected at the same time by *E. longipes, Pseudoidium neolycopersici, G. orontii,* and *P. xanthii.*

Selected Reference

Kiss, L., Jankovics, T., Kovács, G. M., and Daughtrey, M. L. 2008. *Oidium longipes,* a new powdery mildew fungus on petunia in the USA: A potential threat to ornamental and vegetable solanaceous crops. Plant Dis. 92:818-825.

(Prepared by A. R. Chase and M. L. Daughtrey)

Verbena Powdery Mildew

Powdery mildew caused by *Podosphaera xanthii* is an occasional problem on vegetatively grown verbena (Fig. 56). Symptoms and signs that begin on the lower foliage of plants grown in hanging baskets often go unnoticed until the disease is well advanced. Powdery mildew injury on verbena often mimics a nutrient deficiency or spider mite infestation; the lower leaves become yellow or purplish before becoming entirely necrotic.

Much of the research on verbena powdery disease has focused on control through the use of resistant cultivars and fungicides. Results of cultivar trials have shown a wide range of susceptibility, and verbenas within a single series often show very different susceptibility (Table 8).

Fig. 56. Powdery mildew on verbena, caused by *Podosphaera xanthii.* (Courtesy M. L. Daughtrey—© APS)

TABLE 8. Severity of Powdery Mildew (*Podosphaera xanthii*) on Verbena Cultivars[a]

Very Low Severity	Low to Moderate Severity	Very High Severity
Aztec Cherry Red	Babylon Neon Rose	Aztec Peach
Aztec Grape Magic	Babylon White	Babylon Light Blue
Aztec Lilac Picotee	Lanai Royal Blue	Babylon Blue
Aztec Magic Purple	Lanai Royal Deep Purple	Babylon Purple
Aztec Red Velvet	Lanai Royal Lavender Star	Babylon Carpet Blue
Aztec Silver Magic	Rapunzel Hot Rose	Babylon Red
Aztec White Trailing	Superbena Blue Purple	Fuego Apricot
Aztec Wild Rose	Superbena Coral Red	Lanai Blue
Babylon Deep Pink	Tapien Blue Violet	Lanai Blush White
Lanai Royal Bright Pink	Tapien Lilac	Napoleon Purple
Lanai Royal Purple with Eye	Temari Patio Blue	Napoleon Red
Napoleon Blue	Temari Patio Hot Pink	Quartz Blue
Rapunzel Hot Rose		Quartz Burgundy with Eye
Rapunzel Orchid		Quartz Magenta
Superbena Dark Blue		Sparkler Deep Blue/White
Superbena Large Lilac Blue		Sparkler Purple/White
Superbena Pink Shades		Sparkler Red/White
Superbena Purple		Sparkler Sky Blue/Red
Temari Patio Red		Sparkler Violet/White
		Spitfire Violet/White
		Superbena Coral Red
		Temari Burgundy Improved
		Tukana Scarlet
		Tukana White
		Wildfire Purple Improved

[a] Courtesy M. L. Daughtrey and M. K. Hausbeck—© APS.

TABLE 9. Effectiveness of Fungicides on Powdery Mildew of Verbena and Gerber Daisy[a]

Fungicide (FRAC Group)[b]	Verbena	Gerber Daisy
Azoxystrobin (11)	Some	Very good to excellent
Bacillus subtilis (NC)	Excellent	Good
Copper salts of rosin and fatty acids (M1)	Some	Some
Copper pentahydrate (M1)	Excellent	Some
Extract from giant knotweed (NC)	Not tested	Very good
Fenhexamid (17)	Very good	Very good
Fludioxonil and cyprodinil (12 and 9)	Not tested	Excellent
Hydrogen dioxide, peroxy-acetate (NC)	Some	Very good
Myclobutanil (3)	Excellent	Excellent
Piperalin (5)	Not tested	Excellent
Potassium bicarbonate (NC)	Not tested	Excellent
Propiconazole (3)	Very good	Excellent
Pyraclostrobin (11)	Excellent	Excellent
Triadimefon (3)	Very good	Not tested
Trifloxystrobin (11)	Very good to excellent	Excellent
Triflumizole (3)	Excellent	Very good to excellent
Triticonazole (3)	Some	Very good to excellent

[a] Courtesy A. R. Chase and M. L. Daughtrey—© APS.
[b] NC = Not classified.

Fungicide trials on verbena and gerber daisy have shown that products in FRAC groups 3 and 11 are generally most effective in managing powdery mildew (Table 9). Chemical control may vary among the different genera of powdery mildew fungi, so both fungicide rotation and grower evaluations after beginning a treatment protocol are recommended.

Selected References

Daughtrey, M. L., and Hausbeck, M. K. 2011. Verbena Cultivar Sensitivity to Powdery Mildew. Rep. No. 133. AFE Research Reports—Disease Management (Ser. 101). American Floral Endowment, Alexandria, VA.

Sconyers, L. E., and Hausbeck, M. K. 2005. Evaluation of Verbena cultivars for resistance to powdery mildew, 2004. Biol. Cult. Tests 20:O015.

(Prepared by A. R. Chase and M. L. Daughtrey)

Diseases Caused by *Rhizoctonia solani*

Rhizoctonia solani has a very wide host range and occurs worldwide on many crops, including crops produced from seeds, cuttings, and bulbs. This fungus causes damping-off, root rot, crown and stem rot, as well as leaf spot and aerial or web blight (Fig. 57). *R. solani* diseases occur on most bedding plants from plug production through landscape use. This fungus is one of the most common plant pathogens encountered in some locales.

Symptoms

Damping-off caused by *R. solani* causes stems to become constricted, giving them a shrunken, wiry appearance that is sometimes termed "wirestem." Development of shrunken, brown cankers that only partially girdle stems is also common, especially on older plants (Fig. 58). When conditions are warm and very humid, the fungus can grow over an entire plug or cutting flat (Fig. 59). The light-brown mycelium produced by *R. solani* is fairly distinctive, but it can sometimes be mistaken for mycelia of other blighting pathogens, such as *Rhizopus* spp. Rhizoctonia damping-off usually develops in a circular or arc-shaped pattern within a flat or bed of plants. Host plants particularly susceptible to Rhizoctonia damping-off include *Catharanthus roseus* (annual vinca), *Antirrhinum majus* (snapdragon), and *Dianthus chinensis* (China pink), along with species of *Begonia, Impatiens, Petunia, Verbena,* and *Zinnia*.

R. solani also causes a web blight on a wide range of woody, herbaceous, tropical, and bedding crops in warm, wet climates and conditions. The mycelium appears as a brown, spiderweb-like growth on leaves. Discrete leaf spots commonly coalesce to form large areas of blight on stems and leaves.

Root rots caused by *R. solani* are characterized by discrete, brown lesions and a firm rot of cortical tissues. Root infection in the absence of other symptoms is not common on bedding plants.

Bedding plants propagated as stem cuttings—such as *Osteospermum* spp., *Pelargonium* spp. (geranium), and *I. hawkeri* (New Guinea impatiens)—are very susceptible to Rhizoctonia cutting rot. Disease is usually apparent as a dry, brown basal rot (Fig. 60). On geranium, the brown coloration helps to distinguish disease caused by *R. solani* from the more common Pythium black leg, which generally results in a black canker at the stem base.

Crown rot can result from progression of the pathogen from the roots into the crown or from direct fungal penetration of the crown tissue. On bedding plants, *R. solani* commonly causes crown or stem rot in the absence of root rot (Fig. 61). On *Begonia* spp., *Impatiens* spp., and many other hosts, *R. solani* causes a brown crown or stem canker. Longitudinal cracking and a dry appearance of the rotted crown tissue often develops on older plants. Chlorosis, wilting, loss of lower leaves, and at times, stunting of the entire plant may result. Crown rot often develops after plants have been transplanted into the landscape, and plant death commonly occurs following attack by *R. solani*.

Causal Organism

Definitive morphological characteristics of *R. solani* (syn. *Thanatephorus cucumeris*) include the presence of multinucleate cells in young vegetative hyphae, branching near the distal septum and constriction at the base of the branch, the formation

Fig. 59. Rhizoctonia cutting rot (*Rhizoctonia solani*) on osteospermum cuttings. (Courtesy A. R. Chase—© APS)

Fig. 57. Rhizoctonia damping-off (*Rhizoctonia solani*) on seedlings of gomphrena. (Courtesy A. R. Chase—© APS)

Fig. 60. Rhizoctonia stem rot (*Rhizoctonia solani*) on portulaca. (Courtesy A. R. Chase—© APS)

Fig. 58. Rhizoctonia stem rot (*Rhizoctonia solani*) on garden impatiens. (Courtesy M. L. Daughtrey—© APS)

Fig. 61. Rhizoctonia stem rot (*Rhizoctonia solani*) on osteospermum in a garden. (Courtesy A. R. Chase—© APS)

of a septum near the point of branch origin, and the formation of mycelium that is some shade of brown.

Cultures on potato dextrose agar have a fleecy margin and vary in color from light tan to medium brown. Cultures may eventually develop irregularly shaped, brown sclerotia. The following characteristics are usually present, but one or more may be lacking: monilioid cells, sclerotia without a differentiated rind and medulla, hyphae greater than 5 µm in diameter, and a rapid growth rate (often filling a 90-mm-diameter petri dish within 3–4 days).

Not included in the *R. solani* group are *Rhizoctonia*-like organisms with clamp connections, conidia, differentiated sclerotia, rhizomorphs, pigmentation other than brown, and teleomorph stages. The sexual stage of *R. solani, T. cucumeris,* is not commonly observed on infected bedding plants.

Anastomosis grouping has indicated relationships within groups of isolates, allowing separation of *R. solani* into genetically independent entities. The reported isolates of *R. solani* from nearly all bedding plants belong to anastomosis group 4 (AG4).

Host Range and Epidemiology

Rhizoctonia root rot occurs on many bedding plants and causes considerable economic losses. Root rot, crown rot, and/or web blight are commonly encountered on snapdragon, *Begonia* spp., *Plectranthus scutellarioides* (coleus), annual vinca, *Gerbera jamesonii* (gerber daisy), China pink, *I. walleriana* (garden impatiens), New Guinea impatiens, *Pelargonium* × *hortorum* (florist's geranium), *Petunia hybrida, Verbena hybrida,* and *Zinnia elegans.*

A study conducted in the early 1980s in Ohio found that soil (even when steam treated) was a source of *Pythium* spp. but not *Rhizoctonia* spp. None of the other components of the potting medium were infested with *Rhizoctonia* spp., but dust and soil mix samples from walkways, floors, beds, and used flats were contaminated with both *Pythium* and *Rhizoctonia* spp. Even a small amount of dust or soil from one of these areas produced damping-off when sprinkled onto freshly seeded flats.

R. solani often attacks plants at the soil line. Disease incidence and severity depend on the host, fungal strain, and environmental conditions. *R. solani* generally grows best in soils that are evenly moist and warm. Rhizoctonia root rot increases at soil temperatures of 17–26°C and in soil with a moisture-holding capacity of less than 40%. *R. solani* grows best at low carbon dioxide and high oxygen levels.

Management

Sanitation practices should be initiated at propagation time and continued through the crop cycle. Soiled hands and tools should not come in contact with soilless media. In general, cleanliness of the plant-growing area is recommended. Diseased plants and plant debris should be removed as they appear on the bench, space between plants should be increased to improve aeration, and excessive splashing during watering should be avoided.

Differences in susceptibility to *R. solani* infection among New Guinea impatiens cultivars have been observed. In an inoculation experiment with cultivars Astro, Aurora, Columbia, Corona, Cosmos, Equinox, Gemini, Milky Way, Nova, Red Planet, Sunset, Twilight, and Twinkle, Gemini and Milky Way grew best and were more resistant to infection than the others. Cultivars Astro, Aurora, Nova, and Sunset were the most susceptible.

A study conducted on cultivars of *B. semperflorens* (wax begonia) in Texas showed the importance of choosing the right cultivar of a bedding plant for the landscape. The cultivars most resistant to *R. solani* infection were the Stara colors and Party Rose Pink. Testing on vinca and verbena cultivars has also been pursued. However, because of the rapidly changing nature of the bedding plant market, by the time a resistant cultivar is identified, it may have been replaced by a new cultivar with unknown levels of resistance to this common pathogen. For this and other reasons, resistance is not likely a practical means of controlling *R. solani* in bedding plants. Poor performance in annual flower trials is often associated with Rhizoctonia disease, so selection away from the most susceptible cultivars tends to occur without experimentation using inoculated plants.

Binucleate *Rhizoctonia* spp. (BNR) have been researched for their potential to control pre-emergence damping-off of impatiens caused by *R. solani.* Amendment of a soilless potting mix 3 days prior to seeding and infesting did not improve control compared with amendment 1 day prior to seeding. Control of damping-off was comparable but inconsistent between formulations of BNR fungi and thiophanate-methyl (FRAC 1). Damping-off was better controlled with formulations of BNR fungi than with *Trichoderma virens.* The shelf life of BNR formulations significantly affected control of pre-emergence damping-off; a controlled atmosphere enhanced BNR survival at 4°C, which was significantly better for BNR activity than storage at 25°C.

Controlling Rhizoctonia disease once it has been established in a landscape bed is a serious challenge. Solarization trials in Florida (United States) evaluated the effectiveness of single and combination treatments with the biological control agents *Streptomyces lydicus* and *Pseudomonas chlororaphis* (syn. *P. aureofaciens*) and the reduced-risk fungicide fludioxonil (FRAC 12) for managing soilborne pathogens of impatiens. Naturally infested soil was solarized for 47 or 48 days using two layers of 25-µm, clear, low-density polyethylene mulch, separated by an air space of up to 7.5 cm. Solarization decreased the final incidence and progress of Rhizoctonia crown rot and blight and root discoloration and resulted in increased shoot biomass in both experiments. Using this technique also consistently reduced root-knot nematode severity and population densities of parasitic nematodes (*Meloidogyne incognita, Dolichodorus heterocephalus, Paratrichodorus minor,* and *Criconemella* spp.). (See also "Diseases Caused by Nematodes" later in Part I.) The incidence of crown rot and blight was reduced by treatment with fludioxonil but not with the biological control agents.

Fungicides such as fludioxonil, iprodione (FRAC 2), thiophanate-methyl, strobilurins (FRAC 11), and triflumizole (FRAC 3) are effective against *R. solani* in foliar and drench treatments. When soil applications of fungicides are used to control root and crown rots, the applications must be made preventively, before infection occurs. A number of different fungicides are registered in the United States for spray application to control Rhizoctonia web blight on bedding plants. Growers should check fungicide labels for legal uses (site, application method, rate, and interval).

During production, fungicides are commonly used on plants likely to be affected by *R. solani.* Although biological controls have been tested rather extensively, they are often not as effective as standard fungicides against this fast-moving pathogen and require preventive rather than responsive treatment. The type of symptom expressed (foliar versus stem or root) may indicate which product should be used and where it should be applied. In general, root diseases should be treated with drenches, and foliar diseases (e.g., web blight) should be treated with sprays. However, since the pathogen is soilborne, treating the potting medium will control aerial blight in many cases.

Selected References

Bogran, C. E., Pemberton, B., and Isakeit, T. 2004. Resistance of wax begonia cultivars to *Rhizoctonia solani* in southern landscapes. Proc. South. Nursery Assoc. 49:240-242.

Castillo, S., and Peterson, J. L. 1990. Cause and control of crown rot of New Guinea impatiens. Plant Dis. 74:77-79.

Holcomb, G. E., and Carling, D. E. 2000. First report of leaf blight of *Dianthus chinensis* caused by *Rhizoctonia solani.* Plant Dis. 84:1344.

Holcomb, G. E., and Carling, D. E. 2000. First report of web blight of verbena caused by *Rhizoctonia solani*. Plant Dis. 84:492.
Holcomb, G. E., and Carling, D. E. 2002. First report of web blight caused by *Rhizoctonia solani* on *Catharanthus roseus* in Louisiana. Plant Dis. 86:1272.
Honeycutt, E. W., and Benson, D. M. 2001. Formulation of binucleate *Rhizoctonia* spp. and biocontrol of *Rhizoctonia solani* on impatiens. Plant Dis. 85:1241-1248.
McGovern, R. J., McSorley, R., and Bell, M. L. 2002. Reduction of landscape pathogens in Florida by soil solarization. Plant Dis. 86:1388-1395.
McMillan, R. T., Jr., Johnson, S. V., and Graves, W. R. 2004. Rhizoctonia blight of impatiens and its control. Proc. Fla. State Hortic. Soc. 117:318-320.
Stephens, C. T., Herr, L. J., Schmitthenner, A. F., and Powell, C. C. 1982. Characterization of *Rhizoctonia* isolates associated with damping-off of bedding plants. Plant Dis. 66:700-703.
Stephens, C. T., Herr, L. J., Schmitthenner, A. F., and Powell, C. C. 1983. Source of *Rhizoctonia solani* and *Pythium* spp. in a bedding plant greenhouse. Plant Dis. 67:272-275.
Wright, E. R., Rivera, M. C., Asciutto, K., and Gasoni, L. 2001. First report of *Rhizoctonia solan*i AG-4-HG-II on garden pink in Buenos Aires, Argentina. Plant Dis. 85:1287.
Wright, E. R., Rivera, M. C., Asciutto, K., Gasoni, L., Barrera, V., and Kobayashi, K. 2004. First report of petunia root rot caused by *Rhizoctonia solani* in Argentina. Plant Dis. 88:86.

(Prepared by M. L. Daughtrey, R. L. Wick, and J. L. Peterson; revised by A. R. Chase and M. L. Daughtrey)

Rhizopus Blight of Gerber Daisy and Vinca

Rhizopus blight of *Gerbera jamesonii* (gerber daisy) and *Catharanthus roseus* (annual vinca) is caused by *Rhizopus stolonifer*—a fungus that often occurs as bread mold or as a storage mold on ornamentals, soft fruits, and vegetables. Living plants are attacked by this fungus only in very humid environments.

Symptoms

Under warm, moist growing conditions, the flowers, leaves, and stems are infected. The blighted tissue often becomes covered with webs of mycelium. Sporangia of the fungus are obvious as black specks on the white mycelial mass.

On vinca, infected stem and leaf tissues are water soaked and develop a mushy, soft rot (Fig. 62). Infection of highly susceptible species and those predisposed to infection by wounding or stress often leads to complete plant collapse during a period of only a few days (Fig. 63).

Rhizopus blight of gerber daisy flowers has also been observed in cut-flower production.

Causal Organism

R. stolonifer (syn. *R. nigricans*) is found worldwide. It produces stolons that grow across the substrate and are anchored by rhizoids at each node. (Internodes may be as long as 1–3 cm.) Hyphae are somewhat branched. The sporangia (85–200 µm in diameter) are hemispheric and produced on long (1,000–2,000 µm) sporangiophores, usually clumped into groups of three or more. The irregular-, round-, or oval-shaped sporangiospores are brownish-black, 18 × 7.8 µm on average, and angular with longitudinal striations. The hyphae are wide and coenocytic.

Colonies on potato dextrose agar are cottony and white at first and turn a brownish-black with age.

Host Range and Epidemiology

R. stolonifer has been reported as a pathogen of *Euphorbia pulcherrima* (poinsettia), *Crossandra infundibuliformis* (firecracker flower), gerber daisy, and annual vinca. The fungus also affects a wide range of nonornamental hosts, most commonly causing a postharvest soft rot of fruit.

At high levels of relative humidity (at least 75%) and temperatures of 21–32°C, the pathogen grows very rapidly. *R. stolonifer* is an opportunistic pathogen, affecting crops exposed to stressful conditions, such as wounding, spikes of heat, and sunburn. Flowers are especially likely to be blighted by this fungus.

Management

Spores of *R. stolonifer* are ubiquitous and survive periods of desiccation, so sanitation of the production area following an outbreak of disease is important yet only partially effective. Infected plant debris should be removed.

The crop should be maintained under optimal cultural conditions, because outbreaks of this disease generally follow periods of stress. It is particularly important to maintain ideal conditions for the prompt callusing and rooting of cuttings, since cutting wounds are the major points of entry for the fungus. Delayed symptom expression has been noted after infection that originally occurred during propagation.

No fungicides are registered for use against *R. stolonifer* in the United States. Literature on controlling Rhizopus blight on fruit crops indicates the possible effectiveness of fungicides that contain fludioxonil (FRAC 12).

Fig. 62. Rhizopus blight (*Rhizopus stolonifer*) on annual vinca in a garden. (Courtesy A. R. Chase—© APS)

Fig. 63. Collapse of a petunia plant caused by Rhizopus blight (*Rhizopus stolonifer*). (Courtesy A. R. Chase—© APS)

Selected References

Engelhard, A. W. 1982. Poinsettia Diseases and Their Control. Res. Rep. BRA 1982-21. Agriculture Research Education Center, University of Florida, Institute of Food and Agricultural Sciences, Gainesville.

Harris, D. C., and Davies, D. L. 1987. A disease of *Vinca rosea* caused by *Rhizopus stolonifer*. Plant Pathol. 36:608-609.

(Prepared by M. L. Daughtrey, R. L. Wick, and J. L. Peterson; revised by A. R. Chase and M. L. Daughtrey)

Rust Diseases

Rust diseases affect a wide range of host plants, including many common bedding plants. Rusts are found most commonly on *Bellis perennis* (English daisy), *Chrysanthemum* spp., *Pelargonium* spp. (geranium), *Salvia* spp., and *Antirrhinum majus* (snapdragon). Although the economic impact caused by symptoms of rust is usually relatively small, the added impact of quarantine measures for chrysanthemum white rust, a federally regulated disease, can result in especially costly losses in chrysanthemum crops.

Rust fungi are obligate parasites and are usually host specific or have limited host ranges (affecting plants in the same plant family). Rusts were previously identified by spore characteristics, but sequences of the D1/D2 domain of the large subunit of the ribosomal gene complex are now used to differentiate similar species.

The greatest damage to bedding plants in the landscape results when a type of rust spore called a "urediospore" occurs on plants. A urediospore is a repeating spore stage that allows rapid intensification of disease within a planting.

Bedding plant rust fungi survive as teliospores, as systemic mycelium in dormant plants (including, in some cases, an alternate perennial host other than the bedding plant), and most commonly, as long-lived spores in the greenhouse. A rust infestation often begins with the introduction of infected plants to the greenhouse. Spores are easily spread by splashing irrigation water and may also be moved throughout the greenhouse by air currents from fans or throughout an outdoor planting by wind.

Selected Reference

Cummins, G. B., and Hiratsuka, Y. 2003. Illustrated Genera of Rust Fungi, 3rd ed. American Phytopathological Society, St. Paul, MN.

(Prepared by A. R. Chase and M. L. Daughtrey)

Bellis Rust

English daisy (*Bellis perennis*, family Asteraceae) is used as an ornamental bedding or potted plant in North America, but this European native is also considered a lawn weed in parts of the United States. Rust caused by *Puccinia lagenophorae* has been found throughout the United States, Europe, Africa, and Central and South America. Bellis rust was first seen on English daisy in the United States in 2000, and it has also been reported on this host in Australia and Europe.

Symptoms

Aecia are conspicuous on leaves, as they are both numerous and colorful. Rust spores appear as raised bumps that open to show orange aeciospores surrounded by white peridia (Fig. 64). Severely diseased leaves wilt and die.

Causal Organism

P. lagenophorae is an autoecious rust that forms only aecia (stage I) and (less often) telia (stage III). Aeciospores from infected English daisy in the United States measured 14–18 × 12.5–15.0 µm. The aeciospores serve as the repeating stage for this rust. A single telium observed on English daisy contained single-celled mesospores (22–34 × 13.5–16.0 µm) and two-celled teliospores (31.0–36.5 × 16–19 [to 22] µm).

Host Range and Epidemiology

P. lagenophorae has been reported on 150 host species in the Asteraceae and the Goodeniaceae, including English daisy, *Senecio cruentus* (cineraria), *Calendula officinalis* (calendula or pot marigold), *Dimorphotheca* spp., *Felicia* spp., *Gazania* spp., and *S. vulgaris* (common groundsel). The latter is a European native that is considered weedy or invasive throughout the United States (Fig. 65).

Fig. 64. Rust sporulation produced by *Puccinia lagenophorae* on an English daisy leaf. (Courtesy A. R. Chase—© APS)

Fig. 65. Rust (*Puccinia lagenophorae*) on common groundsel, a weed. (Courtesy A. R. Chase—© APS)

Cross-pathogenicity of different sources of *P. lagenophorae* has been tested repeatedly. In trials in California (United States), 9–14 days after inoculation with a common groundsel isolate, aecia developed on English daisy, cineraria, and common groundsel but did not develop on *S. cineraria* (dusty miller) or pot marigold. The inoculation experiments indicated that the rusts on English daisy, cineraria, and common groundsel are not biologically separated. This means that the widely distributed common groundsel could easily serve as a weed source of the rust for crops of English daisy.

Management

Bellis rust can kill English daisy plants if left unchecked. The best products for control are strobilurins (FRAC 11) (especially with a wetting-agent adjuvant), demethylation inhibitors (DMIs) (FRAC 3), and mancozeb (FRAC M3).

Selected References

Hernández, J. R., Daughtrey, M., and Jens, J. 2003. Occurrence of rust disease caused by *Puccinia lagenophorae* on cineraria in the Americas. Plant Dis. 87:1266.

Koike, S. T., Scholler, M., Herbaria, A., and Herbaria, K. 2001. First occurrence of a rust fungus on English daisy (*Bellis perennis*) in North America. Plant Dis. 85:562.

Scholler, M. 1997. Rust fungi on *Bellis perennis* in central Europe: Delimitation and distribution. Sydowia 49:174-181.

Scholler, M., Lutz, M., Wood, A. R., Hagedorn, G., and Mennicken, M. 2010. Taxonomy and phylogeny of *Puccinia lagenophorae:* A study using rDNA sequence data, morphological and host range features. Mycol. Prog. 10:175-187.

(Prepared by A. R. Chase and M. L. Daughtrey)

Brown Rust of Chrysanthemum

Brown rust of chrysanthemum (CBR) was first reported from England in 1897 and began to appear throughout Europe and the United States at about the same time. CBR is believed to have occurred in Japan before that time and to have been moved throughout the world on chrysanthemum shipments. Brown rust of chrysanthemum is also known as "chrysanthemum common rust" and "chrysanthemum black rust."

Symptoms

The first symptom of CBR is the development of small, yellow, irregularly shaped patches on the upper leaf surface; the patches eventually turn brown (Fig. 66). Chocolate-brown rust pustules (primarily, uredinia) form on the lower leaf surface opposite these spots, often creating a "bull's eye" pattern when rings of secondary pustules form around the first. Rust lesions can cover the leaves on a highly susceptible cultivar, while other cultivars growing nearby may show few or no pustules. Stems and involucres may also support sporulation at times. Leaves may be killed and plants defoliated. The causal fungus may overwinter on the leaves of basal shoots as mycelium.

Fig. 66. Brown rust (*Puccinia chrysanthemi*) on chrysanthemum leaves. (Courtesy A. R. Chase—© APS)

CBR (caused by *Puccinia chrysanthemi*) does not affect the host as severely as chrysanthemum white rust (caused by *P. horiana*), except in Japan, where CBR commonly causes losses. Losses from CBR were also reportedly significant in Germany in 1926–1927. (See also the following section, "Chrysanthemum White Rust.")

Causal Organism

The CBR pathogen, *P. chrysanthemi* (syn. *P. tanaceti*), is described in Japan as autoecious with a brachy-form life cycle that includes stages 0, II, III, and IV. The rust has been said to have a reduced life cycle outside Japan, including urediospores but not teliospores. Teliospores have, however, been observed in North America and Sweden, as well as Japan. Urediospores are echinulate and rich brown, and they measure 24–52 × 17–27 μm. They have been observed to function as overwintering spores at times.

Teliospores of brown rust are chestnut brown and thus strikingly different from those of white rust. CBR teliospores are oblong to ellipsoid and measure 35–57 × 20–25 μm; a single teliospore has hardly any constriction at the septum, and it has hyaline, persistent pedicels 35–60 μm long. Telia and uredia of CBR may form on the same leaves with telia of chrysanthemum white rust, making field identification challenging.

Host Range and Epidemiology

CBR is hosted primarily by species of *Chrysanthemum,* but there have also been a few reports of the disease on *Achillea, Artemisia,* and *Tanacetum* spp. The optimal temperature range for spore germination is 15–21°C, and infection can occur from 16 to 27°C. Free water on leaves is required for spore germination.

The disease appears to be common only in coastal locations because of a high degree of sensitivity to temperatures higher than approximately 32°C. A well-established rust infection will be killed when temperatures are higher than 38°C for 8 h, and a younger infection is even more easily halted by high temperatures. This sensitivity to high temperatures has been identified as the reason that CBR has never been a significant problem in inland areas of the United States.

Management

Although some research on chrysanthemum cultivar response has shown differences in susceptibility, cultivar selection will not likely be used as a means of reducing CBR (except when extraordinarily susceptible cultivars may fall out of favor with growers). Other more serious diseases of chrysanthemum, as well as market demands, will dictate which cultivars are produced.

The groups of fungicides that control rust diseases in general (including FRAC groups M3, 3, and 11) have been effective in controlling CBR. Monitoring for the first appearance of the disease is important for CBR management.

Selected References

Alaei, H., DeBacker, M., Nuytinck, J., Maes, M., Hofte, M., and Heugens, K. 2009. Phylogenetic relationships of *Puccinia horiana* and other rust pathogens of *Chrysanthemum* × *hortorum* based on rDNA ITS sequence analysis. Mycol. Res. 113:668-683.

Campbell, C. E., and Dimock, A. W. 1955. Temperature and the geographical distribution of chrysanthemum rust. Phytopathology 45:644-648.

Engelhard, A. W., and Magie, R. O. 1972. Occurrence and susceptibility of chrysanthemums to leaf rust in Florida. Plant Dis. Rep. 56:939-941.

Punithalingham, E. 1968. *Puccinia chrysanthemi.* CMI Descriptions of Pathogenic Fungi and Bacteria, No. 175, Set 18. CABI, Kew, Surrey, UK.

(Prepared by A. R. Chase and M. L. Daughtrey)

Chrysanthemum White Rust

Plants with chrysanthemum white rust (CWR) are intercepted at ports of entry across the United States each year, and CWR is occasionally found in outdoor garden chrysanthemums and cut-flower production, especially in the Northeast and along the West Coast. The disease was first described in Japan in 1895 and has been reported from China since 1922. CWR began to be problematic elsewhere in the world in 1963, and it has since become widespread in the Far East, Australia, Africa, Europe, and South and Central America.

CWR was found in both commercial nurseries and gardens in California in 1991, 1998, and 2002. Outbreaks also occurred in commercial nurseries in New Jersey, Oregon, and Washington State in 1995–1997 and in dooryard and hobbyist plantings in New York and New Jersey in 1997. Additional instances were reported in nurseries in Hawaii, Rhode Island, Pennsylvania, Maryland, and Delaware in 2004 and again in Pennsylvania in 2006. In 2007, CWR was found in surveys or trace-forwards/trace-backs in Connecticut, New York, Pennsylvania, Rhode Island, Maine, Delaware, Massachusetts, Maryland, and North Carolina. Since 2007, the disease has been seen sporadically in the northeastern United States. However, the federal quarantine regulations that apply to CWR have led to underreporting of outbreaks, so the actual number of cases is not known.

Symptoms

The first symptom of CWR is the development of yellow spots of various sizes (up to 4 mm in diameter) on the upper leaf surface (Fig. 67). These spots may be mistaken as symptoms or signs of many other diseases and disorders, including spray injury, insect feeding, and virus infection.

The subsequent development of pustules on the lower leaf surface is diagnostic of a rust disease. Pustules may be pinkish or buff colored but mature to a waxy white (Fig. 68); they occur most commonly on the younger leaves and bracts. Pustules may also develop on stems, bracts, and even flowers. Under dry conditions, severely infected leaves desiccate and cling to the stem.

Causal Organism

Puccinia horiana is an autoecious rust fungus with raised, whitish or buff, telial sori that are most often produced on the lower surfaces of leaves. No other spore stages are produced. The two-celled teliospores have pale-yellow walls; they are oblong to oblong–clavate, measure 32–45 × 12–18 μm, and are slightly constricted (Fig. 69). The hyaline pedicel is persistent and may be up to 45 μm long. Teliospores germinate from one or both cells to produce the infective basidiospores.

P. horiana has been shown to establish systemic infections within chrysanthemum plants.

Host Range and Epidemiology

The host range of CWR includes a number of *Chrysanthemum* spp. and their close relatives, including florist's chrysanthemum (*C. morifolium*), the perennial garden plant known as "silver and gold chrysanthemum" (*C. pacificum* [syn. *Ajania pacifica*]), and the Nippon or Montauk daisy (*Nipponanthemum nipponicum*).

Teliospores germinate at 4–23°C to produce basidiospores, which are the means by which the rust spreads to new plants via wind, wind-driven rain, and splashing irrigation water.

Fig. 68. Pustules on the lower surface of a chrysanthemum leaf, symptomatic of white rust (*Puccinia horiana*). (Courtesy M. L. Daughtrey—© APS)

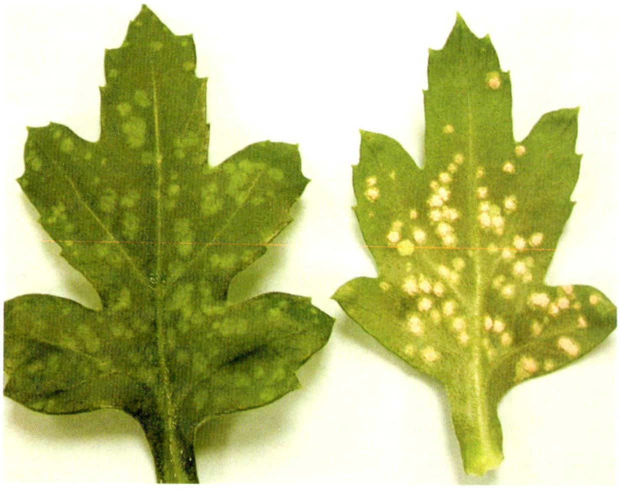

Fig. 67. Signs of white rust (*Puccinia horiana*) on the upper (left) and lower (right) surfaces of chrysanthemum leaves. (Courtesy M. L. Daughtrey—© APS)

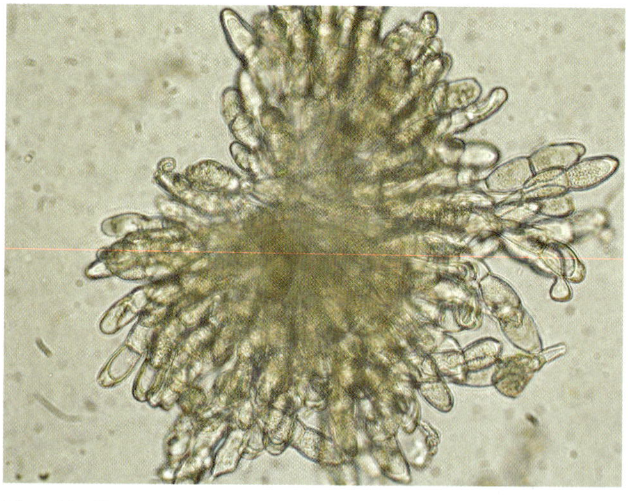

Fig. 69. Two-celled teliospores of *Puccinia horiana,* the fungus that causes chrysanthemum white rust. (Courtesy M. L. Daughtrey—© APS)

Basidiospores are released from pustules after as little as 3 h of high relative humidity (96–100%) when temperatures are between 40 and 73°F (optimum 63°F). Spores that land on a chrysanthemum can germinate and penetrate in as little as 2 h at an optimal temperature (63–75°F). A film of free water is required for infection.

Following infection, the fungus grows in the plant for 5–14 days before causing overt symptoms. Pustules produce teliospores that remain attached to the leaves. They can germinate in place to produce another generation of basidiospores when temperature and humidity conditions are favorable.

The long-distance dispersal of CWR depends on the movement of infected plant materials, which can be latently infected. Basidiospores can be carried short distances (up to 700 m) by wind. However, they are so ephemeral that even under conditions of 100% humidity and cool temperatures, they rapidly lose viability because of desiccation and generally remain viable for no more than 1 h. High relative humidity (90% or greater) is needed to keep basidiospores from drying out. If the humidity level is 80% or less, basidiospores will survive for only 5 min. In one study, CWR teliospores survived associated with dried plant debris for as long as 2 weeks when buried in air-dried soil and for 8 weeks in pustules on detached leaves at 50% relative humidity or lower. In another study, teliospores maintained viability for only 35 days in a growth chamber set to simulate winter temperatures in the northeastern United States. These short periods of teliospore survival are not sufficient to carry the rust fungus through the winter as teliospores; thus, it is presumed that systemic infections reported in the root crowns of chrysanthemum plants are the more likely means by which the pathogen can overwinter in temperate climates.

CWR is commonly most severe in the greenhouse environment, but it can also be destructive outdoors when temperature and humidity permit. Research in the Netherlands has identified the use of artificial substrates as a potential cause of ongoing outbreaks of CWR.

Management

The ability of *P. horiana* to overwinter in the northeastern United States within infected perennial plantings of chrysanthemum reduces the effectiveness of quarantine efforts. Consequently, cultural controls and fungicide treatments are important components of an integrated management program for chrysanthemum crops. Reducing periods of leaf wetness by following careful irrigation practices is essential, but natural rainfall will occasionally provide potential periods of infection.

In a study in Poland, chrysanthemums sprayed preventively with azoxystrobin (FRAC 11), myclobutanil (FRAC 3) + mancozeb (FRAC M3), tebuconazole + triadimefon (FRAC 3), or tridemorph + epoxiconazole (FRAC 5 + 3 [phytotoxic to chrysanthemum cultivar Fiji Yellow and not available for use on ornamentals in the United States]) did not develop CWR. In addition, these fungicides provided more than 95% curative control. The curative/eradicative actions of fungicides have been studied a number of times on CWR. Trials conducted in the United States in 1995 found that myclobutanil (FRAC 3) applied 5 days after infection had very strong curative effects that extended from stock plants to subsequent cuttings. This control was deemed superior when treating disease after infection as opposed to before infection. In postinfection treatments, CWR infection is better managed prior to spore production within the pustules.

CWR is a quarantined pest in the United States; thus, a strict response to an outbreak is required. Currently, the federal quarantine requires the following:

- a 6-month postentry quarantine of imported cuttings
- prohibition of importing cuttings from infested countries
- inspection of imported cut chrysanthemums at the port of entry
- eradication when CWR is found

A grower who finds CWR must inform USDA, state, or county officials. Regulatory officials will supervise eradication and treatment programs that may include these measures:

- destruction of infected plants and those immediately surrounding them
- treatment with eradicant fungicides
- a survey of the surrounding premises
- a trace-back to attempt to determine the source and distribution
- a trace-forward to track the incidence in shipments

More extensive actions may be taken if the infection is widespread in an operation.

Official control programs of the USDA–APHIS are subject to revision. The CWR Eradication Protocol is available online (see Selected References).

Selected References

Bonde, M. R., Murphy, C. A., Bauchan, G. R., Luster, D. G., Palmer, C. L., Nester, S. E., Revell, J. M., and Berner, D. K. 2015. Evidence for systemic infection by *Puccinia horiana,* causal agent of chrysanthemum white rust, in chrysanthemum. Phytopathology 105:91-98.

Bonde, M. R., Palmer, C. L., Luster, D. G., Nester, S. E., Revell, J. M., and Berner, D. K. 2013. Sporulation capacity and longevity of *Puccinia horiana* teliospores in infected chrysanthemum leaves. Online. Plant Health Progress. doi:10.1094/PHP-2013-0823-01-RS

Bonde, M. R., Peterson, G. L., Rizvi, S. A., and Smilanick, J. L. 1995. Myclobutanil as a curative agent for chrysanthemum white rust. Plant Dis. 79:500-505.

Gore, M. E. 2008. White rust outbreaks on chrysanthemum caused by *Puccinia horiana* in Turkey. Plant Pathol. 57:786.

O'Keefe, G., and Davis, D. D. 2012. First confirmed report that *Puccinia horiana,* causal agent of chrysanthemum white rust, can overwinter in Pennsylvania. Plant Dis. 96:1381.

Pedley, K. F. 2009. PCR-based assays for the detection of *Puccinia horiana* on chrysanthemums. Plant Dis. 93:1252-1258.

Rattink, H., Zamorski, C., and Dil, M. C. 1985. Spread and control of white rust (*Puccinia horiana*) on chrysanthemums on artificial substrate. Meded. Fac. Landbouwwet. Rijksuniv. Gent 50:1243-1249.

USDA–APHIS. 2017. Chrysanthemum White Rust Eradication Protocol for Nurseries Containing Plants Infected with *Puccinia horiana* Henn. Online. USDA–APHIS, Washington, DC. www.aphis.usda.gov/plant_health/plant_pest_info/ cwr/downloads/cwrplan.pdf

Wojdyla, A. T. 2004. Development of *Puccinia horiana* on chrysanthemum leaves in relation to chemical compounds and time of their application. J. Plant Prot. Res. 44:91-102.

(Prepared by A. R. Chase and M. L. Daughtrey)

Geranium Rust

Geranium rust was first described in 1926 from South Africa. Cultivation spread the disease throughout the Southern Hemisphere, but it was not found in the Northern Hemisphere until the 1960s. It was reported and observed in the continental United States and Canada in 1967, and by the mid-1970s, it was found throughout the United States.

Symptoms

Leaf symptoms begin as pale-white or yellow spots on both the upper and lower surfaces. Spots on the lower leaf surface enlarge into blisters, which mature into recognizable rust pustules (uredinia) by 10–14 days after infection (Fig. 70); the pustules break open to show reddish-brown urediniospores (Fig. 71). Secondary circles of uredinia may develop, centering around the first pustule to mature. Observing yellow spots on the upper leaf surface and pustules filled with spores on the lower leaf surface, directly opposite, allows a diagnosis of rust. At times, uredinia may be seen on the upper leaf surfaces or on stems, petioles, or stipules.

The geranium crop becomes unmarketable, because severely infected leaves become chlorotic and drop from plants.

Causal Organism

Puccinia pelargonii-zonalis is an autoecious rust. Only uredinial and telial stages occur; urediniospores are the repeating stage most often encountered in greenhouses. Urediniospores (18–29 × 17–25 μm) are light brown, finely echinulate, and subglobose or broadly ovate. There are two equatorial germ pores on each spore. The two-celled, ellipsoid or clavate teliospores are rarely seen on geranium; they are light brown and stalked, measure 36–57 × 19–26 μm, and form either in separate telia or mixed with urediniospores in uredia. The apexes of teliospores are rounded, and they are slightly constricted; germ pores are located near the apex and below the septum.

Fig. 70. Initial appearance of rust pustules (*Puccinia pelargonii-zonalis*) on the lower surface of a geranium leaf. (Courtesy M. L. Daughtrey—© APS)

Fig. 71. Sporulation of *Puccinia pelargonii-zonalis* on the lower surface of a geranium leaf. (Courtesy M. L. Daughtrey—© APS)

Host Range and Epidemiology

Geranium rust has been reported on many *Pelargonium* spp., including *P.* × *hortorum* (florist's geranium) and *P. zonale* (zonal geranium), as well as *P. graveolens, P. hybridum, P. odoratissimum, P. roseum, P.* × *inquinans-zonale,* and *P.* × *hortorum-zonale.* Not all *Pelargonium* spp. are susceptible, and susceptibility varies among the many cultivars of florist's geranium. In tests of 77 commercial florist's geranium cultivars in Georgia, 99% were susceptible to rust. Only one tested cultivar appeared to be immune, and six others developed pustules on less than 10% of the leaves. *P. domesticum* (regal or Martha Washington geranium) is not a host for geranium rust, and some cultivars of *P. peltatum* (ivy geranium) are resistant. The disease has not been reported on hardy species of *Geranium* that are grown as herbaceous perennials.

For optimal germination, urediniospores of *P. pelargonii-zonalis* need free water for an approximately 3-h period, along with a temperature of 16–21°C. Disease develops most quickly at 21°C. Penetration of the leaf occurs through the stomates. Following a disease outbreak, urediniospores will remain viable in the greenhouse for at least 3 months.

Management

Incoming cuttings and plugs should be checked for symptoms and signs of rust. Geranium crops should be monitored regularly for symptoms, and the undersurfaces of leaves with yellow spots should be checked for rust pustules. All infected plants should be carefully discarded by gently placing them into plastic bags that have been brought to the area of the diseased plants.

The recommended cultural controls are similar to those used to manage Botrytis blight. Critical attention to managing leaf wetness duration and relative humidity (particularly the uses of ventilation and heat at sunset) will help drive moisture-laden air out of the greenhouse. Any means of improving air movement around plants should be used.

Geranium rust can be introduced with cuttings and plants via latent infections. In a test of hot-water treatment on cuttings, a 90-sec dip in hot water at 50 or 52°C was not completely effective for killing urediniospores; some were able to germinate. In addition, young leaves were damaged by the treatment, although the damage was temporary. Given these outcomes, hot-water treatment does not appear to be an effective option in commercial settings.

Extensive studies on the effectiveness of fungicide applications before and after infection have been conducted on a variety of rust fungi, including *P. pelargonii-zonalis.* In one study, five fungicides registered for use against rusts on ornamental crops were evaluated on geranium: azoxystrobin (FRAC 11); myclobutanil, propiconazole, and triadimefon (FRAC 3); and chlorothalonil (FRAC M5). Four of the fungicides (all but propiconazole) significantly reduced lesion development on geranium when applied preventively up to 15 days before inoculation and infection. Some curative activity was observed for azoxystrobin, myclobutanil, propiconazole, and triadimefon on geranium when applied up to 7 days after infection. In general, preventive fungicide efficacy decreased the longer the time from application to inoculation, and curative activity decreased the longer the time from inoculation to treatment. Chlorothalonil, a contact fungicide, was found effective as a preventive but not as a curative treatment.

Research conducted in vitro evaluated 12 fungicides in seven chemical classes for toxicity to urediniospores of *P. pelargonii-zonalis.* Germination was sharply suppressed during and after exposure to azoxystrobin, chlorothalonil, copper sulfate pentahydrate (FRAC M1), mancozeb (FRAC M3), and trifloxystrobin (FRAC 11); no more than approximately 1% of urediniospores germinated, in contrast to 54–88% germination in controls. Zero to 60% of urediniospores successfully germinated during exposure to fenhexamid (FRAC 17), iprodione (FRAC 2), myclobutanil, propiconazole, thiophanate-methyl (FRAC 1), triadimefon, and triflumizole (FRAC 3), and the percentage of germination usually increased after the fungicide was removed. Germination of urediniospores decreased following a 1-min exposure to azoxystrobin, copper sulfate pentahydrate, or mancozeb. A 30-min exposure to any of these fungicides halted germination almost entirely, while a 4- or 8-h exposure was needed to halt germination using trifloxystrobin or chlorothalonil, respectively.

Less than one lesion per plant developed on geranium seedlings sprayed with azoxystrobin, chlorothalonil, copper sulfate pentahydrate, or mancozeb before they were inoculated with urediniospores of *P. pelargonii-zonalis*; control seedlings developed 148 lesions per plant. Strobilurins (azoxystrobin and trifloxystrobin), broad-spectrum protectants (chlorothalonil and mancozeb), and inorganic copper (copper sulfate pentahydrate) were all fungicidal to urediniospores. However, the benzimidazole (FRAC 1—thiophanate-methyl), dicarboximide (FRAC 2—iprodione), hydroxyanilide (FRAC 17—fenhexamid), and demethylation-inhibiting (DMI) (FRAC 3—myclobutanil, propiconazole, triadimefon, and triflumizole) fungicides were only fungistatic to rust urediniospores.

Drench applications of fungicides to prevent geranium rust have shown excellent control with azoxystrobin up to 3 weeks after a single application. Myclobutanil provided excellent control at 1 week but lost some efficacy at the 3-week rating.

Little research has been done on the biological control of geranium rust. A 1989 study examined *Bacillus* isolates from florist's geranium. A strain of *B. subtilis* (either as washed cells or a cell-free culture filtrate) was found to inhibit urediniospore germination consistently and reduced disease severity on treated plants. Further development of commercial biocontrol products may eventually lead to an effective tool for geranium rust management.

Selected References

Buck, J. W. 2007. Potential risk of commercial geranium to infection by *Puccinia pelargonii-zonalis*. Online. Plant Health Progress. doi:10.1094/PHP-2007-1031-02-RS

Harwood, C. A., and Raabe, R. D. 1979. The disease cycle and control of geranium rust. Phytopathology 69:923-927.

McCoy, R. E. 1975. Susceptibility of *Pelargonium* species to geranium rust. Plant Dis. Rep. 59:618-620.

Mueller, D. S., Jeffers, S. N., and Buck, J. W. 2004. Effect of timing of fungicide applications on development of rusts of daylily, geranium, and sunflower. Plant Dis. 88:657-661.

Mueller, D. S., Jeffers, S. N., and Buck, J. W. 2005. Toxicity of fungicides to urediniospores of six rust fungi that occur in ornamental crops. Plant Dis. 89:255-261.

Phillips, D. J., and McCain, A. H. 1973. Hot-water therapy for geranium rust control. Phytopathology 63:273-275.

Rytter, J. L., Lukezic, F. L., Craig, R., and Moorman, G. W. 1989. Biological control of geranium rust by *Bacillus subtilis*. Phytopathology 79:367-370.

Scocco, E. A., Buck, J. W., and Jeffers, S. N. 2012. Drench applications of fungicides for management of geranium rust, 2011. Plant Dis. Manag. Rep. 6:OT018.

(Prepared by A. R. Chase and M. L. Daughtrey)

Salvia Rusts

Many species of *Puccinia* have been reported worldwide from various species of *Salvia* (Fig. 72), as well as species of *Aecidium, Coleosporium,* and *Uredo*. Salvia rusts have been reported numerous times in the United States since 1900.

Fig. 72. Rust (*Puccinia* sp.) on blue salvia. (Courtesy A. R. Chase—© APS)

In 1995, a salvia rust was described from Louisiana that had previously been reported in the United States around 1900 in Florida and Texas. Telia of *P. salviicola* were seen on stems of *S. coccinea* (Texas sage) cultivars Bicolor, Lactea, and Punicea. Although *S. leucantha, S. guaranitica,* and *S. elegans* were growing in the same garden, they did not become infected with the rust. Leaves of Texas sage showed cinnamon-brown uredinia and, less often, telia. Urediniospores (19–25 × 21–27 μm) were prominently echinulate and had two equatorial pores. Telia were blackish-brown and had puccinioid, smooth-walled teliospores (18–28 × 35–50 μm); the pore of the upper cell at the apex and the pore of the lower cell were equatorial. This rust has also been reported from Mexico and Cuba.

In 2008, *S. greggii* (autumn sage) plants in a container nursery in California were found to have moderate defoliation associated with heavy coverings of rust pustules on older leaves. Only the cultivars with solid red or pink flowers were affected; for example, autumn sage cultivar Hotlips, a popular bicolor, was not affected. Round, cinnamon-brown uredinia appeared on both the upper and lower leaf surfaces (sometimes, with yellow halos), and some pustules were found on stems, as well. The echinulate urediniospores (22–27 × 24–32 μm; average 24.9 × 26.9 μm) were broadly obovoid and subglobose to broadly ellipsoid; a single urediniospore had one apical and two or three equatorial pores. Telia were found (primarily in stem lesions) associated with the same disease a few years later. Teliospores were chocolate brown and two celled (25–31 × 32–40 μm; average 28.6 × 35.3 μm). The rust fungus was identified as *P. ballotiflora* (syn. *P. ballotaeflora*). It had previously been reported from Texas, Mexico, and Colombia on other *Salvia* spp. but never seen before on autumn sage.

P. ballotiflora is now well established in horticultural trade in California; frequent fungicide applications are used to maintain plant salability. As the line between plants used for annual bedding and those marketed as perennials continues to become blurred, many more rust diseases will have the potential to become problematic.

Selected References

Blomquist, C. L., Scheck, H. J., Woods, P. W., and Bischoff, J. F. 2014. First detection of *Puccinia ballotiflora* on *Salvia greggii*. Plant Dis. 98:1270.

Holcomb, G. E., and R. A. Valverde. 1995. First report of rust on *Salvia coccinea* in Louisiana. Plant Dis. 79:426.

(Prepared by A. R. Chase and M. L. Daughtrey)

Snapdragon Rust

The serious rust disease of snapdragon (*Antirrhinum majus*), caused by the fungus *Puccinia antirrhini,* was first reported in 1879 in California and then in 1909 in Oregon. It has since been prevalent in the United States, as well as Canada and Bermuda, for many years.

Outside these areas, the disease was practically unknown until 1933, when it was noted to have spread quite rapidly. The initial report was made from England in 1933, and by 1934, the disease was present in 28 countries in Europe. By 1935, it was reported in Scotland, Ireland, Holland, Denmark, Germany, Czechoslovakia, Italy, and Austria. The disease had become well established in Eastern Cape Province of South Africa and in Egypt by 1956.

Snapdragon is native to the Mediterranean region, where the rust was previously unknown. Thus, it seems probable that the rust was indigenous to California on a wild relative of snapdragon and somehow introduced to areas in which snapdragon was cultivated.

Symptoms

An early symptom of snapdragon rust is the development of small, round (0.16 cm diameter), light-colored flecks, which occur first on the lower leaf surface and then on the opposite upper leaf surface. The spots soon produce pustules (uredinia) in great profusion on the lower surfaces of leaves (Fig. 73). Concentric rings of pustules filled with dark-brown spores may develop around the initial spots (Fig. 74).

When rust pustules appear on stems, they are somewhat elongated; when stems become girdled, branches may die. Petioles and calyces may also bear sporulation. Plants may be stunted or killed by this fungus infection, especially those subjected to rainy conditions in the landscape.

Causal Organism

Snapdragon rust is caused by *P. antirrhini.* The spores most commonly seen during the growing season are urediniospores (22–30 × 21–25 µm), produced in uredinia. Urediniospores are yellowish-brown, spherical to elliptical, and echinulate; a single urediniospore has two or three pores. Cold temperatures and/or day-length effects on outdoor plantings may result in the appearance of brown teliospores (36–50 × 17–26 µm); a single teliospore has a pointed, rounded, or truncated apex and a slight constriction between the two cells.

Host Range and Epidemiology

P. antirrhini has been reported from a number of wild species of *Antirrhinum, Cordylanthus,* and *Linaria* in California and from a *Fuchsia* sp. in Mexico. However, the economic effects caused by this fungus have all come from its infection of snapdragon, a valuable cut-flower and bedding plant. As early as 1937, the existence of at least two races of *P. antirrhini* had been postulated, based on reactions of resistant and susceptible snapdragon cultivars.

P. antirrhini is a short-cycle rust. It is autoecious and has only uredial and telial stages, both of which occur on snapdragon and presumably the other known hosts. *P. antirrhini* survives in plant debris, rather than in soil.

Disease development is favored by cool nights (10–13°C) and warm days (21–24°C)—conditions that provide abundant dew and thus facilitate infection. Spore germination can occur at temperatures as low as 5°C; the optimum is 10°C and the maximum, 20°C. Disease severity is much reduced when inoculations are made at 15 or 18°C compared with 10°C. Urediniospores are killed by extended exposure to temperatures above 34°C.

On the West Coast of the United States, snapdragon rust is usually problematic in the low desert areas during early to late spring and then develops in areas of higher elevation during the summer. In coastal areas of California, the rust is found year-round, but in the U.S. Northeast, it is a summer disease, affecting plants primarily in the landscape. In drier climates, the rust appears to injure its host through the desiccation of infected leaves, whereas in wetter climates, the attack of secondary fungi, such as *Fusarium* spp., has been observed to cause infected leaves to deteriorate.

Management

In the late 1990s, researchers studied the susceptibility of snapdragon cultivars to rust over three seasons in the southeastern United States. Disease severity from year to year was extremely variable; ratings ranged from 1 (no rust) to 6 (76–100% of foliage had pustules). Based on the extreme variability of single cultivars to rust infection, determining the possibility of races of *P. antirrhini* and even choosing resistant plants for commercial use seems daunting. Ten *Antirrhinum* spp. were identified as being resistant to rust: *A. asarina, A. calycinum, A. charidemi, A. chrysothales, A. glandulosum, A. glutinosum, A. ibanjezii, A. maurandioides, A. orontium,* and *A. siculum.* However, in the 1950s, these species were deemed too diffi-

Fig. 73. Spores produced by *Puccinia antirrhini,* the pathogen that causes snapdragon rust, occur primarily on the lower leaf surface. (Courtesy M. L. Daughtrey—© APS)

Fig. 74. Concentric rings of pustules may develop on leaves and stems with snapdragon rust (*Puccinia antirrhini*). (Courtesy A. R. Chase—© APS)

cult to cross with snapdragon (*A. majus*). Perhaps more modern plant-breeding techniques will help solve this problem.

Like many other rusts, snapdragon rust can be effectively controlled by applying fungicides to healthy plants: namely, chlorothalonil (FRAC M5); mancozeb (FRAC M3); mancozeb + thiophanate-methyl (FRAC M3 + FRAC 1); myclobutanil, triadimefon, or triforine (FRAC 3); and strobilurins (FRAC 11). Neem oil and horticultural spray oil have also been used to treat rust with some success.

Infected plants in greenhouses, nurseries, and landscapes should be removed and destroyed. Growers should always monitor plants carefully for the first symptoms, so that large amounts of inoculum will not already have been produced by the time symptoms are observed. In greenhouses, growers should prevent moisture condensation and avoid wetting foliage. Plants should be spaced to allow good air circulation.

Resistance is available in some cultivars, but effectiveness may vary with location.

Selected References

Doran, W. L. 1921. Rust of *Antirrhinum*. Mass. Exp. Stn. Bull. 202:39-66.

Emsweller, S. M., and Jones, H. A. 1934. The inheritance of resistance to rust in snapdragon. Hilgardia 8(7):197-211.

Holcomb, G. E., Buras, H., and Cox, P. 1998. Reaction of snapdragon cultivars to leaf rust, 1997. Biol. Cult. Tests 13:70.

Holcomb, G. E., Buras, H., and Cox, P. 1999. Reaction of snapdragon cultivars to leaf rust, 1998. Biol. Cult. Tests 14:73.

Holcomb, G. E., Buras, H., and Cox, P. 2000. Reaction of snapdragon cultivars to leaf rust, 1999. Biol. Cult. Tests 15:76.

McClellan, W. D. 1953. Rust and other disorders of snapdragon. Pages 568-572 in: Plant Diseases: The Yearbook of Agriculture 1953. U.S. Department of Agriculture, Washington, DC.

Peltier, G. L. 1919. Snapdragon rust. Univ. Ill. Agric. Exp. Stn. Bull. 221:534-548.

Yarwood, C. E. 1937. Physiologic races of snapdragon rust. Phytopathology 27:113-115.

(Prepared by A. R. Chase and M. L. Daughtrey)

Disease Caused by *Sclerotinia sclerotiorum*

Sclerotinia sclerotiorum infects more than 360 plant species, including important vegetable and ornamental crops. Sclerotinia blight is a serious disease seen occasionally in the landscape but more commonly in the outdoor production of bedding plants and occasionally in the greenhouse. The genus *Sclerotinia* is related to the genus *Botrytis,* and many of the plants that are attacked by *B. cinerea* are also highly susceptible to *S. sclerotiorum.*

Sclerotinia blight appears to be distributed worldwide on bedding plants. Lisianthus (*Eustoma grandiflorum*), *Petunia hybrida,* primrose (*Primula* spp.), and sweet alyssum (*Lobularia maritima*) are common hosts of *S. sclerotiorum.*

Symptoms

Flower or leaf spots and stem and crown rot are symptoms of Sclerotinia infection. Sclerotinia blight most often develops as a stem or crown rot that appears suddenly and proceeds to engulf entire plants or flats of plants (Fig. 75). The disease is sometimes called "white mold" in reference to the massive, white growth of mycelium that covers the stems and sometimes other parts of infected crops (Fig. 76). Infected tissue turns brown, and the final outcome is rapid and complete collapse of the infected plant.

White, fanlike growth of mycelium across the surface of the potting medium is often seen and somewhat diagnostic. Black sclerotia form at the edges of a blighted area; these structures resemble mouse or rat droppings. Sclerotinia blight is often misdiagnosed as infection with a *Rhizoctonia, Phytophthora,* or even *Pythium* sp.

Causal Organism

The conspicuous mycelium of *S. sclerotiorum,* an ascomycete, is white and cottony; the fungus has no asexual spore form. The sexual spore structures often go unnoticed, but they may be important for disease spread. Apothecia are 2–4 mm across and borne on stalks 2.0–2.5 cm long. Sclerotia are seen frequently and form conspicuous lumps within or on top of mycelium; they vary in size but are commonly 5–30 × 3–10 mm. A single sclerotium has a dark-brown to black rind surrounding a white mass of pseudoparenchymatous cells. Sclerotia also often form inside or on rotted stems or under the leading edge of a melted-out area in a flat (Fig. 77). They are effective survival structures and known to retain viability for years.

Host Range and Epidemiology

A number of bedding plants are common hosts of Sclerotinia blight, including sweet alyssum, begonia (especially wax begonia [*Begonia semperflorens*]), gerber daisy (*Gerbera jamesonii*), larkspur (*Delphinium* spp.), lisianthus, lobelia (*Lobelia erinus* and others), pansy (*Viola* × *wittrockiana*), petunia, primrose, garden stock (*Matthiola incana*), and *Zinnia elegans.* Cultivar resistance has not been reported in any bedding plant.

Fig. 75. Sclerotinia blight (*Sclerotinia sclerotiorum*) on a flat of blue salvia. (Courtesy A. R. Chase—© APS)

Fig. 76. White mycelium produced by *Sclerotinia sclerotiorum* on stems of stock (*Matthiola* sp.) plants, characteristic of Sclerotinia blight. (Courtesy A. R. Chase—© APS)

The primary source of infection is sclerotia from soil and plant debris. Sclerotia are quite resistant to environmental extremes, including drought and cold. The optimal temperature for germination of sclerotia is 13–15°C, but a few will germinate at the extremes of the range of 4–26°C. In the landscape, the soil must be very wet for an extended period for sclerotia to germinate (reportedly, as long as 10 days).

Sclerotia may either germinate to infect plants directly, or more importantly, they may produce one or more (as many as 100) of the sexual fruiting structures (apothecia) that disseminate ascospores. Inoculum production from apothecia greatly magnifies the inoculum potential of sclerotia. In one laboratory study, an average of 2.3×10^6 ascospores were produced per single apothecium. Apothecia do not usually form in the greenhouse; the absence of ultraviolet (UV) light in a greenhouse deters the formation of sclerotia. In one study, the production of apothecia under a standard agricultural vinyl film was compared with that under a UV-absorbing film; apothecium formation was found to be significantly or entirely inhibited under the UV-absorbing vinyl film. This light effect may be important to growers that use plastic-covered structures (high tunnels).

On the West Coast of the United States, Sclerotinia blight often starts at the end of spring, when temperatures increase. Disease development is promoted by temperatures around 10–24°C, damage to the crop (phytotoxicity and wounding), and poor air circulation. Symptoms appear within 10 days of infection, depending on the environmental conditions.

Management

A 1982 report evaluated the susceptibility of a variety of bedding plants to *S. sclerotiorum*. Of the 25 seed-propagated and 22 cutting-propagated geraniums (*Pelargonium* spp.) tested, none was susceptible. In the same trial, all 25 of the pansy cultivars tested were also resistant to infection. Among the 16 petunias tested, most were susceptible to some degree, as were the zinnia, *Dahlia* spp., *Calendula* spp., and strawflower (*Helichrysum bracteatum*) that were tested. Among the marigolds (*Tagetes* spp.) tested, the most resistant cultivars were Dwarf French Gold, Harmony Boy, and Red Wheels; the most susceptible were Viking and Pumpkin Crush. Although the popular cultivars have changed since this study was conducted, it is helpful to know that cultivar resistance is possible.

Fig. 77. Sclerotia formed by *Sclerotinia sclerotiorum* in a zinnia stem, symptomatic of Sclerotinia blight. (Courtesy A. R. Chase—© APS)

Most recently, studies were reported on four genera of bedding plants tested under field conditions. Two moss rose cultivars (*Portulaca grandiflora* 'Sundial Scarlet' and 'Sundial White'), a star flower cultivar (*Pentas lanceolata* 'Graffiti Pink'), and a fan flower cultivar (*Scaevola aemula* 'Whirlwind White') were highly susceptible to Sclerotinia blight in a controlled environment but had significantly lower levels of disease incidence and severity than the zinnia hybrid (*Z. elegans* × *Z. angustifolia* 'Profusion White'), the susceptible control, under field conditions. New Guinea impatiens (*Impatiens hawkeri*) cultivars Sonic Red, Sonic Amethyst, and Sonic White and garden impatiens (*I. walleriana*) cultivars Blitz 3000 Red, Blitz 3000 White, and Blitz 3000 Rose displayed abscission of diseased plant tissue in an atypical resistance response. Plants from all four genera evaluated became infected with *S. sclerotiorum* to a lesser extent than susceptible control plants under field conditions and could be used as part of an integrated disease management program for Sclerotinia blight in ornamental flowerbeds planted year after year.

Fungicides with the highest efficacy in preventing Sclerotinia blight include those that contain chlorothalonil (FRAC M5), the combination of fludioxonil and cyprodinil (FRAC 12 and 9), and the combination of pyraclostrobin and boscalid (FRAC 11 and 7). The latter premix has been very effective in eradicating Sclerotinia blight on bedding plants.

Selected References

Garibaldi, A., Minuto, A., and Gullino, M. L. 2001. First report of *Sclerotinia sclerotiorum* on *Calendula officinalis* in Italy. Plant Dis. 85:446.

Grabowski, M. A., and Malvik, D. K. 2015. Evaluation of annual bedding plants for resistance to white mold. HortScience 50:259-262.

Holcomb, G. E. 2001. First report of occurrence of Sclerotinia blight on petunia in Louisiana. Plant Dis. 85:95.

Honda, Y., and Yunoki, T. 1977. Control of Sclerotinia disease of greenhouse eggplant and cucumber by inhibition of development of apothecia. Plant Dis. Rep. 61:1036-1040.

Schwartz, H. F., and Steadman, J. R. 1978. Factors affecting sclerotium populations of, and apothecium production by, *Sclerotinia sclerotiorum*. Phytopathology 68:383-388.

Wegulo, S. N., and Counsell, J. 2007. Evaluation of fungicides for control of Sclerotinia rot of stock, 2006. Plant Dis. Manag. Rep. 1:OT02.

Wright, E. R., and Palmucci, H. E. 2003. Occurrence of stem rot of chrysanthemum caused by *Sclerotinia sclerotiorum* in Argentina. Plant Dis. 87:98.

(Prepared by A. R. Chase and M. L. Daughtrey)

Southern Blight

Southern blight occurs on many agronomic, vegetable, and ornamental crops in the southern United States, in Central and South America, and in Africa, southern Asia, and Australasia. Reports have also been made from other areas of the world, such as the United Kingdom, central and southern Europe up to southern Sweden, and the northeastern United States and California. Even so, the disease is most common in areas with tropical and subtropical climates.

Symptoms

The pathogen, *Athelia rolfsii,* attacks all portions of the plant but most commonly affects the stems and leaves. Initially, symptoms on stems are confined to water-soaked, necrotic lesions at or near the soil line.

Round sclerotia form almost anywhere on the affected portions of the plant or on the soil surface (Fig. 78). They are initially white and cottony and approximately the size of a mustard seed. As sclerotia mature, they turn tan and eventually dark brown and also harden. Mycelia and sclerotia generally develop just after rotting of the stem and wilting of the plant, allowing an accurate diagnosis.

Causal Organism

A. rolfsii (syns. *Corticium rolfsii; Pellicularia rolfsii; Sclerotium rolfsii*) has white, septate mycelium with clamp connections. The mycelium forms sclerotia (often, after the host is dead) that are roughly 1–2 mm in diameter and smooth or with a shallowly pitted surface; sclerotia are initially light colored and change to tan and then dark brown. Sclerotia are almost perfectly spherical and have differentiated rinds.

The fungus can live in soil as a saprophyte on crop residues or survive as sclerotia.

Host Range and Epidemiology

A. rolfsii attacks at least 500 species of plants, including some ornamental groundcovers, most tropical foliage plants, and many vegetable crops. *A. rolfsii* has been problematic on a few bedding plants. In the United States, *Catharanthus roseus* (annual vinca) was reported as a host in Louisiana in 1999, and *Eustoma grandiflorum* (lisianthus) was described as a host in Florida in 2000. Other bedding plants reported as hosts include *Ageratum houstonianum, Antirrhinum majus* (snapdragon), *Calendula officinalis, Campanula carpatica, Centaurea cyanus* (cornflower), *Gerbera jamesonii* (gerber daisy), *Matthiola incana* (garden stock), and *Petunia hybrida,* as well as species of *Aster, Begonia, Chrysanthemum, Coreopsis, Cosmos, Dianthus, Impatiens, Ranunculus, Salvia,* and *Zinnia*.

A. rolfsii can survive as sclerotia up to 5 years in plant debris and soil. Disease development is favored by a high level of soil moisture, a pH between 3 and 5, and temperatures between 25 and 35°C. Symptoms on lisianthus developed 6 days after infection, whereas those on vinca developed after only 3 days. The fungus can be distributed as a contaminant in seed.

Management

Following good sanitation practices is the primary strategy for managing southern blight. In landscape plantings, growers should promptly remove diseased plants and surface soil in patches where southern blight has occurred. Replacing plants with less susceptible ornamentals is recommended. Shading, crowding, and practices that keep the soil surface moist should be avoided, because all favor the development of disease.

Chemical control of southern blight is difficult, and product registrations for this use are limited. The only registered fungicide for control of southern blight on bedding plants is fludioxonil (FRAC 12). *Trichoderma* spp. have been studied as potential bioantagonists.

Fig. 78. Sclerotia produced by *Athelia rolfsii,* the fungus that causes southern blight, at the base of a New Guinea impatiens stem in the landscape. (Courtesy A. R. Chase—© APS)

Selected References

Holcomb, G. E. 2000. First report of *Sclerotium rolfsii* on *Catharanthus roseus*. Plant Dis. 84:200.

McGovern, R. J., Bouzar, H., and Harbaugh, B. K. 2000. Stem blight of *Eustoma grandiflora* caused by *Sclerotium rolfsii*. Plant Dis. 84:490.

Mordue, J. E. M. 1974. *Corticium rolfsii*. CMI Descriptions of Pathogenic Fungi and Bacteria, No. 410. CABI, Kew, Surrey, UK.

(Prepared by A. R. Chase and M. L. Daughtrey)

Disease Caused by *Thielaviopsis basicola*

Thielaviopsis basicola is a widely distributed root pathogen and has been reported from at least a dozen plant families, including the Fabaceae, Solanaceae, and Cucurbitaceae. The disease caused by this fungus is black root rot, and many ornamental plants are susceptible to it, including both woody and herbaceous species. In the bedding plant industry, *T. basicola* is most commonly encountered on plug-produced crops that are grown in reused flats or under conditions of stressful temperatures.

Symptoms

Aboveground symptoms of *T. basicola* infection are not unique; they are common to other causes of root disease and to unfavorable growing conditions. Depending on the susceptibility of the host and the degree of root rot, wilting may or may not occur. Infected plants are significantly stunted, often chlorotic, and sometimes purplish in color. The portions of the roots affected by *T. basicola* darken to a very dark brown or black—hence, the common name "black root rot" (Fig. 79).

In some plants, such as pansy (*Viola* × *wittrockiana*) and viola (*V. cornuta*), distinct blackening occurs when abundant

Fig. 79. Black root rot on pansy, caused by *Thielaviopsis basicola*. (Courtesy M. L. Daughtrey—© APS)

chlamydospores and pseudoparenchymatous stromatic tissues develop on the surfaces of root lesions. However, when these darkly pigmented structures do not form in abundance or form deeper in the root cortex, lesions are more brown than black and may be more difficult to distinguish from those caused by other root pathogens, such as *Rhizoctonia solani*.

The most commonly affected bedding plants are pansy and viola, followed by *Calibrachoa hybrida* (Fig. 80), annual vinca (*Catharanthus roseus*) (Fig. 81), *Salvia* spp., and *Petunia hybrida* (Fig. 82). Plants show wilting, chlorosis, and stunting, along with the diagnostic blackened root system. An irregular stand of plants, with variable amounts of stunting, is a key diagnostic feature, and deaths of some seedlings in plug trays and packs is common. Black root rot can be distinguished from a nutritional problem, because the latter is more likely to occur uniformly throughout the crop, rather than in the more random pattern typical of a contagious disease. A pictorial severity key has been developed for pansy/viola and black root rot.

On *Cyclamen persicum, T. basicola* causes black spotting and discoloration of roots; the small feeder roots are most commonly infected. Severely affected plants grow slowly, but the fungus generally does not invade cyclamen corms or leaf petioles.

T. basicola causes severe root rot in gerber daisy (*Gerbera jamesonii*) in both Europe and the United States. Leaves of affected plants are stunted and distorted and develop marginal chlorosis. Infected roots are at first spotted and become black with age.

Causal Organism

T. basicola (syn. *Chalara elegans*) is a dematiaceous hyphomycete that can be identified by cylindrical endoconidia and dark-brown chains of chlamydospores (aleuriospores). Conidia are 7.5–19.0 × 3–5 µm and extruded in long chains from the conidiophore; they are unicellular, hyaline, and cylindrical to doliform and have truncate or obtuse ends. Chlamydospores (also called "aleuriospores") are 6.5–14.0 × 9–13 µm and form in series of 5–7 loosely connected spore units (Fig. 83); they are dark brown and unicellular, have thin outer and thicker inner walls (1–2 µm), and have transverse germ pores at both ends. Chlamydospores are short and cylindrical, except for the apical chlamydospore, which is conoid. Chlamydospores are distinctive when viewed under magnification.

The fungus can often be seen within infected root tissue using a compound microscope, but direct isolation of the fungus on culture media can be challenging. *T. basicola* can be baited from infected tissue by spreading a thin layer of soil or colonized root tissue over 5-mm-thick carrot root disks in sterile petri dishes. Sterile water is then atomized over the disks to moisten the soil or roots; after 2–4 days at room temperature, the incubated disks are rinsed with water and incubated again. If the fungus is present, endoconidia form in about 6 days, and the chlamydospores form thereafter. The long chains of hyaline endoconidia can ordinarily be identified with a dissecting microscope. Isolates on potato dextrose agar are slow growing

Fig. 80. Black root rot on calibrachoa, caused by *Thielaviopsis basicola.* (Courtesy M. L. Daughtrey—© APS)

Fig. 82. Irregular stand of petunias, some with stunted growth, caused by black root rot (*Thielaviopsis basicola*). (Courtesy M. L. Daughtrey—© APS)

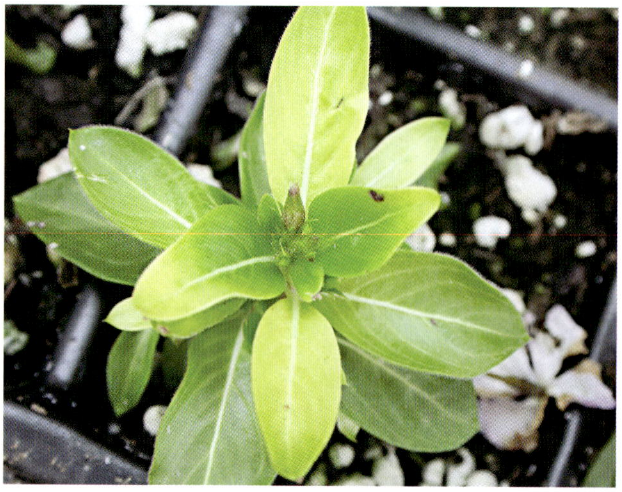

Fig. 81. Severe chlorosis of annual vinca leaves, symptomatic of black root rot (*Thielaviopsis basicola*). (Courtesy M. L. Daughtrey—© APS)

Fig. 83. Chlamydospores and an endoconidiophore of *Thielaviopsis basicola,* the fungus that causes black root rot. (Courtesy M. L. Daughtrey—© APS)

and usually develop a brown pigment, but sectoring to a less-pigmented mycelium has been observed in cultures maintained over time.

Host Range and Epidemiology

Bedding plant hosts of *T. basicola* include snapdragon (*Antirrhinum majus*), *Begonia* spp., *Calibrachoa* hybrids, annual vinca, cyclamen, *Diascia* hybrids, gerber daisy, florist's geranium (*Pelargonium* × *hortorum*), pansy, petunia, *Phlox* spp., *Verbena* spp., and viola.

T. basicola forms chlamydospores, which allows long-term survival in the soil. Chlamydospores are stimulated to germinate by root exudates over a range of pH values (5.0–8.5).

Root rot is most severe at soil temperatures of 13–17°C, but some researchers have reported optimal temperatures as high as 25°C. Moderate disease development may occur when the moisture-holding capacity (MHC) of the soil is at 36%, but the disease is significantly more serious at MHCs of 70% and higher. Root rot is also more serious in plants grown in neutral and alkaline soils. The disease is greatly reduced at pH levels below 5.5.

There are reports of *T. basicola* contamination of commercial peat moss. In Florida (United States) citrus crops, three of 14 samples of unused peat-based potting media and two of 12 Canadian bales of sphagnum peat were found to be infested with low levels of *T. basicola*. Isolates of *T. basicola* from these sources were able to cause significant black root rot on several species of citrus tested.

The reuse of trays and pots contaminated with chlamydospores is a problem within the bedding plant industry. Crops highly susceptible to black root rot should not be grown in reused pots unless the pots are first thoroughly washed and disinfested.

Both fungus gnats (*Bradysia* spp.) and shore flies (*Scatella stagnalis*) may serve as vectors of the fungus. Chlamydospores of *T. basicola* were consistently observed in frass excreted by adults and larvae of shore flies collected in the immediate vicinity of naturally infected corn salad (*Valerianella locusta*) plants from a commercial greenhouse production facility. Approximately 95% of the adult flies and 85% of the larvae were internally infested with the pathogen. Pathogen-free adult shore flies were subsequently shown to acquire the pathogen by ingestion when feeding on naturally infected plants. Viable propagules of the pathogen were excreted by these internally infested adults; thus, they were able to transmit the pathogen to healthy seedlings, which subsequently became infected.

Management

Cultivars of pansy have been tested for susceptibility to *T. basicola;* the greatest resistance was seen in the cultivars Fama Silver Blue, Clear Sky White, Crown Golden, Fama Blue Angel, Fama Dark-Eyed White, and Happy Face Y1/Blotch. In general, cultivar resistance is not a practical method of control, given the rapidly changing spectrum of bedding plant cultivars.

Several factors—including a pH below 5.5 and low phosphorus, high ammonium, and high aluminum levels—have been reported to reduce the severity of black root rot. Research in the mid-1990s determined the effect of nitrogen source and ratio and pH on the severity of black root rot on pansy. The ratio of nitrate to ammonium was the most critical factor in determining black root rot severity. Disease incidence was lowest with high ammonium and low potassium levels. Disease incidence in sand culture was not affected by pH, probably because of the low cation-exchange capacity in that medium. The effect of pH has been demonstrated primarily in soil and in media containing soil. Spore viability has been shown to be reduced with a high level of ammonium, and aluminum has been shown to inhibit germination.

Chlorine dioxide is a disinfestant used to control pathogens in water. Survival of chlamydospores of *T. basicola* was evaluated under different conditions. To kill 50% of the spores (LD_{50}), a higher concentration of chlorine dioxide was needed at pH 8 than at pH 5. The concentration had to be increased to maintain LD_{50} when a divalent metal ion solution was added that contained copper, iron, manganese, and zinc. The presence of nitrogen and hard water also affected the LD_{50}, and this effect varied with pH. Chlorine dioxide doses resulting in 50% mortality ranged from 0.5 to 11.9 mg/L for conidia and from 15.0 to 45.5 mg/L for chlamydospores of *T. basicola*.

Many fungicides have been tested against black root rot on pansy, vinca, and petunia. The most consistently effective products are in FRAC group 1 (notably, thiophanate-methyl), and somewhat lower levels of control have been achieved with products containing polyoxin D (FRAC 19). Good results are sometimes seen with strobilurins (FRAC 11), triflumizole (FRAC 3), and fludioxonil (FRAC 12). To obtain acceptable results with fungicides, growers must first minimize stress to the host, avoid high soil/potting medium pH levels, and control fungus gnat and shore fly vectors. Moreover, growers should use products preventively, rather than after symptoms have appeared in the crop.

Sanitation practices should be initiated at propagation time and continued through the crop cycle. Soiled hands and tools should not come in contact with soilless media. Cleanliness of the plant-growing area is recommended, and insect vectors should be kept under control. Diseased plants and plant debris should be removed as they appear on the bench, and excessive splashing during watering should be avoided. *T. basicola* chlamydospores can persist in organic debris from year to year, making the reuse of trays and pots especially risky for crops prone to black root rot. Plug trays should be treated before reusing them. A spray of 27% hydrogen dioxide (2.5 fluid oz/gal) or a dip for 10 min in 5.25% sodium hypochlorite (12.8 fluid oz/gal) has been proven effective for killing *T. basicola* chlamydospores.

Selected References

Benson, D. M., and Parker, K. C. 2000. Evaluation of pansy for resistance to black root rot, 1999. Biol. Cult. Tests 15:71.

Copes, W. E., and Hendrix, F. F. 1996. Influence of NO_3/NH_4 ratio, N, K, and pH on root rot of *Viola* × *wittrockiana* caused by *Thielaviopsis basicola*. Plant Dis. 80:879-884.

Copes, W. E., and Stevenson, K. L. 2008. A pictorial disease severity key and the relationship between severity and incidence for black root rot of pansy caused by *Thielaviopsis basicola*. Plant Dis. 92:1394-1399.

Copes, W. E., Chastagner, G. A., and Hummel, R. L. 2004. Activity of chlorine dioxide in a solution of ions and pH against *Thielaviopsis basicola* and *Fusarium oxysporum*. Plant Dis. 88:188-194.

Graham, J. H., and Timmer, N. H. 1991. Peat-based media as a source of *Thielaviopsis basicola* causing black root rot on citrus seedlings. Plant Dis. 75:1246-1249.

Hood, M. E., and Shew, H. D. 1997. The influence of nutrients on development, resting hyphae and aleuriospores induction of *Thielaviopsis basicola*. Mycologia 89:793-800.

Keller, J. R., and Potter, H. S. 1954. *Thielaviopsis* associated with root rots of some ornamental plants. Plant Dis. Rep. 38:354-355.

McGovern, R. J., and Seijo, T. E. 1999. Outbreak of black root rot on *Catharanthus roseus* caused by *Thielaviopsis basicola*. Plant Dis. 83:396.

Stanghellini, M. E., Rasmussen, S. L., and Kim, D. H. 1999. Aerial transmission of *Thielaviopsis basicola*, a pathogen of corn-salad, by adult shore flies. Phytopathology 89:476-479.

Warfield, C. Y., and Konczal, K. M. 2003. Survival of *Thielaviopsis* spores on re-used plug trays and efficacy of disinfestants on spore viability. SNA Res. Conf. 48:545-547.

(Prepared by M. L. Daughtrey, R. L. Wick,
and J. L. Peterson; revised by
A. R. Chase and M. L. Daughtrey)

Verticillium Wilt

Verticillium wilt, which is caused by two species of *Verticillium,* affects a broad range of plants, including many of those grown as bedding plants. In 1962, the disease was reported to be spread via infected cuttings of florist's geranium (*Pelargonium* × *hortorum*). Today, clean stock production through the use of culture indexing, tissue culture, and strict sanitation has largely eliminated cuttings as a source of initial infection for geranium. The widespread adoption of soilless growing media since the 1980s has greatly reduced the likelihood of introducing the pathogen to bedding plants during propagation and greenhouse culture. Verticillium wilt is now most commonly seen in the landscape, where soil contamination with the pathogen is always a possibility.

Symptoms

Symptoms of Verticillium wilt are essentially indications of water stress: stunting, wilting, chlorotic foliage, leaf scorch, and leaf drop. On infected *Impatiens* spp., the initial symptoms are yellowing and dropping of the lower leaves, as well as plant wilting (Fig. 84). Another characteristic of Verticillium wilt (along with other vascular wilts) is that symptoms may develop on one side of the plant or sometimes on one side of the midveins on leaves of an infected plant.

Vascular discoloration may occur but is not always obvious (Fig. 85). In florist's geranium, the discoloration is light brown and hard to distinguish. In *Impatiens* spp., coleus (*Plectranthus scutellarioides*), and snapdragon (*Antirrhinum majus*), in contrast, the xylem may be blackened, making the discoloration visible externally. Garden impatiens (*I. walleriana*) with Verticillium wilt show wilting, stunting, leaf drop, stem dieback, and black streaks in the vascular system. Plant death has been observed in the landscape when plants were grown under water-stressed conditions but not in greenhouse trials when water was applied as needed.

Fig. 84. Verticillium wilt (*Verticillium* sp.) on garden impatiens in the landscape. (Courtesy A. R. Chase—© APS)

Fig. 85. Black, vascular streaking in garden impatiens, symptomatic of Verticillium wilt (*Verticillium* sp.). (Courtesy A. R. Chase—© APS)

Wilting of a few leaves followed by general chlorosis occurs on some crops. Yellow, V-shaped wedges may develop in geranium leaves, which gradually turn brown and dry before abscising. For many hosts, infected plants grow slowly and are stunted. In the landscape, stunting may be the only symptom. In addition, infection may be latent for several months before symptom development.

Symptoms of Verticillium wilt on geranium can be similar to those caused by systemic invasion of *Xanthomonas hortorum* pv. *pelargonii.* Laboratory diagnosis is recommended to properly identify either disease.

Causal Organisms

Two cosmopolitan species of *Verticillium, V. albo-atrum* and *V. dahliae,* cause vascular wilt on bedding plants. The species can be distinguished easily in the laboratory: *V. dahliae* produces microsclerotia in culture, whereas *V. albo-atrum* produces only a dark, resting mycelium with hyphae 3–7 µm in diameter after 10–15 days in agar culture. *V. albo-atrum* can also be distinguished from *V. dahliae* because it is unable to grow at 30°C. Additionally, the conidiophore bases of *V. albo-atrum* on plant tissue may be brownish, whereas those of *V. dahliae* are always hyaline.

Both species have verticillately branched conidiophores; there are 2–4 phialides per node for *V. albo-atrum* and 3–4 for *V. dahliae.* Phialides are mostly 20–30 (to 50) × 1.4–3.2 µm for *V. albo-atrum* and 16–35 × 1.0–2.5 µm for *V. dahliae.* The mostly one-celled, hyaline conidia are produced singly at the tips of the phialides and are ellipsoidal to irregularly subcylindrical. *V. albo-atrum* conidia measure 3.5–10.5 (to 12.5) × 2–4 µm, and those of *V. dahliae* measure 2.5–8.0 × 1.4–3.2 µm.

Host Range and Epidemiology

Reported bedding plant hosts of *V. albo-atrum* include species in these genera: *Antirrhinum, Aster, Begonia, Browallia, Calceolaria, Coreopsis, Dahlia, Delphinium, Chrysanthemum, Gerbera, Impatiens, Matthiola, Pelargonium, Phlox, Portulaca,* and *Tagetes. V. dahliae* has been reported from species in these genera: *Begonia, Coleus, Cosmos, Dianthus, Echinacea, Impatiens, Osteospermum, Rudbeckia,* and *Tagetes.* Host specialization has been seen in some but not all isolates of *V. dahliae.* Since the 1990s, most reports have been from garden settings or in-ground plantings, rather than from greenhouse production, where soilless mixes are widely used.

Verticillium spp. are soilborne and can survive in the soil for many years in the absence of a host. During composting, a temperature of at least 65°C is necessary to eliminate *Verticillium* propagules. Thermal therapy of geranium for *Verticillium* spp. was found to be ineffective, because the plant's heat sensitivity was greater than that of the fungus.

Greenhouse insect pests may play a role in crop contamination and dissemination of *Verticillium* spp. The fungus gnat *Bradysia impatiens* was shown to be a vector of *V. albo-atrum* during greenhouse culture of alfalfa (*Medicago sativa*). The insect introduced the fungus into feeding wounds on the roots, stems, or leaves.

Management

Culture-indexed plants should be purchased when available, and symptomatic plants should be removed and destroyed as soon as they are detected. Populations of insect pests, such as

TABLE 10. Relative Susceptibilities of Bedding Plants to Verticillium Wilt[a]

Host Scientific Name	Host Common Name(s)	Susceptibility
Ageratum houstonianum	Ageratum, floss flower	Resistant
Antirrhinum majus	Snapdragon	Susceptible
Aster spp.	Aster	Susceptible
Begonia semperflorens	Wax begonia	Variable
Bellis perennis	English daisy	Resistant
Browallia speciosa	Browallia, amethyst flower	Variable
Calceolaria crenatiflora	Slipperwort, calceolaria	Susceptible
Calendula officinalis	Calendula, pot marigold	Resistant
Celosia argentea var. *cristata*	Cockscomb	Susceptible
Chrysanthemum morifolium	Chrysanthemum	Susceptible
Coreopsis lanceolata	Lanceleaf tickseed	Susceptible
Dahlia hybrids	Dahlia	Susceptible
Dianthus spp.	Pink, Sweet William	Variable
Gerbera jamesonii	Gerber daisy, gerbera daisy, Transvaal daisy	Susceptible
Impatiens walleriana	Garden impatiens	Susceptible
Lobularia maritima	Sweet alyssum	Resistant
Matthiola incana	Garden stock, hoary stock, ten-weeks stock	Susceptible
Osteospermum fruticosum	Trailing African daisy, osteospermum	Susceptible
Pelargonium × hortorum	Florist's geranium	Susceptible
Petunia hybrida	Petunia	Susceptible
Phlox spp.	Phlox	Susceptible
Plectranthus scutellarioides	Coleus	Susceptible
Portulaca grandiflora	Moss rose, portulaca	Variable
Primula spp.	Primrose	Resistant
Salvia farinacea	Blue salvia, mealycup sage	Susceptible
Tagetes spp.	Marigold	Susceptible
Torenia fournieri	Torenia, wishbone flower	Resistant
Verbena hybrida	Garden verbena	Resistant
Viola spp.	Pansy, viola	Resistant
Zinnia spp.	Zinnia	Resistant

[a] Courtesy A. R. Chase and M. L. Daughtrey—© APS.

fungus gnats and shore flies (*Scatella stagnalis*), that might serve as vectors should be monitored and managed. In landscape situations, growers should try to use plants believed to be resistant to Verticillium wilt if planting in an area known to be contaminated with one of the causal pathogens (Table 10).

In general, fungicides do not give adequate protection against Verticillium diseases. Drenches with benzimidazoles (FRAC 1), such as thiophanate-methyl, have shown statistically significant disease reduction but should not be relied on for complete control. Steam pasteurization is more effective and safer than fumigants for treatment of the soil to eliminate *Verticillium* spp.

Selected References

Blomquist, C. L., Rooney-Latham, S., Haynes, J. L., and Scheck, H. J. 2011. First report of Verticillium wilt on field-grown cosmos caused by *Verticillium dahliae* in California. Plant Dis. 95:361.

Garibaldi, A., Bertetti, D., Poli, A., and Gullino, M. L. 2011. First report of Verticillium wilt caused by *Verticillium dahliae* on *Coleus verschaffeltii* in Italy. Plant Dis. 95:878.

Garibaldi, A., Minuto, A., and Gullino, M. L. 2005. Verticillium wilt of African daisy (*Osteospermum* sp.) in Italy caused by *Verticillium dahliae*. Plant Dis. 89:688.

Kalb, D. W., and Millar, R. L. 1986. Dispersal of *Verticillium albo-atrum* by the fungus gnat (*Bradysia impatiens*). Plant Dis. 70:752-753.

McCain, A. H., Raabe, R. D., and Wilhelm, S. 1981. Plants Resistant to or Susceptible to Verticillium Wilt. Leaflet 2703. Division of Agricultural Sciences, University of California, Berkeley.

McWhorter, R. P. 1962. Diverse symptoms in *Pelargonium* infected with *Verticillium*. Plant Dis. Rep. 46:349-353.

Taylor, N. J. 1993. First report of Verticillium wilt of *Impatiens walleriana* caused by *Verticillium dahliae* in the United States. Plant Dis. 77:429.

(Prepared by M. L. Daughtrey, R. L. Wick, and J. L. Peterson; revised by A. R. Chase and M. L. Daughtrey)

White Smut

White smut, which can be caused by any of six *Entyloma* spp., has been reported on a few bedding plants, as well as wildflower and weed species, in many locations worldwide. White smut is seen occasionally on ornamental crops, but losses are rarely extensive.

Symptoms

Symptoms include the formation of circular to irregularly shaped leaf spots that measure 0.63–1.27 cm. The spots are pale greenish-yellow to yellow to brown and sometimes have darker-brown borders (Fig. 86); they are initially light colored and become brown with age (Fig. 87). The spots are somewhat thick and evident on both sides of the leaf.

On *Dahlia* spp., one researcher reported that white smut caused round, black spots that coalesced. When spots coalesce,

Fig. 86. Early leaf spots caused by white smut (*Entyloma gaillardianum*) on gaillardia. (Courtesy A. R. Chase—© APS)

Fig. 87. Leaf spots caused by white smut (*Entyloma* sp.) on calendula. (Courtesy A. R. Chase—© APS)

the leaves may become necrotic. Older, lower leaves are often more severely affected.

Causal Organisms

Entyloma spp. are members of the Tilletiaceae, a family known as "white smuts" because of the pale lesions they produce on infected foliage. *Entyloma* spp. occur primarily (85%) on members of the Asteraceae and Ranunculales. They grow and produce teliospores intercellularly, and in some cases, they also produce conidia on the surfaces of lesions.

The best studied of the white smuts on ornamentals is *E. dahliae* (syn. *E. calendulae* f. sp. *dahliae*). It produces acicular conidia (called "sporidia") that form frosty coatings on the surfaces of lesions and spherical teliospores within leaf tissue of dahlia. The pale-green to brown teliospores of *E. gaillardianum* within leaves of *Gaillardia* × *grandiflora* were described as round and ranged in size from 11.0 (to 10.5) to 14.5 (to 16.0) µm in diameter; teliospores had single or double walls that ranged in thickness from 1.0 (to 0.5) to 2.5 µm. Other *Entyloma* spp. have similar spores.

Host Range and Epidemiology

E. microsporum occurs on *Ranunculus asiaticus* and has been widely reported on many other *Ranunculus* spp. *E. polysporum* has been reported on *Calendula officinalis* from Canada and was also seen in a 2002 outbreak in Virginia (United States) on *G.* × *grandiflora*. *E. calendulae* has also been reported on calendula. Gaillardia was found to be the host of *E. gaillardianum* in a commercial nursery in California in 2009.

In 1997, *E. bellidis* was observed for the first time in the United Kingdom on *Bellis* cultivars. White smut was also seen on wild *B. perennis* (English daisy) and on calendula in Shropshire, United Kingdom. In 1940, a large planting of dahlias in Oregon (United States) was found to be heavily infected by *E. dahliae,* and white smut has been seen occasionally since.

The majority of research on white smut has concentrated on taxonomic issues, rather than control strategies. Comparison of internal transcribed spacer (ITS) sequences has helped with identification of *Entyloma* spp. on gaillardia, which is susceptible to more than one species.

Entyloma spores are windborne and rainborne. Most outbreaks of disease occur in nurseries in which overhead irrigation is used; white smuts appear sporadically during production. The fungi survive on hosts and host refuse, and closely related weed hosts might be inoculum sources in some instances. Overwintering in the soil as teliospores has been reported for *E. dahliae* in the United Kingdom, but teliospores may not be able to survive in northern regions without protection from the winter cold.

Management

Crops should be protected from rain, and irrigation practices should minimize the length of time that foliage is wet. Treating foliage with broad-spectrum fungicides or materials that are highly effective on rust fungi may limit losses to disease.

Studies in the United Kingdom have shown variations in susceptibility of *Dahlia* cultivars to *E. dahliae,* suggesting that plant health-targeted breeding could improve the resistance of dahlias to white smut.

Selected References

Begerow, D., Lutz, M., and Oberwinkler, F. 2009. Implications of molecular characters for the phylogeny of the genus *Entyloma*. Mycol. Res. 106:1392-1399.

Glawe, D. A., Barlow, T., and Koike, S. T. 2010. First report of leaf smut of *Gaillardia* × *grandiflora* caused by *Entyloma gaillardianum* in North America. Online. Plant Health Progress. doi:10.1094/PHP-2010-0428-01-BR

Hong, C. X., and Banko, T. J. 2003. First report on white smut of *Gaillardia* × *grandiflora* caused by *Entyloma polysporum* in Virginia. Plant Dis. 87:313.

McWhorter, F. P. 1940. Brief notes on plant diseases. Plant Dis. Rep. 24:442-443.

Okaisabor, E. K. 1969. The epidemiology of leaf smut disease of dahlia caused by *Entyloma calendulae* f. sp. *dahliae*. Mycopathol. Mycology. Applic. 39:145-154.

Piepenbring, M. 1996. Two new *Entyloma* species (Ustilaginales) in Central America. Sydowia 48:241-249.

Preece, T. F., and Clement, J. A. 1997. *Entyloma calendulae* f. *bellidis* causing smut disease of cultivated *Bellis* in Britain. Mycologist 12:64-66.

Vánky, V. 2013. Illustrated Genera of Smut Fungi, 3rd ed. American Phytopathological Society, St. Paul, MN.

(Prepared by A. R. Chase and M. L. Daughtrey)

Diseases Caused by Oomycetes

The oomycetes are one of the most important groups of plant pathogens, because they include *Pythium* spp., *Phytophthora* spp., and the downy mildews. These organisms are not fungi but are fungus-like in that they are filamentous, microscopic organisms. Unlike fungi, both *Pythium* and *Phytophthora* spp. and some downy mildew pathogens produce zoospores with two types of flagella, and their cell walls are predominantly cellulose (in contrast to the chitin walls of fungi). The oomycetes also include species of *Pustula, Wilsoniana,* and *Albugo,* which cause diseases known as "white blister rusts."

The common denominator for all of these pathogens is a strong tie to water for dissemination. In some cases, oomycete diseases are host specific (certain downy mildews), while in other cases, the host range is very large. (*Pythium* and *Phytophthora* spp. cause many diseases.) Finally, as a group, the oomycetes share sensitivity to certain classes of pesticides.

(Prepared by A. R. Chase and M. L. Daughtrey)

Downy Mildews

The downy mildew genera *Basidiophora, Hyaloperonospora, Peronospora,* and *Plasmopara* have been reported on the bedding plants covered in this compendium (Table 11). Downy mildew diseases are not only becoming more common but are also causing more damage on bedding plants.

The characteristic that most clearly distinguishes the downy mildew pathogens from other oomycetes is that they are obligate parasites that obtain nutrients from their hosts with haustoria. The downy mildews also have determinate sporangiophores, whereas *Phytophthora* and *Pythium* spp. have indeterminate sporangiophores. The genera of downy mildews were originally distinguished largely by the morphology of their sporangia. Downy mildew pathogens in the genus *Plasmopara* produce zoospores, whereas those in the genera *Hyaloperonospora* and *Peronospora* rarely if ever form zoospores.

TABLE 11. Downy Mildews That Affect Bedding Plants[a]

Host(s)	Pathogen(s)
Antirrhinum majus (snapdragon)	*Peronospora antirrhini*
Callistephus chinensis (China aster)	*Basidiophora entospora*
Cleome houtteana (spider flower)	*Hyaloperonospora* sp.
Eustoma grandiflorum (lisianthus)	*Peronospora chlorae*
Gerbera jamesonii (gerber daisy)	*Plasmopara* sp.
Impatiens spp.	*Plasmopara constantinescui*, *P. obducens*
Lobularia maritima (sweet alyssum)	*Hyaloperonospora lobulariae*, *H. parasitica*
Matthiola incana (garden stock)	*Hyaloperonospora parasitica*, *Peronospora matthiolae*
Plectranthus scutellarioides (coleus)	*Peronospora* sp.
Primula spp. (primrose)	*Peronospora oerteliana*
Salvia officinalis (common sage), *S. splendens* (red salvia), *S. coccinea* (Texas sage), and certain other *Salvia* spp.	*Peronospora lamii*
Salvia officinalis (common sage) and certain other *Salvia* spp.	*Peronospora swinglei*
Salvia officinalis (common sage)	*Peronospora salviae-officinalis*
Viola spp. (viola and pansy)	*Peronospora violae*

[a] Courtesy A. R. Chase and M. L. Daughtrey—© APS.

(The sporangia, called "conidia," germinate directly in these two genera.) In the past, some downy mildew species (e.g., *Basidiophora entospora*) were thought to have broad host ranges across entire plant families. However, based on phylogenetic analysis, *B. entospora* and a number of other downy mildew pathogens previously thought to have broad host ranges are now considered to be composed of separate species with similar morphologies that are closely adapted to their host genera.

In rare cases, downy mildew growth can be so dense on leaves that it is confused with powdery mildew. Microscopic examination of the hyphae quickly differentiates the two.

Selected References

Söküсü, A., and Thines, M. 2014. A molecular phylogeny of *Basidiophora* reveals several apparently host-specific lineages on Astereae. Mycol. Prog. 13:1137-1143.

Spencer, D. M. 1981. The Downy Mildews. Academic Press, London.

(Prepared by A. R. Chase and M. L. Daughtrey)

Downy Mildew on Sweet Alyssum and Stock

Downy mildew on members of the Brassicaceae—the plant family that includes sweet alyssum (*Lobularia maritima*) and garden stock (*Matthiola incana*)—is caused by *Hyaloperonospora parasitica* and other *Hyaloperonospora* and *Peronospora* spp.

Symptoms

Typical symptoms of downy mildew on stock are leaves with diffuse patches of chlorosis, necrotic blotching, and some twisting; these symptoms appear only after masses of white sporulation have become evident on the undersurfaces of leaves (Fig. 88). Under optimal conditions, downy mildew on stock can cause extensive aerial blighting that looks very similar to frost damage (Fig. 89). In an epidemic on stock reported from New Zealand in the 1950s, damage was most severe on young plants; sporulation appeared on the cotyledons, as well as the upper and lower surfaces of leaves, and seedlings were defoliated or killed.

Symptoms on sweet alyssum infected with *H. lobulariae* are pale-yellow areas on leaves and white sporulation on the undersurfaces opposite the yellowing. As infection progresses, cotyledons and leaves wilt and fall from alyssum stems.

Causal Organisms

The aggregate group that comprises *Hyaloperonospora* was formed in 2002 and includes the earlier name *Peronospora parasitica* along with five other species—all of which have conidiospores that are hyaline, have recurved branch tips, and have similar ITS1, ITS2, and 5.8S rDNA sequences. The haustoria for species in this genus are globose to lobate.

H. parasitica (syn. *P. parasitica*) was first seen on stock in New Zealand in the late 1950s and quickly developed into an epidemic both in greenhouses and outdoor landscapes. *H. parasitica* was documented on stock in California (United States) in 2000; the disease had been observed there for years, however, on bedding plants, as well as cut flowers. Conidiophores of *H. parasitica* have slender, curved tips that are 2.8 µm long. Conidia are hyaline and ovoid and measure 22–25 × 19–22 µm.

A second downy mildew pathogen on stock is *P. matthiolae*. It has been reported from Europe, central Asia, and Uzbekistan.

On sweet alyssum, *H. lobulariae* and *H. parasitica* cause downy mildew. *H. lobulariae* has been reported from Germany, Hungary, Japan, Poland, and Ukraine but not the United

Fig. 88. Patches of white sporangia produced by *Hyaloperonospora parasitica* on stock leaves, characteristic of downy mildew. (Courtesy A. R. Chase—© APS)

Fig. 89. Stock plants with severe leaf collapse, caused by downy mildew (*Hyaloperonospora parasitica*). (Courtesy A. R. Chase—© APS)

States. The ovoid to ellipsoid, hyaline conidia are 17.5–30.0 × 14.2–21.7 μm (average 22.9 × 16.9 μm) and borne on highly branched conidiophores 300–480 μm long.

Host Range and Epidemiology

H. parasitica has been reported on stock and also on *Arabidopsis thaliana* and many other plants in the Brassicaceae. However, there is a high degree of host specialization within this pathogen species. Strains that affect *A. thaliana,* for example, have not been found to affect other cruciferous plants. Consequently, the downy mildew pathogen on stock, which studies in New Zealand have shown to be host specific, may be renamed in the future once mycologists agree on what constitutes a species within *H. parasitica*.

Infection with *H. parasitica* developed well on stock at temperatures of 15.5–21.0°C; lesions developed after 5–6 days under favorable conditions. Conidia germinated best at 15.5°C and not at all at temperatures lower than 4.5°C or at or higher than 27°C.

Additional hosts for *P. matthiolae* are *M. chenopodifolia* and *M. longipetala* subsp. *bicornis*. This downy mildew pathogen also appears to have a narrow host range.

Management

Cultivars of both stock and alyssum react quite differently to their downy mildew pathogens. However, given the constant influx of bedding cultivars that are not tested in advance for their reactions to downy mildew, it is nearly impossible to use cultivars with lower susceptibility as part of an integrated pest management (IPM) program.

The use of fungicides to manage downy mildew is common, because many bedding crops in southern areas of the United States are produced without a controlled environment and thus are vulnerable to weather conditions. Sweet alyssum is especially sensitive to many fungicides, including most that contain copper. The resulting phytotoxicity is often confused with continued development of downy mildew. Stock plants are also sensitive to copper products at times. In California trials, the products with the best effectiveness against downy mildew contained azoxystrobin, fenamidone, and pyraclostrobin (all FRAC 11) and fosetyl aluminum (FRAC 33). Alternating products from different FRAC groups is critical for resistance management.

Selected References

Jafar, H. 1963. Studies on downy mildew (*Peronospora matthiolae*) (Roumeguerre) Gaumann on stock (*Matthiola incana* R. Br.). N.Z. J. Agric. Res. 6(1-2):70-82. doi:10.1080/00288233.1963.1041 9322

Koike, S. T. 2000. Downy mildew of stock, caused by *Peronospora parasitica,* in California. Plant Dis. 84:103. doi:10.1094/ PDIS.2000.84.1.103E

Satou, M., Chikuo, Y., and Matsushita, Y. 2015. Downy mildew of alyssum caused by *Hyaloperonospora lobulariae* in Japan. J. Gen. Plant Pathol. 81:83-86.

(Prepared by A. R. Chase and M. L. Daughtrey)

Downy Mildew on Coleus

Coleus (*Plectranthus scutellarioides* [syns. *Coleus blumei; Solenostemon scutellarioides*]) is a popular bedding plant widely used in landscapes. Coleus plants exhibiting symptoms of downy mildew disease were first observed in New York and Louisiana (United States) in 2005, and the pathogen was determined to be a *Peronospora* sp. In 2006, a downy mildew disease affected coleus at wholesale container and retail nurseries in Florida. The Florida outbreak began in November and continued to affect new foliage throughout the year during cool, foggy conditions. This new downy mildew disease has since spread throughout the United States and may be found wherever coleus is produced or used in landscapes in mass plantings.

Symptoms

The primary symptoms of downy mildew on coleus are chlorotic and necrotic leaf spotting, sometimes accompanied by leaf curling (Figs. 90 and 91). Stunting is apparent in plants that are infected when young. Infection results in defoliation and death of young seedlings and some defoliation of more mature plants of highly susceptible cultivars. Leaf lesions are sometimes vein bounded and sometimes more round or irregular in shape.

Under suboptimal conditions, sporulation is sparse. Under humid conditions, the undersides of affected leaves may show extensive gray to violet-brown sporulation (Fig. 92).

Causal Organism

The *Peronospora* sp. on coleus described from Florida had hyaline conidiophores that measured 343.5–561.0 × 9.5–15.0 μm; they branched dichotomously, ending in slender, curved branchlets, and the angle between ends was usually a right angle. Conidia were light brown, ovoid to ellipsoid, and mostly nonpapillate; they measured 17.0–24.5 × 16.5–25.5 μm. Oospores were not observed in plant tissue. Internal transcribed spacer (ITS) sequence data clearly distinguished this

Fig. 90. Brown spots on the lower surface of a coleus leaf, indicating downy mildew infection (*Peronospora* sp.). (Courtesy A. R. Chase—© APS)

Fig. 91. Brownish-green, vein-bounded spots on coleus with downy mildew (*Peronospora* sp.). (Courtesy A. R. Chase—© APS)

Peronospora sp. from *P. lamii,* which is morphologically almost identical. The Florida results agreed with the description of the *Peronospora* sp. on coleus first reported from New York and Louisiana.

Downy mildew on basil (*Ocimum basilicum*) was described in Switzerland as being caused by *P. belbahrii,* which is morphologically similar but not identical to the *Peronospora* sp. that causes downy mildew on coleus. Conidial shape, color, and dimensions are slightly different. The morphological differences and the unique host ranges of the coleus and basil pathogens suggest that further studies will establish them as being two separate species.

Host Range and Epidemiology

Epidemiological studies at Michigan State University identified the optimal environmental conditions for sporangial release in the greenhouse. Spore trapping in a greenhouse with downy mildew-infected coleus showed that a high level of relative humidity (>95%) followed by a reduction in humidity prompted the release of a large number of sporangia.

Experimentally, development of disease has been found greatest at temperatures of 15–20°C and has been considerably reduced at 25–30°C. For the coleus cultivar Volcano, 15 and 20°C were most favorable for disease development when humidity was also favorable; only leaf spotting was seen at 25°C, and plants remained healthy when grown at a constant 30°C.

Management

Choosing less susceptible cultivars can allow growers and landscape gardeners to better manage coleus downy mildew. A total of 147 cultivars were tested at Michigan State University and the Cornell University Long Island Horticultural Research and Extension Center between 2006 and 2013. Of these cultivars, 27.2% were rated as highly (H) susceptible to downy mildew, 5.4% had medium-high (M-H) susceptibility, 29.3% had medium (M) susceptibility, 6.1% had low-medium (L-M) susceptibility, and 32.0% had low (L) susceptibility. The following cultivars had the lowest susceptibility: Beauty, Big Blond, Black Ducksfoot, Cranberry Bog, Etna, Fairway Lemon, Fairway Orange, Fairway Red Velvet, Fairway Rose, Fairway Salmon Rose, Florida Sun Lava, Florida Sun Rose, Freckles, Fright Night, Gator Glory, Giant Palisandra, Glory of Luxembourg, Gold Edge, Golden Dream, Green Autry, Harlequin, Henna, Hunky Dory, Indian Summer, Keystone Kopper, Kiwi Fern, Midway Curly Magenta, Night and Glow, Oompah, Pegasus, Redhead, Ruby Dream, Russet, Saturn, Smoldering, Solar Furnace, Spumoni, Tapestry, Tilt a Whirl, Trailing Garnet Rose, Twist of Lime, Under the Sea Bone Fish, Under the Sea Electric Coral, Under the Sea Langostino, Under the Sea Lime Shrimp, Versa Lime, and Wild Streak.

Fig. 92. Gray-brown sporulation of a *Peronospora* sp. on the underside of a coleus leaf, characteristic of downy mildew. (Courtesy M. L. Daughtrey—© APS)

Reduced-risk fungicides that are effective against downy mildew on coleus include fenamidone (FRAC 11), mandipropamid (FRAC 40), and azoxystrobin (FRAC 11). In trials, other effective products contained fluopicolide (FRAC 43), dimethomorph (FRAC 40), mefenoxam (FRAC 4), and pyraclostrobin (FRAC 11). Overall, some of the best-performing fungicides for downy mildew on coleus are fenamidone, dimethomorph, and mefenoxam. Growers should rotate products among FRAC groups to delay development of resistance to fungicides.

In addition, growers should promptly remove diseased plants from the greenhouse and keep periods of high humidity and leaf wetness to a minimum in coleus production.

Selected References

Daughtrey, M., Harlan, B., Linderman, S., and Hausbeck, M. K. 2014. Coleus Cultivars and Downy Mildew. Spec. Res. Rep. No. 136. Online. American Floral Endowment. http://endowment.org/wp-content/uploads/2013/03/136-ColeusDM-Cv-2014.pdf

Daughtrey, M. L., Holcomb, G. E., Eshenaur, B., Pal, M. E., Belbahri, L., and Lefort, F. 2006. First report of downy mildew on greenhouse and landscape coleus caused by a *Peronospora* sp. in Louisiana and New York. Plant Dis. 90:1111.

Harlan, B. R., and Hausbeck, M. K. 2011. Epidemiology and management of downy mildew, a new pathogen of coleus in the United States. Acta Hortic. 952:813-818.

Harlan, B. R., and Hausbeck, M. K. 2013. Understanding Coleus Downy Mildew. Spec. Res. Rep. No. 134. Online. American Floral Endowment. http://endowment.org/wp-content/uploads/2014/03/134ColeusDMDisease2013.pdf

Palmateer, A. J., Harmon, P. F., and Schubert, T. S. 2008. Downy mildew of coleus (*Solenostemon scutellarioides*) caused by *Peronospora* sp. in Florida. Plant Pathol. 57:372.

Thines, M., Telle, S., Ploch, S., and Runge, F. 2008. Identity of the downy mildew pathogens of basil, coleus, and sage with implications for quarantine measures. Mycol. Res. 113:532-540.

(Prepared by A. R. Chase and M. L. Daughtrey)

Downy Mildew on Impatiens

Impatiens downy mildew, caused by *Plasmopara obducens,* was first reported in 1877 from Germany on *Impatiens noli-tangere.* It has been present in the United States since at least 1881, when it was collected in Woods Hole, Massachusetts, on the native *I. capensis* (syn. *I. fulva*), known as "common jewelweed." Subsequently, numerous reports of the disease on wild impatiens have come from both Europe and North America. Both *I. pallida* and *I. capensis* can host *P. obducens,* and both species grow as wildflowers over a large portion of North America.

Downy mildew caused by *P. obducens* on garden impatiens (*I. walleriana*) was reported from the United Kingdom in 2003. In February 2004, double-flowered ("double") garden impatiens infected with *P. obducens* were received by a commercial grower in southern California, and later that year, the same downy mildew was reported in the greenhouse trade in Tennessee and New York. By 2009, the disease had been found in landscapes in at least 11 states, and by the end of the 2013 growing season, 38 states had reported impatiens downy mildew. The disease continues to be reported throughout the United States and Europe.

Symptoms

Garden impatiens plants are destroyed by downy mildew infection in greenhouse production and in landscapes. The upper surfaces of systemically infected leaves become pale yellow, but distinct lesions do not form; downward curling of leaves is common (Fig. 93). Stunting and chlorotic growth are typically seen in systemic infections with downy mildew. When environmental conditions are favorable, the undersurfaces of leaves and even flower buds can become partially or completely covered with the white sporangia of the pathogen (Fig. 94).

As the disease progresses, first the flowers and then the leaves abscise; thus, it is common to see bare stems throughout an infected landscape planting (Fig. 95). With favorable temperature and moisture conditions, plants may be killed within 4–8 weeks.

Causal Organisms

Sporangiophores of *P. obducens* are hyaline and thin walled, emerge from stomata, and have slightly swollen bases. Sporangiophore branching is monopodial, and smaller sporangiophore branches are arranged at right angles. Tips of branches are 8–14 µm long. Sporangia are ovoid and hyaline; each sporangium has a single pore on the distal end, which may have a slight papilla but is more often flat. Short pedicels attach sporangia to sporangiophores. Sporangia measure 19.4–22.2 (to 25.0) × 13.9–16.7 (to 19.4) µm and will germinate to produce zoospores if conditions allow. Oospores have commonly been found in infected *Impatiens* stem tissues in temperate areas of the United States.

P. constantinescui (syn. *Bremiella sphaerosperma*) is another downy mildew pathogen on impatiens. It occurs in some of the same areas as *P. obducens*—namely, the United States, eastern Canada, and far-eastern Russia—on *I. capensis, I. noli-tangere,* and *I. pallida*. This pathogen was discovered in 1991 during examination of old herbarium specimens labeled "*P. obducens*." It causes pale-yellow to ochre spots, and some have reddish-brown margins.

Care should be taken to avoid confusing *P. constantinescui* with *P. obducens,* which widely affects garden impatiens. *P. constantinescui* has not been seen on garden impatiens.

Host Range and Epidemiology

P. obducens is pathogenic on a number of species of *Impatiens*—some native to North America (e.g., jewelweed) and Europe and some plant collectors' specimens originating from Asia and Africa. However, the pathogen has not been seen to cause disease on landscape plantings of New Guinea impatiens (*I. hawkeri*), another species of *Impatiens* that is widely cultivated as a bedding plant.

A number of other *Impatiens* spp. have also been found to be strongly resistant, but all of the garden impatiens cultivars and hybrids tested have shown significant susceptibility. Pathogenicity tests were performed on double-impatiens cultivars Fiesta, Tioga Red, and Tioga Cherry Red and on single cultivars Cajun Watermelon and Accent Lilac. Symptoms appeared about 12 days after infection at temperatures of 22–24°C. Sporulation typically appears at temperatures of 15–22°C. Extraordinarily severe symptoms were produced on garden impatiens, but the other *Impatiens* spp. showed more localized infections (in most cases, only leaf spots) under the same conditions that caused collapse of the popular bedding plant.

Greenhouse conditions are generally conducive to the development of infection and disease, so the presence of even a few infected plants can lead to a widespread epiphytotic. Sporangia are easily disseminated by air movement from fans, and wet plant surfaces allow infection to occur.

Oospores released from decaying plant tissue in the garden provide a means for the pathogen's year-to-year survival in the landscape, but details of how long oospores might survive and what conditions allow them to germinate are still being studied. Environmental conditions vary from year to year, making it difficult to anticipate how early in the growing season impatiens downy mildew will appear in the landscape.

Management

Prevention is critical for control of impatiens downy mildew. Even if disease symptoms are minimal during production, when infected plants are sold, the disease will explode and destroy landscape plantings once environmental conditions are conducive to development of an epidemic. Control strategies

Fig. 93. Chlorosis and downward curling of leaves of garden impatiens infected with downy mildew (*Plasmopara obducens*). (Courtesy A. R. Chase—© APS)

Fig. 94. White sporulation produced by *Plasmopara obducens* covering the undersurfaces of leaves of garden impatiens systemically infected with downy mildew. (Courtesy A. R. Chase—© APS)

Fig. 95. Severe defoliation of garden impatiens plants with downy mildew (*Plasmopara obducens*). (Courtesy A. R. Chase—© APS)

must be routinely implemented during production to prevent downy mildew from becoming established in the crop.

In the landscape, growing alternative crops, such as New Guinea impatiens and its hybrids, is the safest way to avoid the disease. Seed transmission is not known, so producing one's own crop from seed is another alternative.

Fungicide applications will protect impatiens against infection during production, and some products have long-lasting effects that carry over into the landscape. Resistance to mefenoxam (FRAC 4) has been reported from Europe and from Florida landscapes, probably because of excessive use without options for fungicide rotation. A large number of fungicide management trials have been conducted under both greenhouse and landscape conditions. When the disease first began to affect crops in the United States, the recommendation was to drench incoming plug trays and crops nearing sale with a combination of products—often, FRAC 4 (mefenoxam) and FRAC 43 (fluopicolide). This approach is still effective in areas in which FRAC 4 resistance is not a factor. Other currently effective materials for use while finishing plants for final sale include cyazofamid (FRAC C4), dimethomorph and mandipropamid (FRAC 40), fluxapyroxad (FRAC 7), mancozeb (FRAC M3), oxathiapiprolin (FRAC 49), and phosphorous acids (FRAC 33); strobilurins (FRAC 11) are also helpful. Label restrictions for some of these products require tank mixing with another product effective against downy mildew, and additional restrictions may apply, as well.

Fungicide use in the landscape is impractical, and U.S. product labels do not allow effective rotation. Thus, protecting impatiens during production remains the priority.

Selected References

Conners, I. L. 1967. An Annotated Index of Plant Diseases in Canada and Fungi Recorded on Plants in Alaska, Canada, and Greenland. Publ. 1251. Research Branch, Canada Department of Agriculture, Ottawa.

Constantinescu, O. 1991. *Bremiella sphaerosperma* sp. nov. and *Plasmopara borreriae* comb. Nov. Mycolog. 83(4):473-479.

Saccardo, P. A. 1888. *Plasmopara obducens* Schroet. Sylloge Fung. 7:242-243.

Vajna, L. 2011. First report of *Plasmopara obducens* on impatiens (*Impatiens walleriana*) in Hungary. New Dis. Rep. 24:13. doi:10.5197/j.2044-0588.2011.024.01.3

Voglmayr, H., and Thines, M. 2007. Phylogenetic relationships and nomenclature of *Bremiella sphaerosperma* (Chromista, Peronosporales). Mycotaxon 100:11-20.

Warfield, C. 2011. Downy mildew of impatiens. GrowerTalks 79(10).

(Prepared by A. R. Chase and M. L. Daughtrey)

Downy Mildew on Pansy and Viola

Downy mildew on pansy (*Viola* × *wittrockiana*) was first described in 1886, and since then, it has been problematic only periodically in ornamentals production and in landscape ornamentals. Little research outside taxonomic studies of the pathogen has been published on this disease, although it has caused significant losses on pansy in the northwestern United States. Pansy downy mildew is more common along the U.S. Pacific Coast than in areas of North America with warmer, drier climates.

Symptoms

One of the earliest signs of downy mildew infection on pansy or viola (*V. cornuta*) caused by *Peronospora violae* is the development of chlorotic, distorted leaves; usually, the newest leaves on the plant are affected. Plants appear nutrient deficient and at times show interveinal chlorosis. Sporulation of the pathogen appears readily on lower leaf surfaces; it is thick and often lavender to violet, making the disease easy to identify (Fig. 96).

A second downy mildew, *Plasmopara megasperma,* produces faintly chlorotic lesions on viola. They are more or less circular, from 8 to 16 mm wide, and without defined margins.

Causal Organisms

Two downy mildews are reported to cause disease on *Viola* spp.: *Peronospora violae* and *Plasmopara megasperma* (syn. *Bremiella megasperma*).

P. violae sporangiophores are not swollen at the bases; they are 250–350 µm long and branch dichotomously (three to five times). The slightly reflexed tips bear ellipsoid, lilac-colored, nonpapillate sporangia that measure 24 × 16 µm. Oospores formed in leaves are 22–28 µm and have thick, smooth walls. Since the 1990s, *P. violae* has been the downy mildew pathogen found most frequently on bedding plants, based on examinations of isolates from California and Texas.

Sporangiophores of *Plasmopara megasperma* emerge through stomata and are hyaline, straight or slightly curved, and tapered toward the tips. A single sporangiophore is 175–275 µm long, is dichotomously branched, and has a swollen apex and a 2.5–4.0 µm vesicle. Sporangia are hyaline, ellipsoid, and 30–65 × 23–32 µm; oospores are 30–40 µm and have smooth walls.

Host Range and Epidemiology

Both viola and pansy are susceptible to downy mildews. *P. violae* has been reported on pansy, Johnny jump-up (*V. tricolor*), and wild violas. *P. megasperma* is known on various wild and cultivated *Viola* spp. but has not been officially reported from pansy.

Downy mildew on *Viola* spp. has been associated with foggy, cool coastal conditions, but little research has been conducted on its epidemiology. Spore trapping at an Australian nursery revealed that conidia were released beginning at 7:00 A.M.; the release level peaked at 9:00 A.M. and tapered to 0 at 7:00 P.M. No spores were trapped during the night over a 9-day period.

Management

Recognizing variations in cultivar susceptibility may allow growers to avoid serious problems with downy mildew. In the early 2000s, a few trials evaluated pansy cultivars for susceptibility to downy mildew. The cultivars with the lowest incidence of downy mildew were as follows: Dancer Beaconsfield, Delta Red with Blotch, Crown Rose, Bingo Light Blue, Crystal Bowl Supreme True Blue, Colossus Deep Blue, Bingo with Blotch, Delta Red and Yellow, Majestic Giant 2 Yellow with Blotch, Bingo Blue Frost, Crown Orange, Bingo White with Blotch, Pure Red, and Maxima Orange.

Fig. 96. Purplish-gray downy mildew sporulation produced by *Peronospora violae* on the underside of a pansy leaf. (Courtesy A. R. Chase—© APS)

Fungicides are routinely used for control of downy mildew on pansy and viola in production but not in the landscape. Trials conducted in the early 2000s indicated optimal control with phosphonates (FRAC 33) and dimethomorph (FRAC 40). Strobilurins (FRAC 11) were slightly less effective, and fenamidone (FRAC 11) was significantly more effective than strobilurins. Copper (FRAC M1) products were also somewhat effective. In managing all downy mildews, fungicides from different FRAC groups should be used in rotation to slow the development of fungicide resistance.

Selected References

Boudier, B. 1987. Le mildiou de la pensee. Methode de lutte et sensibilities carietales. (Pansy mildew. Methods of control and varietal susceptibility.) Hortic. Fr. 189:7-8.

Constantinescu, O. 1979. Revision of *Bremiella* (Peronosporales). Trans. Brit. Mycol. Soc. 72:510-515.

Hall, G. 1989. *Peronospora violae*. CMI Descriptions of Pathogenic Fungi and Bacteria, No. 977. Mycopathologia 106:199-201.

Minchinton, E., Hepworth, G., et al. 1998. Control of Downy Mildews in Nursery Seedlings. Publ. NY406. Report of the Horticultural Research and Development Corporation, Gordon, New South Wales, Australia.

Wilson, G. W. 1914. Studies in North American Peronosporales—VI. Notes on miscellaneous species. Mycologia 6:192-210.

(Prepared by A. R. Chase and M. L. Daughtrey)

Downy Mildew on *Salvia* spp.

In the United States, downy mildew on common sage (*Salvia officinalis*), red salvia or red sage (*S. splendens*), and Texas sage (*S. coccinea*) is caused by *Peronospora lamii*. Downy mildew is also often seen on blue salvia or mealycup sage (*S. farinacea*), which develops purple leaf spots. The disease has been reported from throughout the southeastern United States and on the West Coast.

Symptoms

The initial symptom of downy mildew is the development of angular, pale-green to yellow patches between leaf veins; the patches may become tan to black but retain their angular shape. In severe infections of young plants, complete collapse can occur (Fig. 97). Violet-colored sporulation on the lower leaf surfaces is often present, but sometimes, the only symptom is the development of black, greasy, angular lesions on the upper leaf surfaces (Fig. 98). These symptoms are easily confused with those of bacterial leaf spot, caused by a *Xanthomonas* sp. on the same plants.

Fig. 97. Collapse of *Salvia* sp. plants in a plug tray, caused by infection with downy mildew (*Peronospora lamii*). (Courtesy A. R. Chase—© APS)

Fig. 98. Black, angular lesions on *Salvia* sp. plants with downy mildew (*Peronospora lamii*). (Courtesy A. R. Chase—© APS)

Downy mildew also causes leaf abscission on some *Salvia* spp. Death of young seedlings can occur, as well.

Causal Organisms

Conidiophores of *P. lamii* have been variously reported as 440–560 µm and as 250–600 × 8 µm. They have dichotomous branching (five to seven times); branch ends are 8–16 × 2–3 µm, and each gently tapers to a point. Pale-brown conidia are 18–26 × 16–22 µm and ovoid to ellipsoid.

P. swinglei has been reported on various sage species around the world and was collected in the Czech Republic from 1984 to 1995 on *S. officinalis, S. pratensis,* and *S. verticillata.* However, recent studies have indicated that this pathogen is restricted to one host, *S. reflexa.* Another pathogen, *P. salviae-plebeiae,* occurs on *S. plebeia,* and *P. salviae-officinalis* occurs on *S. officinalis.*

Host Range and Epidemiology

Disease caused by *P. lamii* developed in 5 days when the pathogen was incubated in a growth chamber and in 6–8 days under greenhouse conditions over a temperature range of 10–23°C. With additional research, a number of more host-specific species may be identified within what has been considered *P. lamii.* Careful morphological studies and DNA sequence information from European research have provided new insights into the identities of downy mildews that were once thought to be *P. lamii.* For example, downy mildew of common sage has recently been described as *P. salviae-officinalis.*

Management

Growers should reduce the greenhouse humidity level as much as possible to avoid the development of a downy mildew epidemic. One trial from 2003 showed better control of the disease with phosphonates (FRAC 33) than with copper (M1) fungicides.

Selected References

Choi, Y.-J., Shin, H.-D., and Thines, M. 2009. Two novel *Peronospora* species are associated with recent reports of downy mildew on sages. Mycol. Res. 113(12):1340-1350.

Holcomb, G. E. 2000. First report of downy mildew caused by *Peronospora lamii* on *Salvia splendens* and *Salvia coccinea.* Plant Dis. 84:1154.

Humphreys-Jones, D. R., Barnes, A. V., and Lane, C. R. 2008. First report of the downy mildew *Peronospora lamii* on *Salvia officinalis* and *Rosmarinus officinalis* in the UK. Plant Pathol. 57:372.

McMillan, R. T., Jr., and Graves, W. R. 1994. First report of downy mildew of *Salvia* in Florida. Plant Dis. 78:317.

Thines, M., Telle, S., Ploch, S., and Runge, F. 2008. Identity of the downy mildew pathogens of basil, coleus, and sage with implications for quarantine measures. Mycol. Res. 113:532-540.

Wood, T. F., McMillan, Jr., R. T., and Graves, W. R. 2001. Downy mildew of *Salvia splendens* caused by *Peronospora lamii*. Proc. Fla. State Hortic. Soc. 114:237-238.

(Prepared by A. R. Chase and M. L. Daughtrey)

Downy Mildew on Snapdragon

Downy mildew caused by *Peronospora antirrhini* was originally described on a wild snapdragon in Germany in 1874. However, it was not reported on cultivated snapdragon (*Antirrhinum majus*) until 1936, when a serious outbreak began in Ireland; within 10 years, the disease had spread rapidly throughout the United Kingdom and Europe and into the United States. A full description of the disease was provided in California in 1947. Downy mildew has been found on *A. majus; A. nuttallianum,* which is native to California; and *Misopates orontium* (syn. *A. orontium*), which is native to Europe and naturalized in North America.

By the late 1970s, downy mildew on snapdragon was not felt to be a common threat, but it still causes occasional losses in greenhouse cut flowers and landscapes. Bedding plants largely escape this downy mildew.

Symptoms

Downy mildew on snapdragon presents itself in both systemic and localized fashions. A localized infection is typically observed as the development of chlorotic patches on the upper leaf surface, accompanied by tan to gray or violet sporulation on the lower leaf surface (Fig. 99). Plants with darker (e.g., red or purple) flowers may show purplish rather than chlorotic spotting. A systemic infection causes shoots to have chlorotic, downward-curling, stunted, and distorted leaves (Fig. 100); these symptoms closely mimic an aphid infestation. Secondary shoots may be produced from the base of the stem. Systemically infected seedlings do not reach full development and may die while still in the plug stage.

The pathogen is thought to overseason in the soil via oospores.

Causal Organism

Under optimal conditions for development, *P. antirrhini* forms intracellular haustoria with four to eight fingerlike branches. The dichotomously branched conidiophores, which emerge from stomates, are 350–700 µm long. Conidia are ovoid and 14–17 × 21–29 µm; they germinate best at 13°C if free water is available. No sporulation occurs at temperatures lower than 7°C or higher than 22°C. Oospores, which measure 30–38 µm, form abundantly in petioles, stems, and roots of infected plants.

Host Range and Epidemiology

Snapdragon downy mildew affects wild and cultivated plants in the genera *Antirrhinum* and *Misopates*. In studies performed in a field of cut-flower snapdragons in Palm Beach County, Florida (United States), airborne conidia were released according to a diurnal pattern; peak spore release occurred from the hours of 0500 to 1200. A similar daily cycle might be expected in bedding plant production under similar environmental conditions. In the same studies, temperatures less than 10°C reduced conidial production or release, and temperatures higher than 30°C also limited the presence of conidia in the air. Large spore releases occurred during dew periods that lasted longer than 6 h. Dry weather with shorter periods of leaf wetness slows the development of an epidemic.

Management

Maintaining good sanitation and providing adequate ventilation may help manage downy mildew in production areas. However, landscape plantings cannot be shielded from the weather.

Utilizing differences in cultivar susceptibility is not easy because of the constant introduction of new cultivars into the industry and the mixing of relatively susceptible and relatively resistant plants within a single series. Even so, cultivar comparisons have shown wide variations in disease susceptibility. In a trial that included three members of the Rocket series, Lemon and Orchid were less susceptible than White.

Tank mixes of fungicides are more effective on snapdragon downy mildew than some other treatments (Table 12). Results demonstrated optimal control with tank mixing of phosphonate (FRAC 33) and fenamidone (FRAC 11); phosphonate and mancozeb (FRAC M2); and fenamidone and mancozeb.

Fig. 99. Gray downy mildew sporulation produced by *Peronospora antirrhini* on the lower surface of a snapdragon leaf. (Courtesy A. R. Chase—© APS)

Fig. 100. Systemic downy mildew infection (*Peronospora antirrhini*), causing chlorosis and distortion of snapdragon leaves. (Courtesy M. L. Daughtrey—© APS)

TABLE 12. Results of Fungicide Trials on Snapdragon Downy Mildew[a]

Product	FRAC Group	Damage Caused	Effectiveness
Azoxystrobin	11	Slight damage on some cultivars	Excellent
Cupric hydroxide	M1	No damage to some damage	None
Cupric hydroxide + mancozeb	M1/M2	No damage	None
Fenamidone	11	No damage	Good
Fenamidone + mancozeb	11/M2	No damage	Excellent
Fluopicolide	43	No damage	Very good to excellent
Fosetyl aluminum	33	Slight damage in one of four trials	Good to excellent (mainly excellent)
Fosetyl aluminum + mancozeb	33/M2	No damage	Excellent
Kresoxim methyl	11	No damage	Excellent
Mancozeb	M2	No damage	Very good to excellent
Mandipropamid	40	No damage	Very good to excellent
Potassium bicarbonate	Not classified	Moderate damage	Some
Pyraclostrobin	11	No damage	Excellent
Trifloxystrobin	11	No damage	Some

[a] Courtesy A. R. Chase and M. L. Daughtrey—© APS.

Selected References

Byrne, J. M., Hausbeck, M. K., and Sconyers, L. E. 2005. Influence of environment on atmospheric concentrations of *Peronospora antirrhini* sporangia in field-grown snapdragon. Plant Dis. 89:1060-1066.

Byrne, J. M., Sconyers, L. E., and Hausbeck, M. K. 2003. Evaluation of snapdragon cultivars for resistance to downy mildew, 2000 and 2001. Biol. Cult. Tests 19:O010.

Green, D. E. 1938. Downy mildew on *Antirrhinum*. J. Roy. Hortic. Soc. 63:159-165.

Harris, M. R. 1939. Downy mildew on snapdragon in California. Plant Dis. Rep. 23:16.

Kirby, R. S. 1945. Downy mildew on snapdragon in Pennsylvania. Plant Dis. Rep. 29:371.

Yarwood, C. E. 1947. Snapdragon downy mildew. Hilgardia 17:241-249.

(Prepared by A. R. Chase and M. L. Daughtrey)

Downy Mildew on Verbena

Downy mildew on *Verbena officinalis* (common verbena or common vervain) was reported from Germany in 2009 and from the Czech Republic in 2010. This relative of *V. hybrida* (garden verbena), which is commonly grown as a bedding plant, is native to Europe and is a medicinal herb used occasionally as an ornamental.

The cause of downy mildew on common verbena is *Peronospora verbenae*. Symptoms include the development of yellowish-green, later brownish, and sometimes purplish-violet discolorations or spots on the upper leaf surface. Downy, gray to brown growth of the pathogen may be visibile on the lower leaf surface. The spots are often vein limited and sometimes coalesce; the leaves become necrotic and dry.

Conidiophores of *P. verbenae* are solitary and emerge through stomates; a single conidiophore is 200–350 µm long and has a straight trunk (110–220 × 8–10 µm). The branches are straight to slightly curved and arborescent. Ultimate branchlets are more or less conical and straight to curved; a single branchlet is 5–20 µm long and 2–4 µm wide at the base and has an obtuse to subacute tip. Conidia are narrowly to broadly ovoid, broadly ellipsoid–ovoid, and rarely subglobose; a single conidium measures 25–35 × 15–25 µm and has a length/width ratio of 1.2/1.7. The pale-brown (when mature) conidia have broadly rounded ends and granular contents. Conidia vary from being almost smooth to verruculose; there is no pedicel, and scars are inconspicuous. Oospores have not been observed.

Selected References

Braun, U., Jage, H., Richter, U., and Zimmerman, H. 2009. *Peronospora verbenae* sp. nov.—A new downy mildew on *Verbena officinalis*. Schlechtendalia 19:77-80.

Choi, Y. J., Lebeda, A., Sedlarova, M., and Shin, H. D. 2010. First report of downy mildew caused by *Peronospora verbenae* on verbena in the Czech Republic. Plant Pathol. 59:1166.

(Prepared by A. R. Chase and M. L. Daughtrey)

Phytophthora Diseases

In 2000, the genus *Phytophthora* was considered to have 60 species, including many important plant pathogens. Since then, many new species have been described; there are now 123 formally described species, and the number is ever increasing. At least 12 *Phytophthora* spp. have been reported on the bedding plants covered in this compendium.

In general, *Phytophthora* spp. are more often pathogenic than *Pythium* spp., but as with any *Pythium* sp., the pathogenicity and virulence of a given *Phytophthora* sp. depends on many factors, including the environmental conditions, host, and species.

Some species of *Phytophthora* have relatively narrow host ranges. *P. capsici* has a relatively narrow host range but attacks plants in the Solanaceae, Fabaceae, and Cucurbitaceae. *P. nicotianae* (tobacco), in contrast, is very common on bedding plants and attacks many different genera. Additionally, *P. cryptogea, P. drechsleri, P. palmivora,* and *P. tropicalis* (the latter, previously within *P. capsici*) have been reported occasionally, as well as *P. cactorum, P. cinnamomi, P. colocasiae, P. hibernalis, P. infestans,* and *P. lateralis.*

Phytophthora spp., like *Pythium* spp., produce oospores and zoospores. *Phytophthora* spp. are facultative parasites, stimulated by root exudates, and more virulent under conditions of high soil moisture, low oxygen, and high soluble salt levels. Some *Phytophthora* spp. produce sporangia on aboveground portions of plants and thus can be easily dispersed by splashing water (thus acting like their close relatives, the downy mildews).

During production, Phytophthora diseases on bedding plants can be managed by implementing sanitation practices and applying fungicides. These diseases are rarely problematic in propagation, but in the landscape, conditions can promote disease development to the point that fungicides are not effective. In the early 1990s, Phytophthora aerial blight on annual vinca (*Catharanthus roseus*) was so severe in the landscape throughout the southern United States that the crop fell out of favor with the gardening public. This development also resulted in breeding vinca for resistance to the pathogen, *P. nicotianae* (syns. *P. parasitica; P. parasitica* var. *nicotianae*), and the later use of improved cultivars.

In the landscape, it often becomes necessary to use other crops to break the disease cycle once a flowerbed becomes in-

TABLE 13. Bedding Plants Showing Some Tolerance of *Phytophthora nicotianae* in Landscape Beds[a]

Good Tolerance	Fair Tolerance	Very Susceptible
Ageratum houstonianum	*Celosia plumosa* 'Apricot Brandy'	*Antirrhinum majus* (snapdragon)
Celosia argentea 'New Look'	*Dahlia* 'Harlequin'	*Begonia semperflorens* 'Olympia White'
Lobularia maritima (sweet alyssum)	*Eustoma grandiflorum* (lisianthus)	*Catharanthus roseus* (annual vinca)
Nicotiana alata 'Dwarf White,' 'Nicki Red'	*Melampodium paludosum* 'Medallion'	*Gazania rigens* 'Mini Star Tangerine'
Portulaca grandiflora	*Nicotiana alata* 'Daylight Mix,' 'Domino Salmon'	*Impatiens walleriana* 'Accent Bright Eye'
Salvia coccinea 'Lady in Red'	*Pelargonium* × *hortorum* (florist's geranium) 'Multibloom Scarlet Eye'	*Salvia splendens* 'Red Hot Sally'
Salvia farinacea 'Victoria Blue'	*Petunia hybrida* 'Polo Salmon,' 'Red Picotee'	*Verbena* 'Imagination,' 'Novalis'
Tagetes 'Gold Fireworks'	*Torenia fournieri* (wishbone flower) 'Clown Mix'	*Verbena speciosa* 'Imagination'
Tagetes erecta 'Inca Orange,' 'Inca Yellow,' 'Moonshot'		*Verbena* × *hybrida* 'Novalis'
Tagetes patula 'Disco Mix,' 'Janie Harmony,' 'Janie Harmony Improved,' 'Scarlet Sophie'		*Viola* × *wittrockiana* (pansy)
Zinnia angustifolia (narrowleaf zinnia)		

[a] Courtesy A. R. Chase and M. L. Daughtrey—© APS.

fested (Table 13). The level of resistance is not always sufficient to warrant switching to another crop, but few other choices are superior to this management strategy. Some tolerance is present across certain genera (e.g., *Tagetes*), while in other genera (e.g., *Salvia, Nicotiana,* and *Celosia*), varying levels of tolerance are associated with particular cultivars or species in that genus.

In production, where the use of fungicides is most common, fungicide resistance can occur. *Phytophthora* isolates were collected from floriculture crops grown in commercial greenhouses in North Carolina (United States) for species identification, compatibility-type determination, and mefenoxam (FRAC 4) sensitivity tests. Isolation from 41 symptomatic plant species at 29 production locations resulted in 483 isolates. *P. cryptogea* (184 isolates) was recovered from dusty miller (*Senecio cineraria*) and gerber daisy (*Gerbera jamesonii*). All the isolates of *P. cryptogea* were insensitive or intermediate in sensitivity to mefenoxam at 1 µg a.i./ml (i.e., active ingredient per milliliter) and were A1 mating compatibility type. *P. nicotianae* (273 isolates) was isolated from African violet (*Saintpaulia ionantha*), lavender (*Lavandula* spp.), pansy (*Viola* × *wittrockiana*), *Petunia hybrida,* and annual vinca. Of these isolates, 21% were insensitive to mefenoxam (1 or 100 µg a.i./ml). Nearly all the isolates of *P. nicotianae* were A2 mating compatibility type, with the exception of those from pansy (one location), which were A1 compatibility type. Mating-type compatibility and mefenoxam sensitivity were uniform among isolates of *P. nicotianae* and *P. palmivora* from a given crop at a given location, suggesting that the epidemics within a location may have originated from a single source of inoculum.

Selected References

Banko, T. J., and Stefani, M. A. 2000. Evaluation of bedding plant varieties for resistance to *Phytophthora*. J. Environ. Hort. 18(1):40-44.

Enzenbacher, T. B., Naegele, R. P., and Hausbeck, M. K. 2015. Susceptibility of greenhouse ornamentals to *Phytophthora capsici* and *P. tropicalis*. Plant Dis. 99:1808-1815.

Gallegly, M. E., and Hong, C. X. 2008. Phytophthora: Identifying Species by Morphology and DNA Fingerprints. American Phytopathological Society, St. Paul, MN.

Hagan, A. K., and Akridge, R. 1999. Survival of selected summary annuals and perennials in beds infested with *Phytophthora parasitica,* 1998. Biol. Cult. Tests 14:74.

Hao, W., Richardson, P. A., and Hong, C. X. 2010. Foliar blight of annual vinca (*Catharanthus roseus*) caused by *Phytophthora tropicalis* in Virginia. Plant Dis. 94:274.

Hong, C. X., Moorman, G. W., Wohanka, W., and Büttner, C., eds. 2014. Biology, Detection, and Management of Plant Pathogens in Irrigation Water. American Phytopathological Society, St. Paul, MN.

Hwang, J., and Benson, D. M. 2005. Identification, mefenoxam sensitivity, and compatibility type of *Phytophthora* spp. attacking floriculture crops in North Carolina. Plant Dis. 89:185-190.

Olson, H. A., Jeffers, S. N., Ivors, K. L., Steddom, K. C., Williams-Woodward, J. L., Mmbaga, M. T., Benson, D. M., and Hong, C. X. 2013. Diversity and mefenoxam sensitivity of *Phytophthora* spp. associated with the ornamental horticulture industry in the southeastern United States. Plant Dis. 97:86-92.

(Prepared by A. R. Chase and M. L. Daughtrey)

Diseases Caused by *Phytophthora cryptogea*

Phytophthora cryptogea was first described in 1919 as a pathogen of *Lycopersicon esculentum* (tomato) and *Petunia hybrida*. Since then, it has been reported worldwide on a number of vegetable and ornamental plants, including some key bedding crops, such as *Gerbera jamesonii* (gerber daisy), *Begonia* spp., and petunia.

Symptoms

P. cryptogea causes root and crown rot, resulting in stunting or wilting of the plant and eventual death (Figs. 101 and 102). On some plants, symptoms include the development of soft, sunken lesions on the leaf or stem surface; dark-brown, soft, necrotic areas (1–8 mm in diameter) develop internally under the lesions. Lesions become sunken and water soaked and sometimes encompass the stem. Infected leaves are also water soaked, dark brown, and limp.

The symptoms caused by *P. cryptogea* are similar to those caused by a number of other pathogens, including *Pythium* spp., other *Phytophthora* spp., and even *Rhizoctonia* spp.

Fig. 101. Wilting of gerber daisy caused by root rot (*Phytophthora cryptogea*). (Courtesy M. L. Daughtrey—© APS)

Causal Organism

P. cryptogea is in Clade 8 of the genus *Phytophthora*, based on sequences of four mitochondrial loci. *P. cryptogea* has coralloid mycelium and sporangia that are ellipsoid, ovoid, or obpyriform with inconspicuous papillae. Sporangia are generally longer than 45 µm and vary in size (20–63 × 15–38 µm); they have a length/breadth ratio of 1.1/2.6. Sporangia are produced by internal proliferation. *P. cryptogea* is typically self-incompatible and has amphigynous antheridia and relatively thick-walled oospores.

P. cryptogea is morphologically similar to *P. drechsleri* and distinguished only on the basis of growth at 35°C. (*P. cryptogea* does not grow at this high temperature.) Isozyme and mitochondrial DNA analyses have made it clear that the two are separate species.

Researchers have proposed that isolates that apparently attack only begonia be considered a *forma specialis: P. cryptogea* f. sp. *begoniae.*

Host Range and Epidemiology

P. cryptogea has many bedding plant hosts: *Begonia* spp., *Calceolaria crenatiflora, Antirrhinum majus* (snapdragon), *Bellis perennis* (English daisy), *Celosia* spp., *Chrysanthemum morifolium, Dahlia* hybrids, *Dianthus barbatus* (Sweet William), gerber daisy, *Matthiola incana* (garden stock), *Osteospermum ecklonis,* petunia, *Salvia officinalis* (common sage), *Senecio cruentus* (dusty miller), *Tagetes erecta* (African marigold), *Verbena hybrida,* and *Zinnia elegans.*

The optimal temperature for growth of *P. cryptogea* is 22–25°C. The pathogen is unable to grow at temperatures as high as 35°C—a trait that separates this species from *P. drechsleri,* which grows optimally at 28–31°C. Zoospore formation of *P. cryptogea* is stimulated by saturated soil, and sporangia form at a very wide range of temperatures (from 5 to nearly 30°C); the optimum is approximately 20°C. As with other *Phytophthora* spp., root exudates stimulate zoospore development. Most typically, the roots are invaded, but the crown of the plant may become colonized, as well.

In one study, unstressed inoculated roots were lightly infected with *P. cryptogea* (20%), whereas salt-stressed roots were more heavily infected (70–88%). The pathogen has been demonstrated to tolerate salt levels that are injurious to plants and to produce sporangia and zoospores under high-salt conditions.

Management

The studies of salt-stressed roots mentioned in the previous section indicate that avoiding overfertilization may help reduce the impact of a *P. cryptogea* infection.

Fig. 102. Crop loss of gerber daisies caused by root rot (*Phytophthora cryptogea*). (Courtesy A. R. Chase—© APS)

Research performed in Italy on cut-flower gerber daisy grown in soilless media evaluated biological control and efficacy of slow-sand filtration on recycled or recirculated water in reducing the severity of *P. cryptogea.* Several strains of *Fusarium* and *Trichoderma* spp. were identified that reduced disease severity. Slow-sand filtration was also found to be effective, and the authors of the study suggested combining the two approaches for optimal results.

A review of fungicide trials on gerber daisy indicated that optimal fungicide control is achieved with etridiazole (FRAC 14), fenamidone (FRAC 11), fluopicolide (FRAC 43), mandipropamid (FRAC 40), and mefenoxam (FRAC 4). (Mefenoxam is ineffective, however, if resistance has developed in the *P. cryptogea* population.) The review also concluded that very good control is achieved with cyazofamid (FRAC 21) and dimethomorph (FRAC 40) (alone and in combination with ametoctradin [FRAC 45]). Finally, some product groups, including strobilurins and phosphonates (FRAC 33), were noted to be excellent sometimes but relatively ineffective other times.

Selected References

Ampuero, J., LaTorre, B. A., Torres, R., and Chávez, E. R. 2008. Identification of *Phytophthora cryptogea* as the cause of rapid decline of petunia (*Petunia* × *hybrida*) in Chile. Plant Dis. 92:1529-1536.

Gill, H. S., Zentmyer, G. A., Ribeiro, O. K., and Klure, L. J. 1976. A Phytophthora disease of African daisies (*Osteospermum* spp.) in California. Plant Dis. Rep. 60:647-649.

Grasso, V., Minuto, A., and Garibaldi, A. 2003. Selected microbial strains suppress *Phytophthora cryptogea* in gerbera crops produced in open and closed soilless systems. Phytopathol. Mediterr. 42:55-64.

Kaewruang, W., Sivasithamparam, K., and Hardy, G. E. 1988. *Phytophthora cryptogea,* an additional pathogen of African daisy in western Australia. Australas. Plant Pathol. 17:67-68.

MacDonald, J. D. 1982. Effect of salinity stress on the development of Phytophthora root rot on chrysanthemum. Phytopathology 72:214-219.

MacDonald, J. D., and Duniway, J. M. 1978. Temperature and water stress effects on sporangium viability and zoospore discharge in *Phytophthora cryptogea* and *P. megasperma.* Phytopathology 68:1449-1455.

Orlikowski, L. 1979. Effect of temperature, pH and peat moisture on sporulation of *Phytophthora cryptogea* from diseased African daisy. Bull. Acad. Pol. Sci. 27:761-767.

Orlikowski, L. 1981. Studies on the biological control of *Phytophthora cryptogea* Pethybr. et Laff., the mycoflora associated with African daisy production in Polish greenhouses and effects of its main components on the development of the pathogen. Prot. Ecol. 2:285-296.

Rattink, H. 1981. Characteristics and pathogenicity of six *Phytophthora* isolates from pot plants. Neth. J. Plant Pathol. 87:83-90.

Stamps, D. J. 1978. *Phytophthora cryptogea.* CMI Descriptions of Pathogenic Fungi and Bacteria, No. 592. CABI, Kew, Surrey, UK.

(Prepared by A. R. Chase and M. L. Daughtrey)

Diseases Caused by *Phytophthora drechsleri*

A number of reports have been made of crown rot on *Calibrachoa* spp. and *Gerbera jamesonii* (gerber daisy) caused by *Phytophthora drechsleri. P. drechsleri* and *P. cryptogea* have indistinguishable morphological traits and must be separated either by DNA sequences or by noting whether the isolate is able to grow at 35°C. (Of the two species, only *P. drechsleri* can grow at this temperature.) Additional bedding plants known to have been infected by *P. drechsleri* include *Celosia argentea, Fuchsia hybrida, Helichrysum bracteatum* (strawflower), *Salvia splendens* (red salvia or sage), and *Senecio cruentus* (dusty miller).

A 2003 study showed a range of *Calibrachoa* cultivar responses to infection, ranging from 20% mortality (Superbells Red) to 100% mortality (Superbells Blue and Superbells Coral Pink). The study also evaluated the application of lime for its effect on disease severity but could not show a correlation.

In one study of gerber daisy, optimal control with fungicides was seen in plants treated with products from FRAC 11 (azoxystrobin and fenamidone), as well as products from FRAC 14, 40, 43, and 4. Results differed in another study; only products from FRAC 43, FRAC 21, and FRAC 11 (fenamidone) and the combination of FRAC 40 and 45 showed excellent control. Results on calibrachoa have been similar.

Selected References

Benson, D. M., and Parker, K. C. 2010. Efficacy of registered and unregistered fungicides for control of *Phytophthora drechsleri* on gerbera, 2009. Plant Dis. Manag. Rep. 4:OT004.

Dicklow, M. B., and Wick, R. L. 2003. Evaluation of lime application on Superbells cultivars and their resistance to *Phytophthora drechsleri* and *Pythium* species, 2002. Biol. Cult. Tests 18:O005.

Hausbeck, M. K., and Glaspie, S. L. 2009. Control of Phytophthora root rot of gerbera daisy with fungicide drenches, 2008. Plant Dis. Manag. Rep. 3:OT007.

Hausbeck, M. K., Woodworth, J. A., and Harlan, B. R. 2004. Evaluation of a biopesticide and fungicides for managing Phytophthora crown rot of calibrachoa, 2003. Fung. Nemat. Tests 59:OT017.

Slinski, S. L., Wick, R. L., and Dicklow, M. B. 2004. Evaluation of fungicides for control of *Phytophthora drechsleri* crown rot of million bells, 2003. Fung. Nemat. Tests 59:OT025.

Slinski, S. L., Wick, R. L., and Dicklow, M. B. 2004. Evaluation of fungicides for rescue management of *Phytophthora drechsleri* crown rot of million bells, 2003. Fung. Nemat. Tests 59:OT026.

(Prepared by A. R. Chase and M. L. Daughtrey)

Diseases Caused by *Phytophthora infestans*

Late blight, which is caused by *Phytophthora infestans,* occurs on solanaceous crops such as potato (*Solanum tuberosum*), tomato (*S. lycopersicum*), and *Petunia hybrida.* Since the first report of petunia late blight in 1856, occurrence of late blight on this flower crop has been only sporadic, leading to the assumption that it is not an important disease in commercial bedding plant production.

However, during a widespread epidemic of tomato late blight in 1998, a commercial production greenhouse in California (United States) lost both petunias and tomatoes to the disease. In 1999, two different genotypes of *P. infestans,* US-8 and US-11, were found on petunia in western Washington State. Losses in a greenhouse petunia crop caused by infection with *P. infestans* resistant to metalaxyl were reported from Maryland in 2001–2002, and in 2003, *P. infestans* was confirmed on petunias shipped into the United States from Israel.

Symptoms

On petunia, late blight appears as a foliar blight and looks much like a downy mildew infection. Patches of necrosis develop on the upper leaf surface. White sporangia form rapidly under humid conditions, appearing on the lower leaf surface and in some cases, the upper surface, as well (Fig. 103).

Artificial inoculation of petunia results in extensive infection under ideal humidity and temperature conditions (Fig. 104). However, infection has been observed to be less extensive on petunia than on tomato grown under the same conditions. (Less leaf area was colonized on inoculated petunias.) Losses may be rapid and severe when the environment favors development of disease.

Causal Organism

The pathogen that causes late blight is *P. infestans,* an oomycete. Sporangia are limoniform to ovoid in shape and caducous.

In a 2003 outbreak on petunia, sporangia measured 39–50 μm (average 45 μm) × 26–28 μm (average 27 μm). Oospores were not found. The isolates were A2 mating type and formed oogonia (36.14 μm) and amphigynous antheridia (16.91 μm). Both A1 and A2 mating types were found on petunia in Washington State but on different plants.

Different cultivars have been shown to have different levels of susceptibility to late blight.

Host Range and Epidemiology

P. infestans has an extensive host range that includes many members of the family Solanaceae, especially in the genus *Solanum.* Potato and tomato are well-known hosts. Other members of the same family have also been identified as hosts, including petunia, *Calibrachoa hybrida,* and *Nicotiana benthamiana* (tobacco). There is only one report of *N. tabacum* as a host (from Mexico); it is not known whether ornamental tobaccos (various *Nicotiana* spp. and hybrids) host the disease.

Most of the petunia cultivars tested have been found susceptible, but they were less severely affected than potato and tomato in the same trial. Two petunia cultivars showed some resistance,

Fig. 103. Sporulation of *Phytophthora infestans,* the late blight pathogen, on a tray of petunia plugs after shipment. (Courtesy M. L. Daughtrey—© APS)

Fig. 104. Necrotic patches on petunias after inoculation with *Phytophthora infestans,* the late blight pathogen. (Courtesy M. Becktell—© APS)

suggesting the presence of *R* genes against *P. infestans*. The hypersensitive response was found in susceptible, partially resistant, and resistant petunia cultivars. Young petunias (3 weeks) were determined to be more susceptible than older petunias (7 weeks).

Ten *Calibrachoa* cultivars were tested; four were found susceptible and six resistant to *P. infestans*. *N. benthamiana* was susceptible to all four *P. infestans* isolates tested in the greenhouse and also developed an epidemic in the field. All solanaceous crops should be considered as potential hosts in a late blight outbreak, regardless of the original host, because *P. infestans* has been demonstrated to move from one host to another.

Late blight-infected petunia produce sporangia that may be dispersed throughout the greenhouse via air currents. Sporulation is high at cooler (18°C) temperatures. One study showed that infected petunias (cultivar White Madness) produced and released fewer sporangia than infected tomatoes (cultivar Sunrise), but infected petunias were seen to release sporangia for twice as long.

The temperature and leaf wetness requirements for pathogen germination, infection, and colonization and the temperature effects on incubation period and sporulation of *P. infestans* on petunia were compared with those on tomato. The environmental parameters were similar on the two hosts and very similar to those reported previously. Temperatures ranging from 13 to 23°C were generally conducive to establishment of infection; very little establishment occurred at 28°C. The minimum leaf wetness period was 2 h; the majority of establishment occurred within 6 h following inoculation. The time from inoculation to lesion development was shortest at 23°C, and the time required for development of sporangia in lesions was shortest at 28°C. Production of sporangia was greatest at 18°C and negligible (0) at 28°C. Although sporulation density at 18°C was slightly less on petunia than tomato, the total lesion area on petunia was only 20% of that on tomato.

Management

Environmental and fungicide methods of managing late blight on petunia are usually confined to production. Methods used to minimize downy mildew diseases will aid in late blight control, as well. Namely, growers should promptly remove diseased plants from the greenhouse or other growing area, provide adequate ventilation, and keep periods of high humidity and leaf wetness to a minimum during production. Irrigation directed to the potting medium can also reduce disease incidence compared with overhead irrigation, which will splash and spread inoculum.

Fungicides in FRAC 40 (dimethomorph), FRAC 11 (azoxystrobin), and FRAC 33 (phosphonates) suppress late blight development. Mefenoxam resistance is characteristic of some genetic groups of *P. infestans*. The plant defense activator acibenzolar-S-methyl (FRAC P) was also effective in tests but is not labeled for use on ornamentals. Another plant defense activator (harpin protein) and several bioantagonists (*Trichoderma harzianum, T. virens* [syn. *Gliocladium virens*], and *Bacillus subtilis*) were ineffective for control of late blight at the rates tested.

Selected References

Becktell, M. C., Daughtrey, M. L., and Fry, W. E. 2005. Epidemiology and management of petunia and tomato late blight in the greenhouse. Plant Dis. 89:1000-1008.

Becktell, M. C., Daughtrey, M. L., and Fry, W. E. 2005. Temperatures and leaf wetness requirements for pathogen establishment, incubation period, and sporulation of *Phytophthora infestans* on *Petunia* × *hybrida*. Plant Dis. 89:975-979.

Becktell, M. C., Smart, C. D., Haney, C. H., and Fry, W. E. 2006. Host–Pathogen interactions between *Phytophthora infestans* and the solanaceous hosts *Calibrachoa* × *hybrida, Petunia* × *hybrida,* and *Nicotiana benthamiana*. Plant Dis. 90:24-32.

Deahl, K. L., and Fravel, D. R. 2003. Occurrence of leaf blight on petunia caused by *Phytophthora infestans* in Maryland. Plant Dis. 87:1004.

(Prepared by A. R. Chase and M. L. Daughtrey)

Diseases Caused by *Phytophthora nicotianae*

Phytophthora nicotianae is the most commonly encountered *Phytophthora* spp. in bedding plants and thus the most extensively researched. Typically, this pathogen invades the roots, crowns, or stems. The disease that results causes extensive losses during production and is especially problematic in the landscape, where *P. nicotianae* is considered the most important soilborne pathogen of *Catharanthus roseus* (annual vinca). In greenhouse production, *P. nicotianae* is seen most commonly on annual vinca and *Viola* × *wittrockiana* (pansy).

Symptoms

Symptoms of the root and crown rot caused by *P. nicotianae* are similar to those caused by other species of *Phytophthora, Rhizoctonia,* and *Pythium* and may also be confused with symptoms of *Impatiens necrotic spot virus* on some hosts (see the later section "Tomato Spotted Wilt, Impatiens Necrotic Spot, and Other Tospovirus Diseases"). Wilting of one or more leaves is often the first symptom.

Roots become water soaked and turn brown to black, and the cortex can be easily removed. While root discoloration and deterioration are often attributed to *Pythium* infection, they are clearly not associated with a single cause. Overfertilization, poor aeration, and attacks by other pathogens (including *Phytophthora* spp.) may all lead to root discoloration and deterioration.

Sometimes, cankers in the form of brown to black lesions develop at the crown. Leaves may also become colonized from contact with the growing medium, soil, or other plants, producing water-soaked lesions that become brown and limp. Early infection of plug seedlings can result in losing them before or after transplanting (Fig. 105).

An infection that starts in propagation or production can be masked by fungicide application, only to appear later in the landscape. On annual vinca, plantings may show stem infections, or disease may appear as an aerial blight (Figs. 106 and 107).

Causal Organism

P. nicotianae (syns. *P. parasitica; P. parasitica* var. *nicotianae*) was first described from *Nicotiana tabacum* (tobacco) in

Fig. 105. Phytophthora crown rot on pansies, caused by infection by *Phytophthora nicotianae.* (Courtesy A. R. Chase—© APS)

1896, and in 1913, the pathogen was described as *P. parasitica*. In 1964, the two species were combined into *P. nicotianae* var. *parasitica* and *P. nicotianae* var. *nicotianae*, based on morphological and pathological characteristics. More recent research on mitochondrial DNA analysis, isozyme analysis, and morphological comparisons supports the concept of a single species in Clade 1 on the phylogenetic tree. There are considerable physiological and pathological variations in isolates of *P. nicotianae*. In this compendium, *P. nicotianae* var. *nicotianae* and *P. nicotianae* var. *parasitica* are referred to as *P. nicotianae*.

P. nicotianae has prominently papillate sporangia and amphigynous antheridia. *P. nicotianae* is a self-incompatible species; however, some isolates form oospores when cultured from plants, presumably because a mixture of both mating types is present. Oospores are typically less than 20 µm in diameter, although they may be larger in some isolates. Sporangia are mostly spherical to ovoid.

P. nicotianae has a fairly high optimal temperature range for growth (25–32°C), which makes it well suited to the greenhouse environment and to landscapes in the southern United States. The pathogen's ability to grow at 35°C and its distinctive patchy growth pattern on corn meal agar help to distinguish it from other *Phytophthora* spp.

Host Range and Epidemiology

P. nicotianae attacks plants in more than 58 different families. Bedding plants reported as hosts include annual vinca, *Begonia* spp., *Gerbera jamesonii* (gerber daisy), and *Petunia hybrida* (Fig. 108). Many other ornamentals are also hosts. Cultivars of *Saintpaulia ionantha* (African violet), *Sinningia speciosa* (gloxinia), and annual vinca vary in susceptibility to *P. nicotianae*.

The most rapid disease development occurs at temperatures higher than 28°C, but asymptomatic colonization can occur at lower temperatures. In saturated soil and artificial growing media, zoospores are released and easily spread by irrigation and rainfall. Splashing of water off sidewalks can help to provide wet foliage, enhancing symptom development.

P. nicotianae is soilborne and can be spread in contaminated soil, water, and plant materials.

Management

The use of solarization to manage pathogen contamination in landscape beds has received some research attention in Florida (United States). Using clear, 25- or 50-mil polyethylene mulch was found to reduce populations of *P. nicotianae* and subsequent disease severity. In the first year, an inoculum load of 53 propagules per gram of soil was reduced to 0 following solarization, and the area under the disease progress curve (AUDPC) was reduced by 86%. In the second year, an inoculum load of 109 was reduced to 15, but the AUDPC was reduced only by 40%. In the third year, an inoculum load of 3,290 was reduced to 0, and the AUDPC was reduced by 95%. A temperature of 45°C was required to achieve these results, and it was reached only in the upper 2 in. of soil in autumn treatments. *P. nicotianae* is one of the more difficult *Phytophthora* spp. to kill using heat.

Studies on the effect of aluminum on damping-off of annual vinca, *Antirrhinum majus* (snapdragon), and petunia have been completed in North Carolina (United States). Aluminum was shown to have direct toxicity for some soilborne fungi, including *P. nicotianae*. Aluminum was tested for control of damping-off in a peat-based, limed medium at several rates. All the rates tested resulted in good management of damping-off on vinca and snapdragon, but only the highest rate was effective on petunia. Aluminum was not phytotoxic to seeds at the rates tested in the limed medium. The rate of aluminum that is effective is directly correlated with the pH of the potting medium; higher rates are needed at higher pH levels.

In a production setting, fungicides from FRAC groups 4, 14, 21, 33, 40, 43, and 45 are equally effective on diseases caused by *P. nicotianae* unless resistance has developed. In one series of studies on azoxystrobin (FRAC 11), varying the rate, interval, and application method demonstrated that even effective products must be used correctly to give optimal results. A

Fig. 107. Mycelial webbing on annual vinca, produced by *Phytophthora nicotianae*. (Courtesy A. R. Chase—© APS)

Fig. 106. Early symptoms of Phytophthora aerial blight (*Phytophthora nicotianae*) on annual vinca in a garden. (Courtesy A. R. Chase—© APS)

Fig. 108. Severe crop loss of petunias caused by Phytophthora crown rot (*Phytophthora nicotianae*). (Courtesy A. R. Chase—© APS)

2-week interval was more effective than a 4-week interval at all the rates tested. Use of the fungicide as a spray, rather than a drench, had little impact on the survival of annual vinca in a landscape trial. In greenhouse studies, a 28-day interval was found to be effective with some fungicides. Control of blight on snapdragon caused by *P. nicotianae* is more challenging than on vinca. The best results have been achieved with mefenoxam (FRAC 4) (as long as resistance has not developed), fenamidone (FRAC 11), mandipropamid (FRAC 40), and fluopicolide (FRAC 43).

The use of soil fungicides in the landscape can affect the pathogen population in the soil. A FRAC 33 fungicide, fosetyl-Al, did not affect the population of *P. nicotianae* when applied on a 4-week interval as a spray but did reduce the population when applied as a drench. In contrast, the FRAC 4 fungicide metalaxyl reduced the population when applied on a 4-week interval as a drench and also improved plant size (as measured by root and shoot weights). Unfortunately, resistance to FRAC 4 fungicides (metalaxyl and mefenoxam) may develop quickly when these very cost effective (and thus often overused) fungicides are not used in rotation or tank mixed.

Biopesticides containing *Streptomyces griseoviridis* strain K61, *Trichoderma virens* GL-21, and *Gliocladium catenulatum* strain J1446, as well as endo- and ectomycorrhizae, have all failed to give significant control of *P. nicotianae*. In contrast, plant extracts and oils have been found to reduce the population of *P. nicotianae* in soil and to improve plant health. The most effective materials tested have been pepper extract–mustard oil, cinnamon oil, clove oil, and cassia extracts.

Selected References

Benson, D. M. 1993. Suppression of *Phytophthora parasitica* on *Catharanthus roseus* with aluminum. Phytopathology 83:1303-1308.

Benson, D. M. 1995. Aluminum amendment of potting mixes for control of Phytophthora damping-off in bedding plants. HortScience 30(7):1413-1416.

Bowers, J. H., and Locke, J. C. 2004. Effect of formulated plant extracts and oils on population density of *Phytophthora nicotianae* in soil and control of Phytophthora blight in the greenhouse. Plant Dis. 88:11-16.

Ferrin, D. M., and Rohde, R. G. 1992. In vivo expression of resistance to metalaxyl by a nursery isolate of *Phytophthora parasitica* from *Catharanthus roseus*. Plant Dis. 76:82-84.

Ferrin, D. M., and Rohde, R. G. 1992. Population dynamics of *Phytophthora parasitica,* the cause of root and crown rot of *Catharanthus roseus,* in relation to fungicide use. Plant Dis. 76:60-63.

Hagan, A. K., and Akridge, R. 1999. Survival of selected summer annuals and perennials in beds infested with *Phytophthora parasitica,* 1998. Biol. Cult. Tests 14:74.

McGovern, R. J., McSorley, R., and Urs, R. R. 2000. Reduction of Phytophthora blight on Madagascar periwinkle by soil solarization in autumn. Plant Dis. 84:185-191.

Oudemans, P., and Coffey, M. D. 1991. A revised systematics of twelve papillate *Phytophthora* species based on isozyme analysis. Mycol. Res. 95:1025-1046.

Rattink, H. 1981. Characteristics and pathogenicity of six *Phytophthora* isolates from pot plants. Neth. J. Plant Pathol. 87:83-90.

Schubert, T. S., and Leahy, R. M. 1989. Phytophthora Blight of *Catharanthus roseus*. Plant Pathol. Circ. 321. Division of Plant Industry, Florida Department of Agricultural and Consumer Services, Gainesville.

Waterhouse, G. M., and Waterston, J. M. 1964. *Phytophthora nicotianae* var. *nicotianae*. CMI Descriptions of Pathogenic Fungi and Bacteria, No. 34. CABI, Kew, Surrey, UK.

Waterhouse, G. M., and Waterston, J. M. 1964. *Phytophthora nicotianae* var. *parasitica*. CMI Descriptions of Pathogenic Fungi and Bacteria, No. 35. CABI, Kew, Surrey, UK.

Yandoc, C. B., Rosskopf, E. N., Shah, D. A., and Albano, J. P. 2007. Effect of fertilization and biopesticides on the infection of *Catharanthus roseus* by *Phytophthora nicotianae*. Plant Dis. 91:1477-1483.

(Prepared by A. R. Chase and M. L. Daughtrey)

Diseases Caused by *Phytophthora tropicalis* and *P. capsici*

Phytophthora capsici is commonly problematic on vegetables in field production and not often associated with ornamentals. Host records from the past may be confusing, however. *P. tropicalis* was considered conspecific with *P. capsici* until 2001, when its differences were deemed substantial enough for a separate species designation. Unlike *P. capsici, P. tropicalis* does not infect *Capsicum annuum* (pepper), and it shows no or minimal growth at 35°C; it also has morphological differences. The decision to separate these species has since been supported by phylogenetic and genetic analyses.

In studies conducted in Michigan following a disease outbreak in New York (both United States), greenhouse flower crops in the Fabaceae and Solanaceae families were checked for their susceptibility to *P. capsici* and *P. tropicalis.* Symptoms appeared beginning 17 and 19 days after inoculations of a *Nicotiana* sp. (tobacco) and a *Lupinus* sp. (lupine) with either *Phytophthora* sp. Plants showed stunting, necrosis, wilting, and sometimes death. *Calibrachoa* hybrid cultivars Callie Gold with Red Eye and Starlette Sunset were somewhat susceptible to *P. capsici* (maximum 25% disease incidence) but very susceptible to *P. tropicalis* (100% disease incidence); plants showed root rot, crown rot, and wilt. In contrast, the *Petunia* sp. tested did not develop symptoms when inoculated with *P. tropicalis* or *P. capsici;* however, the latter was re-isolated at a low percentage from petunia. Both *Phytophthora* spp. were shown to infect some ornamental and some vegetable hosts. Given the possibility of Phytophthora disease moving between ornamental and edible crops, growers should be careful in growing crops in the Fabaceae and Solanaceae families.

Studies from the state of Virginia and from Germany, Italy, and Poland have produced additions to the originally limited host range of *P. tropicalis: Cuphea ignea* (cigar flower), *Cyclamen persicum*, *Gerbera jamesonii* (gerber daisy), *Begonia* spp., *Senecio cineraria* (dusty miller), and lupine. These findings indicate that plants in the families Apocynaceae, Asteraceae, Begoniaceae, Lythraceae, and Primulaceae might also be susceptible.

All methods that assist in controlling soilborne pathogens are helpful in managing this disease. Growers should always use new potting media and new or thoroughly disinfested containers, and when possible, growers should minimize exposure to excess water and overhead irrigation and rainfall.

Selected References

Aragaki, M., and Uchida, J. Y. 2001. Morphological distinctions between *Phytophthora capsici* and *P. tropicalis* sp. nov. Mycologia 93:137-145.

Cacciola, S. O., Spica, D., Cooke, D. E. L., Raudino, F., and Magano di San Lio, G. 2006. Wilt and collapse of *Cuphea ignea* caused by *Phytophthora tropicalis* in Italy. Plant Dis. 90:680.

Enzenbacher, T. B., Naegele, R. P., and Hausbeck, M. K. 2015. Susceptibility of greenhouse ornamentals to *Phytophthora capsici* and *P. tropicalis*. Plant Dis. 99:1808-1815.

Gerlach, W. W. P., and Schubert, R. 2001. A new wilt of cyclamen caused by *Phytophthora tropicalis* in Germany and the Netherlands. Plant Dis. 85:334.

Hong, C. X., Richardson, P. A., and Kong, P. 2008. Pathogenicity to ornamental plants of some existing species and new taxa of *Phytophthora* from irrigation water. Plant Dis. 92:1201-1207.

Orlikowski, L. B., Trzewik, A., Wiejacha, K., and Szkuta, G. 2006. *Phytophthora tropicalis,* a new pathogen of ornamental plants in Poland. J. Plant Prot. Res. 46:103-109.

(Prepared by A. R. Chase and M. L. Daughtrey)

Diseases Caused by *Globisporangium, Phytopythium,* and *Pythium* spp.

The genera *Pythium, Globisporangium,* and *Phytopythium* are oomycetes. These genera, which were previously all included in the genus *Pythium,* include saprophytes and opportunistic plant pathogens associated with damping-off and crown and root rot diseases of a wide range of hosts—among them, bedding plants. The new names of the pathogens reflect differences that are apparent through molecular characterization. However, it will be most practical to refer to the diseases collectively as "Pythium root rot."

Pythium, Globisporangium, and *Phytopythium* spp. are ubiquitous in soils worldwide, surviving easily as saprophytes. Many factors affect the severity of the diseases they cause, including host susceptibility, nutrition, and environmental conditions in the root zone (particularly, moisture and oxygen levels). Specific pathogens may be favored by different temperature conditions and also vary in host preferences; many have extensive host ranges. Table 14 lists some of the *Globisporangium, Phytopythium,* and *Pythium* spp. that cause diseases on bedding plants.

These genera of oomycetes are separated from the closely related genus *Phytophthora* by how they form zoospores. In *Globisporangium, Phytopythium,* and *Pythium* spp., zoospores are formed within a thin-walled vesicle, whereas in *Phytophthora* spp., zoospores are formed directly within the sporangium. The lemon-shaped sporangia of *Phytophthora* spp. may also be more conspicuous under the microscope than the asexual fruiting structures of some *Pythium* spp., in which swollen hyphae may serve the function of producing vesicles where zoospores are formed. In a few *Pythium* spp., the sporangia are indistinguishable from the hyphae.

Symptoms

Damping-off caused by *Pythium, Globisporangium,* and *Phytopythium* spp. can occur on many bedding plants during propagation (Fig. 109). Lab testing is needed to identify which organism is present, as there are dozens of fungi and oomycetes that may kill young seedlings. On older plants, root and stem rot is most often noticed after yellowing or wilting has been observed on foliage, typically beginning with the lowest leaves and progressing up the plant (Fig. 110).

Stunting is also a common symptom of root rot, so irregular growth of a greenhouse crop or garden planting may be the first evidence of disease caused by a *Pythium, Globisporangium,* or *Phytopythium* sp. Close examination of washed roots or roots at the interface with the growing container may show discrete lesions along with lateral root decay or root-tip blackening and decay (Fig. 111). When conditions are prime for disease development, large portions of the root system may appear water soaked and glassy or brown and shriveled. The decayed cortex of the root can be easily stripped off, leaving the vascular core intact.

Symptoms sometimes progress up from the roots into the crown (Fig. 112). If inoculum is distributed above the ground, stem or crown rot or canker may occur without obvious root involvement. The basal portions of unrooted cuttings on some crops, such as *Pelargonium* spp. (geranium), may also rot (called "black leg") (Fig. 113). In many cases, the only symptom of infection is slight to severe stunting (Fig. 114), while in other cases, the disease proceeds up from the roots under wet, humid conditions, so that the foliage is attacked and becomes covered with masses of white hyphae.

TABLE 14. ***Globisporangium, Phytopythium, and Pythium*** **spp. That Cause Diseases on Bedding Plants[a]**

Host	Species
Begonia spp.	*Globisporangium debaryanum, G. intermedium, G. irregulare, G. splendens, G. ultimum*
Calceolaria crenatiflora (slipperwort, calceolaria)	*Globisporangium mastophorum, G. ultimum*
Catharanthus roseus (annual vinca)	*Pythium aphanidermatum*
Dahlia hybrids	*Globisporangium debaryanum, G. ultimum, Phytopythium oedochilum, Pythium acanthicum*
Eustoma grandiflorum (lisianthus)	*Globisporangium spinosum*
Gerbera jamesonii (gerber daisy)	*Globisporangium middletonii*
Impatiens spp.	*Globisporangium debaryanum, G. irregulare, G. paroecandrum, G. spinosum, G. ultimum, Pythium aphanidermatum*
Pelargonium spp. (geranium)	*Globisporangium debaryanum, G. intermedium, G. mamillatum, G. megalacanthum, G. splendens, G. ultimum, Phytopythium vexans, Pythium aphanidermatum*
Primula spp. (primrose)	*Globisporangium irregulare, G. megalacanthum, G. spinosum, G. ultimum*
Ranunculus asiaticus	*Globisporangium debaryanum, G. irregulare, G. ultimum*

[a] Courtesy A. R. Chase and M. L. Daughtrey—© APS.

Fig. 109. Sweet alyssum with Pythium root rot (pathogen not identified). (Courtesy A. R. Chase—© APS)

Fig. 110. Loss of a calibrachoa plant caused by Pythium root rot (pathogen not identified). (Courtesy A. R. Chase—© APS)

None of the symptoms is entirely unique to diseases caused by *Pythium, Globisporangium,* and *Phytopythium* spp. Making mistakes is likely when diagnosing diseases and disorders of bedding plants through symptomology alone.

Causal Organisms

There are hundreds of species in the genera *Pythium, Globisporangium,* and *Phytopythium,* and those that are pathogens of plants generally have broad host ranges. Species such as *G. cryptoirregulare* and *Pythium aphanidermatum* are very common pathogens of bedding plants. *G. irregulare, G. ultimum, Pythium myriotylum,* and others are seen occasionally. Laboratory identificaton is only rarely made at the species level.

As noted earlier, these oomycete genera are closely related to the genus *Phytophthora* but distinguished by the nature of zoospore maturation and sporangium, hyphae, and sex organ morphology. Sporangia may be filamentous (merely swollen hyphal tips) or globose. Since 2010, those with globose sporangia, which were long considered *Pythium* spp., have been considered to belong in the genus *Globisporangium.* Biflagellate, motile zoospores serve to disseminate most *Pythium, Globisporangium,* and *Phytopythium* spp. (Exceptions include some strains of *G. ultimum* that do not produce zoospores.) For *Pythium, Globisporangium,* and *Phytopythium* spp., the protoplasm within the sporangium empties into a thin-walled vesicle and zoospores differentiate within the vesicle. In contrast, zoospores of *Phytophthora* spp. differentiate within the sporangium. *Pythium* and *Globisporangium* spp. have a coenocytic, 5–7 µm (rarely 10 µm) mycelium, which has a uniform diameter. Most species are homothallic, with single colonies producing both oogonia and antheridia, but outcrossing has been demonstrated in many species. Oospores are generally smooth and spherical, although in some species, they are ornamented with spines or blunt projections.

Host Range and Epidemiology

Pythium, Globisporangium, and *Phytopythium* spp. are not considered vigorous soil competitors and survive in soil, usually in a dormant state, in the form of oospores or chlamydospores. The pathogens also survive in dust and soil particles on greenhouse floors and benches and in used containers, such as flats and pots. Occasionally, commercial peat moss and growing media that contains peat moss have been reported as sources of these pathogens.

Oospore germination can be triggered by exudates from roots or germinating seeds. Observations have shown that oospores may germinate within 1.5 h following seed germination. Germ tubes of oospores show positive chemotropism, growing toward the roots of young seedlings. Speed of germination and rapid growth rate are advantageous traits that allow these pathogens to successfully attack plants' roots in advance of other pathogens. *G. ultimum* has been found more aggressive in steam-pasteurized soil and soilless media than in natural field soil because of the populations of competitive microorganisms present in natural soil.

Fig. 111. Pythium root rot at the base of a pansy pot (pathogen not identified). (Courtesy A. R. Chase—© APS)

Fig. 112. Hybrid geranium seedling with Pythium crown rot (pathogen not identified). (Courtesy M. L. Daughtrey—© APS)

Fig. 113. Black leg, caused by a *Pythium* sp., on geranium cuttings. (Courtesy A. R. Chase—© APS)

Fig. 114. Healthy chrysanthemum plant (left) and plant with stunted growth (right), caused by Pythium root rot (pathogen not identified). (Courtesy A. R. Chase—© APS)

Excessive soil moisture has been shown to promote disease development. Similarly, high fertilization rates have been shown to increase the susceptibility of *Pelargonium* × *hortorum* (florist's geranium) and *Euphorbia pulcherrima* (poinsettia) to Pythium root rot. Temperatures that are suboptimal for plant growth also tend to favor development of disease. Some of the species that cause Pythium root rot grow well at relatively high temperatures (greater than 35°C), while others grow well at low temperatures (less than 15°C). These temperature preferences determine which species are problematic in the greenhouse at different times of year.

Pythium, Globisporangium, and *Phytopythium* pathogens are not effectively spread by air currents. Contamination of pots and flats with field soil (via soiled hands, tools, and hose ends) and introduction of contaminated plant materials are the primary means by which oospores and other propagules enter production areas. Zoospores can be moved by flowing and splashing water during irrigation and propagation under mist.

Adult shore flies (*Scatella stagnalis*) can also introduce inoculum into the growing medium. Adult fungus gnats (*Bradysia* spp.) do not efficiently disseminate these pathogens, but larvae ingest and excrete oospores and encysted zoospores, which remain viable in their feces until adulthood. The extent to which these two insects contribute to pathogen distribution is not known. Laboratory trials on the effect of fungus gnat larval feeding and susceptibility of geranium seedlings to subsequent infection by *Pythium aphanidermatum* yielded unexpected results. Seedlings fed on by fungus gnat larvae 24 h before inoculation with this pathogen had almost 50% lower mortality than seedlings inoculated without previous fungus gnat feeding. Mechanically wounded seedlings did not show the same response, however. Researchers postulated that fungus gnat feeding might trigger activation of plant defense systems in seedlings.

Ebb-and-flood subirrigation systems can effectively disseminate *Pythium, Globisporangium,* and *Phytopythium* pathogens when the irrigation solution is heavily contaminated with infested potting medium or plant debris, as may occur after several crop cycles in which pathogens have been present. However, pot-to-pot movement of pathogens in properly operated ebb-and-flood systems may not be any more efficient than in conventional irrigation systems. Crops are successfully produced in ebb-and-flood systems without damaging levels of root rot.

Species that produce spherical oospores can be detected through microscopic examination of symptomatic (discolored, softened) roots. Culture media containing pimaricin may be used to select for some of the pathogens that cause Pythium root rot. Assay kits based on enzyme-linked immunosorbent assay (ELISA) have been developed for onsite diagnosis. Test kits and selective media will detect both pathogens and nonpathogens, so care must be taken not to jump to conclusions when using these detection tools—particularly when water or soil is sampled, rather than symptomatic plants.

Management

Exclusion of *Pythium, Globisporangium,* and *Phytopythium* spp. from growing media is of critical importance. These pathogens are sometimes found in new, bagged, commercial potting media; only the highest-quality media available should be used.

Diseased plants should be removed and discarded as they occur. (Doing so may be particularly important with subirrigation systems.) Occasional inspection of plants' roots may provide advance detection of root rot problems. Growers should prevent contact between soiled tools and hands and growing media and plants, and they should be careful not to drop hose ends on the floor.

The growing medium should be well drained, and to avoid prolonged saturation, irrigation should be applied only as needed. Plants, especially seedlings, should be grown at temperatures conducive to optimum growth and development. Pots and flats that will be reused should be power washed to dislodge clinging growing medium and plant tissue particles; then they should be disinfested in a 0.5% solution of sodium hypochlorite or a greenhouse disinfestant that contains a quaternary ammonium compound or hydrogen dioxide as its active ingredient.

Research has shown that excess levels of certain nutrients enhance Pythium root rot on geranium seedlings. Mortality increased as the levels of nitrogen and phosphorous increased; the increased mortality was not directly associated with increased electrical conductivity of the potting medium. Further research has shown that the use of high levels of fertilizer actually resulted in higher losses of geranium cultivars previously identified as resistant to the disease. Fertilizers should be used according to plants' requirements for adequate growth—never in excess.

Control of shore flies and fungus gnats is also necessary, because they have been demonstrated capable of ingesting and excreting *P. aphanidermatum.* Biological and chemical methods are available for control of these insects (see Part III, "Arthropod Pests"). Sticky cards can be used to monitor these insects' population levels in the greenhouse.

If field soil is used as a component of the growing medium, it must be treated to eliminate pests and pathogens. Steam is the least expensive, safest, and most effective method of treatment. In treating soil with steam, the whole soil mass must reach a temperature of 82°C for at least 30 min. The oomycetes that cause root rot will be killed at lower temperatures, but this higher temperature will provide control of a wide spectrum of soilborne pests. Various fumigants may also be used, but they can be hazardous and must be handled cautiously. In addition, fumigants may not completely penetrate large pieces of organic material that harbor plant pathogens, and residual fumigants in the treated medium may be phytotoxic. Fumigation of container media is rare in the United States, because limited products are available for this use.

Resistant cultivars are sometimes available; for example, extensive research on geranium and *Viola* × *wittrockiana* (pansy) has been reported. Even so, the use of cultivar information to reduce the severity of Pythium root rot is challenging because of the ever-changing list of available cultivars of bedding plants. Research on the method of resistance was studied in *Antirrhinum majus* (snapdragon) for Pythium damping-off caused by *G. ultimum.* Increased resistance apparently resulted from physiological rather than anatomical characteristics, such as lignification.

Biological control of these root-colonizing oomycetes is feasible, although limited products are commercially available. *Bacillus, Trichoderma,* and *Streptomyces* spp. are being used as biocontrols for agents of Pythium root rot, both alone and in combination. In one study, geraniums treated with microbial inoculants showed better root growth and lower colonization with *G. ultimum.* The use of properly composted pine bark at 20% or more in potting mixes has been shown to provide some control of Pythium and Phytophthora root rots. The use of municipal sludge at 2.5% has also been found to suppress Pythium root rot. There appears to be some interaction of the potting mix with the microbial treatment, so experimentation with each combination of microbial agent and potting medium may be needed to optimize benefits. Growing bedding plants organically has also gained some support in the past few years.

A review was conducted of more than 20 years of trials on bedding plants—specifically, on the common hosts *Impatiens hawkeri* (New Guinea impatiens), geranium, *Catharanthus roseus* (annual vinca), *Gerbera jamesonii* (gerber daisy), *Eustoma grandiflorum* (lisianthus), and snapdragon. Among all the trials, the best control of Pythium root rot (and black leg) with fungicides occurred with etridiazole (FRAC 14), cyazofamid (FRAC 21), and mefenoxam (FRAC 4) (unless resistance had occurred). Mefenoxam resistance is commonly found in surveys of *Pythium* and *Globisporangium* spp. that affect greenhouse crops in the United States. Strobilurins (e.g., azoxystrobin [FRAC 11]), fenamidone (FRAC 11), and phosphonates (e.g., fosetyl-aluminum [FRAC 33]) are sometimes partially effective. Using a sublethal dose of mefenoxam has been shown to enhance Pythium root rot on geranium caused by *G. ultimum,* particularly if the isolate is resistant to mefenoxam. For crown and stem rots, a foliar spray may be as effective as a drench application of an effective product. Growers should consult labels before using products in an ebb-and-flood irrigation system; not all fungicides are specifically labeled for this use. The off-label use of mefenoxam in this manner has led to resistance development and crop failure.

One of the most interesting aspects of managing Pythium root rot is that many plants can recover from an initial root loss if conditions improve or effective fungicides are applied. Sometimes, plants simply outgrow the damage. It is also possible for a crop to experience damage from one of these oomycete pathogens without showing any overt symptoms, especially if a slight reduction in plant size is the only result of infection.

The use of leaf temperature as a nondestructive means of detecting root rot stress has been demonstrated. It might be an excellent means of monitoring bedding plants in the greenhouse and in the landscape, where root examination is more difficult.

Selected References

Bolton, A. T. 1981. Specificity among isolates of *Pythium splendens* from geranium, chrysanthemum and Rieger begonia. Can. J. Plant Pathol. 3:177-179.

Braun, S. E., Sanderson, J. P., Nelson, E. B., Daughtrey, M. L., and Wraight, S. P. 2009. Fungus gnat feeding and mechanical wounding inhibit *Pythium aphanidermatum* infection of geranium seedlings. Phytopathology 99:1421-1428.

Burns, J. R., and Benson, D. M. 2000. Biocontrol of damping-off of *Catharanthus roseus* caused by *Pythium ultimum* with *Trichoderma virens* and binucleate *Rhizoctonia* fungi. Plant Dis. 84:644-648.

Castillo, S., and Peterson, J. L. 1990. Cause and control of crown rot of New Guinea impatiens. Plant Dis. 74:77-79.

Chagnon, M.-C., and Belanger, R. R. 1991. Tolerance in greenhouse geraniums to *Pythium ultimum.* Plant Dis. 75:820-823.

Cohen, Y., and Coffey, M. D. 1986. Systemic fungicides and the control of oomycetes. Annu. Rev. Phytopathol. 24:311-338.

Del Castillo Múnera, J., and Hausbeck, M. K. 2016. Characterization of *Pythium* species associated with greenhouse floriculture crops in Michigan. Plant Dis. 100:569-576.

Dick, M. W. 1990. Keys to *Pythium.* University of Reading, Reading, Berkshire, UK.

Garzón, C. D., Molineros, J. E., Yánez, J. M., Flores, F. J., Jiménez-Gasco, M. M., and Moorman, G. W. 2011. Sublethal doses of mefenoxam enhance Pythium damping-off of geranium. Plant Dis. 95:1233-1238.

Gladstone, L. A., and Moorman, G. W. 1989. Pythium root rot of seedling geraniums associated with various concentrations of nitrogen, phosphorus, and sodium chloride. Plant Dis. 73:733-736.

Gladstone, L. A., and Moorman, G. W. 1990. Pythium root rot of seedling geraniums associated with high levels of nutrients. HortScience 25(8):982.

Goldberg, N. P., and Stanghellini, M. E. 1990. Ingestion-Egestion and aerial transmission of *Pythium aphanidermatum* by shore flies (Ephydrinae: *Scatella stagnalis*). Phytopathology 80:1244-1246.

Gravel, V., Menard, C., and Dorals, M. 2009. Pythium root rot and growth responses of organically grown geranium plants to beneficial microorganisms. HortScience 44(6):1622-1627.

Hausbeck, M. K., Stephens, C. T., and Heins, R. D. 1987. Variation in resistance of geranium to *Pythium ultimum* in the presence of absence of silver thiosulphate. HortScience 22(5):940-944.

Hoitink, H. A. J., Inbar, Y., and Boehm, M. J. 1991. Status of compost-amended potting mixes naturally suppressive to soilborne diseases of floricultural crops. Plant Dis. 75:869-873.

Jarvis, W. R., Shipp, J. L., and Gardiner, R. B. 1993. Transmission of *Pythium aphanidermatum* to greenhouse cucumber by the fungus gnat *Bradysia impatiens* (Diptera, Sciaridae). Ann. Appl. Biol. 122:23-29.

Kim, S. H., Forer, L. B., and Longenecker, J. L. 1975. Recovery of plant pathogens from commercial peat products. (Abstr.) Proc. Am. Phytopathol. Soc. 2:124.

Mellano, H. M., Munnecke, D. E., and Endo, R. M. 1970. Relationship of pectic enzyme activity and presence of sterols to pathogenicity of *Pythium ultimum* on roots of *Antirrhinum majus.* Phytopathology 60:943-950.

Mellano, H. M., Munnecke, D. E., and Endo, R. M. 1970. Relationship of seedling age to development of *Pythium ultimum* on roots of *Antirrhinum majus.* Phytopathology 60:935-942.

Middleton, J. D. 1943. The taxonomy, host range, and geographic distribution of the genus *Pythium.* Mem. Torrey Bot. Club 20.

Munera, J. D. C., and Hausbeck, M. K. 2016. Characterization of *Pythium* species associated with greenhouse floriculture crops in Michigan. Plant Dis. 100:569-576.

Omer, M., Locke, J. C., and Frantz, J. M. 2007. Using leaf temperature as a nondestructive procedure to detect root stress in geranium. HortTechnology 17(4):532-536.

Pasura, A., and Elliott, G. 2007. Efficacy of microbial inoculants for control of blackleg disease of geranium in soilless potting mixes, 2006. Plant Dis. Manag. Rep. 1:OT010.

Sanogo, S., and Moorman, G. W. 1993. Transmission and control of *Pythium aphanidermatum* in an ebb-and-flow subirrigation system. Plant Dis. 77:287-290.

Schroeder, K. L., Martin, F. N., de Cock, A. W. A. M., Lévesque, C. A., Spies, C. F. J., Okubara, P. A., and Paulitz, T. C. 2013. Molecular detection and quantification of *Pythium* species: Evolving taxonomy, new tools, and challenges. Plant Dis. 97:4-20.

Stevens, C. T., and Powell, C. C. 1982. *Pythium* species causing damping-off of seedling bedding plants in Ohio greenhouses. Plant Dis. 66:731-733.

Stevens, C. T., Herr, L. J., Schmitthenner, A. F., and Powell, C. C. 1983. Sources of *Rhizoctonia solani* and *Pythium* spp. in a bedding plant greenhouse. Plant Dis. 67:272-275.

Tompkins, C. M. 1975. World Literature on *Pythium* and *Rhizoctonia* Species and the Diseases They Cause. Contr. 24. Reed Herbarium, Baltimore, MD.

Uzuhashi, S., Tojo, M., and Kakishima, M. 2010. Phylogeny of the genus *Pythium* and description of new genera. Mycoscience 51:337-365.

Van Der Plaats-Niterink, A. J. 1981. Monograph of the Genus *Pythium.* Centraalbureau Voor Schimmelcultures and Institute of the Royal Netherlands Academy of Sciences and Letters, Baarn, the Netherlands.

Waterhouse, G. M. 1968. The Genus *Pythium* Pringsheim. CABI, Kew, Surrey, UK.

(Prepared by M. L. Daughtrey, R. L. Wick, and J. L. Peterson; revised by A. R. Chase and M. L. Daughtrey)

White Blister Rusts

White blister rusts are obligate pathogens that are oomycetes. They are not true rusts; they are not even fungi. They are called "white blister rusts" because they form white, rustlike pustules on various plant parts.

White blister rusts are of minor economic importance on bedding plants. They have been reported occasionally on *Senecio cineraria* (dusty miller), *Gerbera jamesonii* (gerber daisy), and *Lunaria annua* (honesty). White blister rust on dusty miller was a consistent problem in the United States during the late 1990s; this disease was also reported from Italy in 2003. White blister rusts have hosts in the Brassicales, Convolvulaceae, Asteranae, and Caryophyllales. Cruciferous vegetables are more commonly affected than flower crops.

Symptoms

Plants infected with white blister rust develop white, crusty masses (pustules with exposed sporangia) on the undersurfaces of leaves (Fig. 115). Symptoms can appear within 10 days of infection, and leaves die after 30 days. Infected plants are very difficult to spot in plug flats.

Causal Organisms

The white blister rust pathogen on honesty, *Lobularia maritima* (sweet alyssum), *Matthiola incana* (garden stock), and *Tropaeolum officinale* (nasturtium) is *Albugo candida*. The pathogen on *Cleome houtteana* (spider flower) is *A. chardonii*.

The genus *Albugo* is a member of the order Albuginaceae, family Albuginales, along with the genera *Pustula* and *Wilsoniana*. Unlike the more familiar oomycetes in the Peronosporales (e.g., *Pythium, Phytophthora,* and *Peronospora* spp.), *Albugo* spp. and other white blister rusts produce chains of sporangia basipetally, beneath the host epidermis; both sporangia and oospores germinate to produce zoospores. The white blister rusts have intercellular hyphae with intracellular haustoria.

A. candida sporangia are primarily subglobose and hyaline to pale yellow; they have verrucose ornamentation and no equatorial wall thickening. Oospores are dark brown. *A. chardonii* is morphologically similar.

Pustula tragopogonis causes white blister rust on dusty miller and gerber daisy. It has chains of hyaline or yellowish, subglobose to cylindrical sporangia. Another species of *Pustula, P. centaurii,* affects *Eustoma grandiflorum* (lisianthus) in Australia. Scanning electron microscopy shows that *Pustula* sporangia are reticulate or striate and that the equatorial portion of the wall is thickened.

Species of the genus *Wilsoniana* cause white blister rust on certain members of the Caryophyllales. They have pyriform to cylindrical sporangia and an equatorial wall thickening; sporangia are verrucose to striate but never reticulate.

Host Range and Epidemiology

The genus *Pustula* is composed of pathogens of the superorder Asteranae. Some common hosts of *P. tragopogonis* (syn. *Albugo tragopogonis*) include *Tragopogon* spp. (salsify), *Helianthus annuus* (sunflower), and several familiar weeds (including species of *Ambrosia, Antennaria, Artemisia, Cirsium, Matricaria,* and *Parthenium*). However, further study may reveal that multiple pathogens affect this broad host range. Another *Pustula* species affects two species of coneflower: *Echinacea purpurea* and *E. angustifolia*. White blister rust on gerber daisy caused by *P. tragopogonis* is common in Australia and New Zealand and known in Africa, Europe, and Cuba. White blister rust was seen on gerber daisy in Mexico in 1995, but it has not been reported in the United States.

White blister rust caused by *P. tragopogonis* occasionally affects greenhouse crops of dusty miller, but it is not common on the related *S. cruentus* (cineraria) and has not been reported on cineraria in the United States. The occurrence of *P. tragopogonis* on dusty miller is fairly rare. White blister rust was found on dusty miller in the U.S. greenhouse industry for 3–4 consecutive years in the late 1990s, possibly indicating seed contamination as a source.

Fig. 115. White blister rust *(Albugo* sp.) on dusty miller plugs. (Courtesy A. R. Chase—© APS)

Another white blister rust species, *A. candida,* has an unusually wide host range, affecting cruciferous flowers and vegetables, as well as members of the Convolvulaceae. Local introduction from weed hosts is possible for both this pathogen and *P. tragopogonis*.

Infection of host plants by white blister rusts usually occurs via sporangia (the disseminative spore form) during cool, wet conditions. The sporangia produce motile zoospores that encyst, germinate, and penetrate the host epidermis via a germ tube. An intercellular mycelium forms in the leaf tissue, where the pathogen absorbs nutrients through intracellular, knoblike haustoria. Oospores form in the leaf tissue, allowing *P. tragopogonis* and *A. candida* to persist in spite of adverse environmental factors, such as desiccation and low temperatures.

P. tragopogonis and *A. candida* survive in the greenhouse as sporangia or oospores on dormant plants and debris. During moist, warm conditions in the landscape (and in the greenhouse under the conditions found during plug production), oospores may germinate to form vesicles that contain several zoospores.

As noted earlier, *Wilsoniana* spp. produce white blister rust on certain members of the Caryophyllales, including species of *Alternanthera, Amaranthus,* and *Portulaca*.

Management

To effectively reduce the incidence of disease, diseased plants and their debris should be discarded, and the introduction of diseased plants into the greenhouse should be avoided. The difficulty of finding an early infection must be addressed by at least weekly examination of leaves of known hosts by turning them over. If white blister rust has been found on dusty miller or another host, the grower should use pathogen-free seed and record and track seed lots.

Preventive treatments with products that were effective on downy mildew were effective in controlling white blister rust on dusty miller plugs in a commercial operation. Studies on sunflower from Europe showed a high degree of preventive control with mefenoxam (FRAC 4) and azoxystrobin (FRAC 11).

Selected References

Choi, Y.-J., Shin, H.-D., and Thines, M. 2009. The host range of *Albugo candida* extends from Brassicaceae through Cleomaceae to Capparaceae. Mycol. Prog. 8:329-335.

Doepel, R. F. 1965. White rust of African daisies. J. Agric. West. Aust. Ser. 46(7):439.

du Toit, L. J., and Derie, M. L. 2014. White rust of *Echinacea angustifolia* and *E. purpurea* in North America caused by a *Pustula* species. Plant Dis. 98:856.

Johnson, D. A., and Gabrielson, R. L. 1990. White rust of seed radish. Wash. State Univ. Ext. Bull. 1570:1-2.

Lava, S. S., Ziooer, R., and Spring, O. 2015. Sunflower white blister rust—Host specificity and fungicide effects on infectivity and early infection stages. Crop Prot. 67:214-222.

Plate, H. P., and Krober, H. 1977. Weisser Rost an Gerbera auf Teneriffa (Erreger: *Albugo tragopogonis* (DC.) S. F. Gray). (White rust on gerbera in Tenerife [Causal agent: *Albugo tragopogonis* (DC.) S. F. Gray].) Nachr. Deut. Pflanzenschutzd. 29(11):169-170.

Ploch, S., Telle, S., Choi, Y.-J., Cunnington, J. H., Priest, M., Rost, C., Shin, H.-D., and Thines, M. 2011. The molecular phylogeny of the white blister rust genus *Pustula* reveals a case of underestimated biodiversity with several undescribed species on ornamentals and crop plants. Fungal Biol. 115:214-219.

Thines, M., and Spring, O. 2005. A revision of *Albugo* (Chromista, Peronosporomycetes). Mycotaxon 92:443-458.

Vasquez, G. L., Aquino, M. J., Norman, M. T., Martinez, F. A., Sandoval, R. V., Corona, R. M. C., and Strider, D. L. 1997. First report of white rust on gerbera caused by *Albugo tragopogonis* in North America. Plant Dis. 81:228.

(Prepared by A. R. Chase and M. L. Daughtrey)

Diseases Caused by Nematodes

Nematodes are small (0.5–5.0 mm in length), nonsegmented roundworms commonly found in soil. Most nematodes feed on other microscopic organisms, such as algae, fungi, bacteria, and other nematodes. Those that are plant parasites are highly specialized, largely obligate parasites. They feed by means of a stylet: a hollow, retractable, spear-shaped mouthpart that punctures plant cells and allows the nematode to acquire nutrients.

Plant-parasitic nematodes primarily parasitize plants' roots, but some species feed on aboveground plant parts. The root parasites, whose feeding may cause aboveground symptoms, are either ectoparasitic or endoparasitic. Ectoparasitic nematodes remain on the outside of the plant root and feed by probing into root hairs or epidermal cells, for the most part. Endoparasitic nematodes enter the root (or aboveground tissue) to feed and thus are potentially more damaging than ectoparasitic nematodes.

Some of the most important plant-parasitic nematodes are root-knot and cyst nematodes. They are sedentary endoparasites and stay localized at one feeding site within the roots throughout adulthood.

Nematodes cause both direct and indirect injury to plants. They feed directly on cells and in some cases also cause injury by burrowing into plant tissue. Symptoms on roots may include inhibited root elongation, swollen (stubby) root tips, galls, and lesions. Secondary infections by fungi that utilize nematode feeding sites as infection courts may increase the impact of a nematode infestation. Damage to the plant's roots affect its nutrient and water uptake, leading to wilting and other indications aboveground, such as stunting and symptoms of nutrient deficiency. The severity of nematode damage correlates directly with the size of the nematode population. Damage is often overlooked until the population becomes sizeable enough to have dramatic effects on plants.

Lapses in sanitation practices that may lead to nematode contamination include improperly treating field soil, allowing flats of plants to contact the ground, and using infested vegetative propagation materials. Contamination of landscape beds with the root-knot nematode is the most common nematode problem with bedding plants; within the United States, this occurs primarily in the South. Foliar nematodes are typically introduced into the greenhouse on cuttings and rooted plants; they do not commonly appear on bedding plants grown from seed unless sanitation practices are lax. The most important plant-pathogenic nematodes of bedding plants are root-knot (*Meloidogyne* spp.) and (to a very limited degree) foliar (*Aphelenchoides* spp.) nematodes.

(Prepared by M. L. Daughtrey, R. L. Wick, and J. L. Peterson; revised by A. R. Chase and M. L. Daughtrey)

Root-Knot Nematodes

Root-knot nematodes (*Meloidogyne* spp.) cause considerable economic losses on food, fiber, and ornamental plants worldwide. The common name is derived from the knotlike growths (galls) these nematodes produce on plants' roots.

Although root-knot nematodes are potentially destructive to crops, they are rarely problematic for propagators of annuals because of their relatively long generation time (about 30 days) and their infrequent association with bedding plants grown in soilless mixes. Challenges with root-knot management more commonly affect gardens in regions with subtropical climates than temperate climates.

The southern root-knot nematode (*M. incognita*) is a common and harmful pest in landscapes in the southern United States. The northern root-knot nematode (*M. hapla*) can survive New England winters and has been problematic in vegetatively propagated, herbaceous perennial ornamentals. Fewer generations of the northern root-knot nematode are produced in the northern United States compared with the southern part of the country, and the galls produced by this nematode are typically less damaging than those produced by the southern root-knot nematode. However, a gradual buildup of nematodes and subsequent damage can occur over time with repeated plantings of annuals in nematode-infested soils, even in temperate climates, and this damage often goes unrecognized.

Symptoms

Diagnosing damage caused by root-knot nematodes is facilitated by the striking symptoms they produce on plants' roots. Galls range in size from little more than root swellings to large, clublike structures 1 cm or more in diameter (Fig. 116). Symptoms vary with the severity of the infestation, the plant's susceptibility, and the species of nematode involved.

The northern root-knot nematode produces galls that are relatively small and have radiating lateral roots, distinguishing them from the galls produced by the southern root-knot nematode and a third species, the peanut root-knot nematode (*M. arenaria*). The peanut root-knot nematode also produces small, beadlike galls, but they do not have the lateral roots typical of galls produced by the northern root-knot nematode. The southern root-knot nematode and a fourth species, the Javanese root-knot nematode (*M. javanica*), may produce discrete galls, but these galls often coalesce into large, clublike structures.

Fig. 116. Galls produced by a root-knot nematode (*Meloidogyne* sp.) on the roots of a coleus plant in a garden. (Courtesy A. R. Chase—© APS)

TABLE 15. Resistance Levels of Some Bedding Plants to Root-Knot Nematodes[a,b]

Very Resistant	Slightly Susceptible	Susceptible	Highly Susceptible
Ageratum houstonianum [5]	*Begonia* spp. [1]	*Begonia* spp. [2]	*Antirrhinum majus* (snapdragon) [5]
Catharanthus roseus (annual vinca) [2]	*Catharanthus roseus* (annual vinca) [3]	*Bellis perennis* (English daisy) [1]	*Begonia* spp. [1]
Coreopsis spp. (tickseed) [1]	*Clarkia amoena* (godetia) [1]	*Celosia* spp. [3]	*Calendula officinalis* [2]
Delphinium spp. (larkspur) [1]	*Dianthus* spp. [2]	*Dahlia* spp. [1]	*Celosia* spp. [1]
Gaillardia × *grandiflora* [2]	*Petunia hybrida* [3]	*Dianthus* spp. [2]	*Delphinium* spp. (larkspur) [1]
Gerbera jamesonii (gerber daisy) [1]	*Portulaca* spp. [1]	*Nicotiana* spp. [1]	*Helianthus annuus* (sunflower) [1]
Lobularia maritima (sweet alyssum) [2]	*Salvia* spp. [3]	*Petunia hybrida* [2]	*Impatiens* spp. [1]
Matthiola spp. (stock) [2]	*Verbena hybrida* [3]	*Plectranthus scutellarioides* (coleus) [2]	*Lobelia erinus* [1]
Phlox spp. [2]		*Viola* × *wittrockiana* (pansy) [3]	*Plectranthus scutellarioides* (coleus) [2]
Salvia spp. [3]			*Tropaeolum officinale* (nasturtium) [2]
Tagetes spp. (marigold) [5]			*Verbena hybrida* [1]
Torenia fournieri [1]			*Viola* × *wittrockiana* (pansy) [3]
Zinnia elegans [4]			

[a] Courtesy A. R. Chase and M. L. Daughtrey—© APS.
[b] Numbers in brackets after host names indicate numbers of trials with these results.

The aboveground symptoms of root-knot nematode infestation include dwarfing, yellowing, and wilting, but these symptoms are, of course, not unique. In some cases, plants infested with root-knot nematodes are taller than usual but have thin, sparse foliage.

Causal Organisms

The genus *Meloidogyne* is represented by approximately 55 species, the most commonly encountered of which are *M. arenaria, M. hapla, M. incognita,* and *M. javanica.* Root-knot nematodes are characterized by sexual dimorphism. Females are distinctively globose (440–1,300 × 300–700 µm), and males and juveniles are vermiform. In addition to morphological features, differential host response can be used to identify these four species (including six races).

Host Range and Epidemiology

Root-knot nematodes affect many species of plants worldwide. Nearly all the plants covered in this compendium are susceptible (Table 15).

Meloidogyne spp. are sedentary endoparasites of plants after the larvae have entered the roots and established feeding locations. The female produces an egg mass that contains up to 1,000 eggs, which are extruded in a gelatinous matrix. The first-instar juvenile undergoes one molt in the egg. Both molting and hatching are stimulated by plant root exudates. The second-stage juvenile emerges from the egg, enters the root, and positions itself next to or within the vascular cylinder; there, its feeding results in the formation of giant cells and swelling of the root. Two more molts occur within the root. The female develops into the characteristically globose adult and is strategically positioned in the gall such that the eggs will be extruded outside the root, thus facilitating dissemination. The male also swells somewhat but becomes vermiform after the last molt and leaves the root.

In the presence of a susceptible host, the nematode's life cycle is influenced by both soil temperature and moisture. The northern root-knot nematode can invade plants' roots at temperatures of 5–35°C (optimum 15–20°C). Growth and reproduction occur at 15–30°C (optimum 20–25°C). The temperature ranges for the Javanese root-knot nematode and related species are approximately 5°C higher than those for the northern root-knot nematode.

Quite a few trials have been performed (primarily in the southern United States) to evaluate the relative resistances of genera of bedding plants to a variety of root-knot nematode species. The results have been reasonably consistent, except for a very few genera. The plants that have generally been very resistant include ageratum, sweet alyssum, gaillardia, marigold, *Salvia* spp., annual vinca, and zinnia. Plants that have been moderately susceptible include begonia, celosia, *Dianthus* spp., and petunia. Finally, the most susceptible genera have been calendula, coleus, pansy, and snapdragon. Interestingly, larkspur and verbena had very inconsistent reactions, ranging from very resistant to highly susceptible.

Management

Recommended management practices include using soilless growing media, steam pasteurization or chemical treatment of field soil, disease-free plant materials, and genetic resistance, as well as avoiding contamination. If plants are to be rooted in nursery soil outdoors, the application of a nematicide or fumigant may be necessary. The use of aerated steam is the safest and most effective way to treat soil to be used in pots or raised beds in the greenhouse. Soilless media is free of plant-parasitic nematodes, but plants potted in soilless mix and placed on groundcover fabric may become infested from the soil below.

Nematodes are killed at a relatively low temperature (about 50°C). Maintaining a soil temperature of 70°C for 30 min will control nematodes and other pests, as well, so using this higher temperature is recommended for soil treatment. Soil fumigation with metam sodium or 1,3-dichloropropene prior to planting is also effective for nematode control. Solarization has been effective in landscapes in some regions in which the length of day and exposure to cloud-free days results in significant heating.

None of these management practices is acceptable for controlling root-knot nematodes in all situations. Thus, using less susceptible plants is the most versatile approach for minimizing landscape problems with root-knot nematodes.

Selected References

Creswell, T. 2009. Nematode management in bedding plants in the landscape. N.C. State Univ. Ornam. Dis. Note 31.

Ismail, W. 1980. Susceptibility of some ornamental plants to the attack of *Meloidogyne incognita*. Indian J. Hortic. 37:326-328.

McSorley, R. 1994. Susceptibility of common bedding plants to root-knot nematodes. Proc. Fla. State Hortic. Soc. 107:430-432.

McSorley, R., and Frederick, J. J. 1994. Response of some common annual bedding plants to three species of *Meloidogyne*. Supp. J. Nematol. 26(4S):773-777.

Schochow, M., Tjosvold, S. A., and Ploeg, A. T. 2004. Host status of *Lisianthus* 'Mariachi Lime Green' for three species of root-knot nematode. HortScience 39:120-123.

Walker, J. T. 1980. Susceptibility of Impatiens cultivars to root-knot nematode, *Meloidogyne arenaria*. Plant Dis. 64:184-185.

Walker, J. T., Melin, J. B., and Davis, J. 1994. Sensitivity of bedding plants to southern root-knot nematode, *Meloidogyne incognita* race 3. Supp. J. Nematol. 26(4S):778-781.

(Prepared by M. L. Daughtrey, R. L. Wick, and J. L. Peterson; revised by A. R. Chase and M. L. Daughtrey)

Foliar Nematodes

Foliar nematodes (*Aphelenchoides* spp.) cause disease by feeding directly on foliage. They are not common on bedding plants, but they can be an important and persistent problem in the landscape, especially with herbaceous perennials. Because of the short crop cycles of bedding plants and their propagation through seeds and cuttings prepared from carefully maintained stock plants, the occurrence of foliar nematodes on bedding species is uncommon.

Symptoms

Depending on the host, foliar nematodes feed ectoparasitically or endoparasitically, causing a variety of symptoms. Endoparasitic feeding on foliage often produces chlorotic or brown to purple or black, water-soaked lesions; they are angular in shape because of delimitation by the leaf veins. The lesions produced on *Begonia* spp. are not angular, however. Also, foliage becomes bronze, yellow, or red, depending on the light intensity, cultivar, and age of the lesion (Fig. 117). Colonized leaves eventually desiccate but remain attached to the plant. Ectoparasitic feeding on some hosts may cause leaf distortion and bud scarring.

Causal Organisms

The genus *Aphelenchoides* contains more than 227 species, some of which are mycophagous and some, predatory. The most common foliar nematodes (also known as "bud" and "leaf" nematodes) are the spring crimp nematode (*A. fragariae*), the chrysanthemum foliar nematode (*A. ritzemabosi*), and the spring dwarf nematode (*A. besseyi*). These three species of foliar nematodes can be distinguished by morphological characteristics.

Fig. 117. Bronzing and necrosis of begonia leaves, caused by infestation with foliar nematodes (*Aphelenchoides* spp.). (Courtesy J. LaMondia—© APS)

Host Range and Epidemiology

The spring crimp nematode attacks more than 250 species of plants in 47 families. It occurs in regions with temperate climates and is distributed in the United States from Massachusetts to Florida; it might be expected to occur in greenhouses worldwide on susceptible hosts. The chrysanthemum foliar nematode has been reported on 190 species of plants throughout regions with temperate climates; it is more likely to be found on plants in the Asteraceae than the spring crimp nematode. The spring dwarf nematode is encountered less frequently than the other two species but has been reported on *Dahlia* hybrids and affects more than 200 other plant species. Bedding plants reported as hosts of various foliar nematodes include ageratum, chrysanthemum, coleus, cyclamen, dahlia, English daisy, marigold, nemesia, phlox, primrose, ranunculus, salvia, snapdragon, torenia, verbena, and zinnia.

Management

Although foliar nematodes can survive in soil for only a few months, adult foliar nematodes can survive desiccation within plant tissue for up to 3 years. Thus, avoiding infested plant materials is the most important method of controlling foliar nematodes. Splashing irrigation water also easily spreads foliar nematodes within a susceptible crop or from one infested crop to another. Vegetative propagation from infested plants will further spread the nematodes. Common hosts should be routinely inspected, especially when new plants are brought into the greenhouse or landscape. Plants with symptoms of foliar nematode infestation should be discarded.

Foliar nematode management is very challenging. Insecticides and miticides have been tested occasionally, particularly since the 1990s, with poor results. In a series of studies conducted on *Begonia* spp. between 1993 and 1995, abamectin, azadirachtin, bifenthrin, carbosulfan, cyromazine, diflubenzuron, fenoxycarb, and imidacloprid were all ineffective in controlling the spring crimp nematode. In a 2002 study, chlorfenapyr and abamectin failed to significantly reduce populations of this nematode on anemone, phlox, and salvia. A study in 2003 found no benefit of applying bifenazate, chlorfenapyr, or peroxyacetate to control foliar nematodes on abelia.

No effective chemicals are registered in the United States for postplant control of foliar nematodes. Sanitation efforts and propagation from clean stock remain the key methods of foliar nematode management.

Selected References

Kohl, L. M. 2011. Foliar nematodes: A summary of biology and control with a compilation of host range. Online. Plant Health Progress. doi:10.1094/PHP-2011-1129-01-RV

LaMondia, J. A. 2004. Evaluation of Avid and Pylon for control of foliar nematodes on Anemone, Phlox, and Salvia, 2002. Fung. Nemat. Tests 59:N006.

McCuiston, J. L., Hudson, L. C., Subbotin, S. A., Davis, E. L., and Warfield, C. Y. 2007. Conventional and PCR detection of *Aphelenchoides fragariae* in diverse ornamental host plant species. J. Nematol. 39:343-355.

Walker, J. T., Oetting, R. D., Johnston-Clark, M., and Melin, J. B. 1997. Evaluation of newer chemicals for control of foliar nematode on begonia. J. Environ. Hort. 15:16-18.

Warfield, C. Y., Parra, G. R., and Hight, P. A. 2004. Evaluation of miticides and a surface disinfestant for control of foliar nematodes on abelia, 2003. Fung. Nemat. Tests 59:OT001.

(Prepared by M. L. Daughtrey, R. L. Wick, and J. L. Peterson; revised by A. R. Chase and M. L. Daughtrey)

Diseases Caused by Viroids

Viroids are the smallest known plant pathogens. A single viroid consists of single-stranded, circular RNA molecules that range in size from 239–401 nucleotides. Viroids do not encode proteins, and they multiply in their plant hosts without so-called helper viruses. There are more than 30 viroid species, and they infect herbaceous and woody plant hosts; several viroid diseases are of considerable economic importance.

Symptoms, Causal Organisms, and Hosts

Despite the many established and potential ornamental host–viroid combinations, symptoms have been reported only occasionally. In ornamentals, the most serious direct effects occur in *Chrysanthemum* spp. and are caused by the *Chrysanthemum stunt viroid* (CSVd) and, to a lesser extent, the *Chrysanthemum chlorotic mottle viroid*. The vast majority of newly discovered viroid infections in ornamental crops are caused by members of the genus *Pospiviroid*. Two pospiviroids—*Potato spindle tuber viroid* (PSTVd) and CSVd—have been responsible for most of the recorded natural infections by members of this genus (11 and 10 plant species, respectively).

Severe symptoms are known for CSVd in chrysanthemum, for which viroid infection may cause growth reductions of up to 65%. In addition, flowers may be bleached, and leaves of some cultivars may develop bright-yellow spots of various sizes. However, the severity of symptoms varies greatly, and an infection may even be symptomless. Therefore, symptoms are often best observed in a crop that has both infected and uninfected plants.

In addition to severely affecting chrysanthemum, CSVd may cause significant yield losses in *Argyranthemum frutescens* because of growth reduction, flower distortion, and leaf necrosis. Symptoms have not been observed, however, in CSVd-infected plants of *Ageratum houstonianum, Petunia hybrida, Senecio cruentus* (cineraria), *Solanum jasminoides* (jasmine nightshade), *Verbena* spp., and *Vinca major* (bigleaf periwinkle).

Chrysanthemum chlorotic mottle viroid (CChMVd) may cause mild mottling or mosaic of young chrysanthemum leaves at elevated temperatures. This symptom may be followed by general chlorosis, some reduced growth of leaves, and a delay in blossom development.

Coleus blumei viroid 1 (CbVd-1) and related variants or species have been reported to cause symptomless infections in *Plectranthus scutellarioides* (coleus). In addition, symptomless infections by *Columnea latent viroid* (CLVd) were reported in *Columnea erythrophaea, Brunfelsia undulata,* and *Nematanthus wettsteinii,* and many more viroid–host combinations have been reported in ornamental crops.

Whether symptomatic or asymptomatic, a viroid infection may also be economically significant by acting as a source of viroid inoculum for other crops that may undergo severe losses when infected, such as *Solanum lycopersicum* (tomato). In addition, both types of infections may result in extra costs (e.g., for certification) when guarantees of viroid status are required for trade and export.

Management

All viroids are mechanically transmissible and spread primarily through vegetative propagation. Several viroids are also seed transmitted in certain hosts, and there is some evidence for insect transmission (mechanical) by pollinators.

Starting a new crop with viroid-free planting materials is very important. Certification and testing are the main tools for guaranteeing viroid-free status.

In addition, introducing viroids via human activities should be prevented. Because viroids are mechanically transmissible, they can be introduced to potential host plants via the hands, clothes, or equipment of people working in or visiting the greenhouse. Cleaning and disinfesting equipment using potassium peroxymonosulfate or sodium hypochlorite may prevent viroid transmission.

Viroid eradication involves destroying viroid-infected plants and thorough cleaning and disinfestation of equipment and greenhouses in which infected plants have been grown. All the infected plants, together with those from an adequate buffer zone, should be destroyed. In the case of a symptomless infection (as commonly observed in ornamentals), all the plants in the respective lot should be destroyed. All pots, as well as plastic used to cover the soil, should also be removed and destroyed.

Selected References

Bostan, H., Nie, X., and Singh, R. P. 2004. An RT-PCR primer pair for the detection of *Pospiviroid* and its application in surveying ornamental plants for viroids. J. Virol. Methods 116:189-193.

EPPO/OEPP. 2002. Certification scheme for chrysanthemum. Bull. OEPP/EPPO 32:105-114.

Hadidi, A., Flores, R., Randles, J. W., and Semancik, J. S., eds. 2003. Viroids. CSIRO, Clayton, Victoria, Australia.

Horst, R. K. 1987. Chrysanthemum chlorotic mottle. Pages 291-295 in: The Viroids. T. O. Diener, ed. Plenum, New York.

Horst, R. K., and Nelson, P. E. 1997. Compendium of Chrysanthemum Diseases. American Phytopathological Society, St. Paul, MN.

Horst, R. K., Langhans, R. W., and Smith, S. H. 1977. Effects of chrysanthemum stunt, chlorotic mottle, aspermy and mosaic on flowering and rooting of chrysanthemums. Phytopathology 67:9-14.

Marais, A., Faure, C., Deogratias, J. M., and Candresse, T. 2011. First report of *Chrysanthemum stunt viroid* in various cultivars of *Argyranthemum frutescens* in France. Plant Dis. 95:1196.

Verhoeven, J. Th. J., Botermans, M., Meekes, E. T. M., and Roenhorst, J. W. 2012. *Tomato apical stunt viroid* in the Netherlands: Most prevalent pospiviroid in ornamentals and first outbreak in tomatoes. Eur. J. Plant Pathol. 133:803-810.

Verhoeven, J. Th. J., Hammond, R. W., and Stancanelli, G. 2017. Economic significance of viroids in ornamental crops. Ch. 3 in: Viroids and Satellites. A. Hadidi, R. Flores, J. Randles, and P. Palukaitis, eds. Academic Press, Boston.

(Prepared by R. Hammond, G. Stancanelli, and J. Th. J. Verhoeven)

Diseases Caused by Viruses

A plant virus consists of a few genes composed of nucleic acid packaged in a protein capsule (capsid). Virus particles are very small (20–300 nm) and visible only with electron microscopy. A virus disrupts a plant's normal functions by altering its structural and chemical processes so as to manufacture more virus particles. Viruses do not usually kill plants, but they dramatically reduce yields of flowers and fruits and destroy the aesthetic value of ornamentals. The name of a virus usually describes one of the symptoms it may cause and includes the name of the first host on which it was described.

Viruses cause a wide array of symptoms, some of which are unique to viruses but many of which mimic those of nutrient disorders, herbicide injuries, insect damage, and other plant pathogens, such as fungi and bacteria. The most common symptom caused by viruses is stunting, which can go undetected unless virus-free plants are present for direct comparison with infected plants. Other symptoms of virus infection include unusual color patterns, such as mosaics and streaks; leaf chlorosis; vein clearing and banding; ring spots; necrotic spots, stem tips, and veins; distorted leaves and fruit; swollen shoots; enations; and flower color breaking (mottling or streaking of color). Infection with a virus can also affect reproduction through reducing the number of flowers or fruits, seed set, or viability of pollen. Only a few virus diseases (e.g., tomato spotted wilt) may cause infected plants to wilt and die.

Diagnostic species (indicator plants) used for virus identification exhibit distinctive local lesions and/or systemic symptoms when inoculated with different viruses. Inclusion bodies, which are aggregates of virus particles or proteins visible with the light microscope, are also helpful for diagnosis. However, because of the availability of many genetic tools, diagnostic species and inclusion bodies are used less commonly than in the past.

Different viruses can be spread in different ways, but all are spread through vegetative propagation. The modern trend of propagating an increasing proportion of bedding plants through tissue cultures and cuttings has increased the possibility of the global spread of viruses. An infection may be latent (showing no symptoms) when the virus titer in the plant is low. Some viruses (notably, *Tobacco mosaic virus*) are spread merely by workers handling plants, whereas some require an insect (or more rarely, a nematode or fungus) vector.

The particular relationship between a virus and its insect vector has important effects on virus transmission. In some cases, the virus is present on the insect's stylet for only a short period and is shed upon molting (nonpersistent transmission). Alternatively, the virus may be held in the insect's foregut, which increases the period of transmission to hours or days (semipersistent transmission). Transmission may also occur when the virus accesses the insect's hemocoel and becomes associated with its salivary glands (persistent or circulative transmission); this form of transmission lasts from days to weeks, and in some cases, the virus can even multiply within its vector. An example of persistent transmission is the spread of *Tomato spotted wilt virus* by the western flower thrips (*Frankliniella occidentalis*); transmission occurs throughout the adult's life following acquisition by the larva that fed on an infected plant.

New viruses are described frequently, and there are many reports of new hosts of previously described viruses. Some of the newly emerging viruses on bedding plants (and their hosts) include *Verbena latent virus* (*Verbena* spp.), *Tobacco ringspot virus* (*Chaenostoma cordatum* [bacopa] and *Portulaca* spp.), *Impatiens flower break virus* (*Impatiens hawkeri* [New Guinea impatiens]), and *Alternanthera mosaic virus* (*Angelonia, Crossandra, Helichrysum, Phlox, Portulaca, Salvia, Scutellaria, Torenia,* and *Zinnia* spp.). The viruses described in the following sections are, in some cases, known to occur only on ornamentals, whereas others affect a wide range of horticultural crops and weed species. Additional viruses that are reported from bedding plants but are generally not of great economic importance on these crops are listed in Table 16.

TABLE 16. Some Viruses of Minor Importance on Certain Bedding Plants[a]

Host	Virus	Symptom(s)
Begonia spp.	*Carnation mottle virus* (CarMV)	Leaf vein clearing; leaf curling; flower deformation; flower color breaking
	Clover yellow mosaic virus (ClYMV)	Bright mosaic; stunting
Coreopsis auriculata 'Nana'	*Lettuce mosaic virus* (LMV)	Chlorotic leaf spots and rings
Cyclamen persicum	*Tomato aspermy virus* (TAV)	Flower malformation
	Potato virus X (PVX) + *Tobacco mosaic virus* (TMV) (co-infection)	Flower distortion; necrotic streaking of petals
Dahlia hybrids	*Potato virus Y* (PVY)	Unknown
Eustoma grandiflorum (lisianthus)	Lisianthus necrosis virus (LNV)	Systemic necrotic leaf spots; rings on leaves and stems; stunting; leaf tip necrosis; flower color breaking
	Tomato mosaic virus (ToMV)	Mosaic; leaf necrosis
	Carnation mottle virus (CarMV)	Systemic necrotic leaf spots
	Turnip mosaic virus (TuMV)	Stunting; systemic yellow leaf spots
	Pepper veinal mottle virus (PVMV)	Yellow leaf spots
	Iris yellow spot virus (IYSV)	Necrotic leaf spots and rings; systemic necrosis
Gerbera jamesonii (gerber daisy)	*Tobacco rattle virus* (TRV)	Black or yellow ring spots on leaves
Impatiens walleriana	*Helenium virus S* (HVS)	Stunting or no symptoms
Petunia hybrida	*Colombian datura virus* (CDV) (syn. *Petunia flower mottle virus* [PetFMV])	Flower mottling; some leaf mottling
	Petunia vein banding virus (PetVBV)	Leaf vein banding
	Petunia vein clearing virus (PVCV)	Leaf vein clearing
	Calibrachoa mottle virus (CbMV)	No symptoms
	Turnip mosaic virus (TuMV)	Flower color breaking
Plectranthus scutellarioides (coleus)	*Tobacco etch virus* (TEV)	Leaf mosaic and vein necrosis
	Coleus vein necrosis virus (CVNV)	Leaf vein necrosis
Primula spp. (primrose)	*Tobacco necrosis virus* (TNV)	Mottling; chlorosis; stunting; necrotic spots along leaf veins
	Tomato bushy stunt virus (TBSV)	Systemic necrosis
	Primula mosaic virus (PrMV) (North America)	Mosaic; green, elevated blisters; upcupping of young leaves; shoestring leaves; petal spots and streaks; stunting
	Primula mosaic virus (PrMV) (Italy)	Leaf chlorosis; flower color breaking
	Primula mottle virus (PrMoV)	Mild leaf mosaic; stunting; flower color breaking
Ranunculus asiaticus	*Ranunculus mosaic virus* (RanMV)	Mosaic
	Ranunculus mild mosaic virus (RanMMV)	Mosaic
Verbena hybrida	*Clover yellow mosaic virus* (ClYMV)	Chlorotic, necrotic flecks on leaves

[a] Courtesy A. R. Chase and M. L. Daughtrey—© APS.

Selected References

Christie, R. G., and Edwardson, J. W. 1977. Light and Electron Microscopy of Plant Virus Inclusions. Fla. Agric. Exp. Stn. Monogr. No. 9.

Martelli, G. P. 1992. Classification and nomenclature of plant viruses: State of the art. Plant Dis. 76:436-442.

Valverde, R. A., Sabanadzovic, S., and Hammond, J. 2012. Viruses that enhance the aesthetics of some ornamental plants: Beauty or beast? Plant Dis. 96:600-611.

(Prepared by M. L. Daughtrey, R. L. Wick, and J. L. Peterson; revised by A. R. Chase and M. L. Daughtrey)

Alternanthera Mosaic

Alternanthera mosaic virus (AltMV), a potexvirus, was first described from Australia in 1998 following its discovery in *Alternanthera pungens* (khaki weed), which also occurs in the southern United States. The experimental host range observed for the virus in the Australian study included plants in nine families. Among those developing a systemic mosaic were *Amaranthus tricolor, Solanum lycopersicum* (tomato), *Spinacia oleracea* (spinach), and *Zinnia elegans.*

Symptoms and Causal Agent

In 1995 in the United States (Kentucky, Kansas, Iowa, and Florida), a potexvirus was noted in ornamental *Portulaca* hybrids. Electron microscopy allowed visualization of flexuous filamentous rods in plants that were stunted and showed chlorotic mottling and irregularly shaped leaf margins (Fig. 118). Growers had reported these symptoms as early as 1993. In serological tests (ELISA), the samples reacted positively to antiserum for *Papaya mosaic virus* (PapMV). Subsequent analysis of isolates disclosed that this was a cross-reaction among potexviruses. The virus in portulaca was later identified as *Alternanthera mosaic virus* (AltMV); it was easily distinguished by molecular sequencing and sufficiently different from PapMV to be considered a separate virus species.

AltMV was also seen in 2001 on *P. grandiflora* (moss rose) in a Ligurian nursery in Italy. The plants showed apical wilting, stunting, and poor flowering. The virus showed high similarity to PapMV, but sequencing showed that the virus had a much higher similarity (94%) to the virus that had recently been described from Australia, which was later given the rank of species and the name AltMV.

Host Range and Epidemiology

AltMV occurs in *Phlox stolonifera* (creeping phlox), *Scutellaria longifolia* (skullcap), *Crossandra infundibuliformis* (firecracker plant), *Angelonia angustifolia,* a *Torenia* sp., a *Helichrysum* sp., *Portulaca* spp., *Salvia splendens* (red salvia), and *Nandina domestica* (heavenly bamboo). Infections of some hosts (e.g., torenia) produce no obvious symptoms.

Fig. 118. Distorted portulaca leaves, symptomatic of infection by *Alternanthera mosaic virus.* (Courtesy M. L. Daughtrey—© APS)

In 2008, stunting, chlorosis, and light-yellow mottling resembling symptoms of nutrient deficiency were reported in angelonia in commercial production in New York (United States). Numerous filamentous particles 520–540 nm long and spherical virus particles 30 nm in diameter were observed by transmission electron microscopy (TEM). Two viruses—the filamentous potexvirus AltMV and the spherical carmovirus *Angelonia flower break virus* (AnFBV)—were subsequently identified on the basis of nucleotide sequence analysis. Mixed infections with two viruses are sometimes found to cause more severe symptoms than infection with either virus alone. The incidence in the horticultural trade of both AltMV and AnFBV appears to have decreased since the 2008 report.

Potexviruses are transmissible without the assistance of an arthropod vector. They are readily spread via contaminated tools and hands through plant handling during vegetative propagation. Potexviruses are not transmitted through seed or pollen. Some have been shown to retain infectivity in plant sap for several months at room temperature.

Management

Roguing and destroying all symptomatic plants is recommended after an outbreak of Alternanthera mosaic. Sanitation is also important to limit spread from infected to healthy plants. Use of 0.5% a.i. (active ingredient) sodium hypochlorite has been suggested for sanitizing cutting tools, pots, and benches that have come into contact with infected plants. A spray of foliar extract of *Mirabilis jalapa* (four o'clock flower) was shown to reduce transmission of AltMV on cutting tools by 83%.

Selected References

Baker, C. A., and Williams, L. 1999. *Alternanthera mosaic virus* in *Portulaca* spp. (rev.). Plant Pathol. Circ. No. 382. Florida Department of Agriculture and Consumer Sciences, Division of Plant Industry, Gainesville.

Baker, C. A., Breman, L., and Jones, L. 2006. *Alternanthera mosaic virus* found in *Scutellaria, Crossandra,* and *Portulaca* in Florida. Plant Dis. 90:833.

Ciuffo, M., and Turina, M. 2004. A potexvirus related to *Papaya mosaic virus* isolated from moss rose (*Portulaca grandiflora*) in Italy. Plant Pathol. 53:515.

Duarte, L. M. L., Toscano, A. N., Alexandre, M. A. V., Rivas, E. B., and Harakava, R. 2008. Identificação e controle do *Alternanthera mosaic virus* isolado de *Torenia* sp. (Scrophulariaceae). (In Portuguese.) Rev. Bras. Hortic. Ornam. 14:59-66.

Eshenaur, B. C., Jariflors, U. E., Kelly, K. A., and O'Mara, J. 1995. Detection of a virus infecting portulaca hybrids in Kentucky and Kansas greenhouses. Phytopathology 85:1171.

Geering, A. D. W., and Thomas, J. E. 1999. Characterisation of a virus from Australia that is closely related to papaya mosaic potexvirus. Arch. Virol. 144:577-592.

Lockhart, B. E., and Daughtrey, M. L. 2008. First report of *Alternanthera mosaic virus* infection in angelonia in the United States. Plant Dis. 92:1473.

(Prepared by A. R. Chase and M. L. Daughtrey)

Angelonia Flower Break

A new carmovirus was isolated from *Angelonia angustifolia* in the United States and Israel; it caused flower color breaking and mild foliar symptoms. *Angelonia flower break virus* (AnFBV) and Angelonia flower mottle virus were proposed in 2006 as appropriate names for this new carmovirus, which causes stunting, mild leaf mottling, flower mottling, and color

Fig. 119. Reddish patterns or mosaic on nemesia leaves, caused by *Angelonia flower break virus.* (Courtesy A. R. Chase—© APS)

breaking. Symptoms have been detected in naturally infected angelonia in the United States, Israel, and Germany. Natural infection of verbena (in the United States and Israel) and phlox (in the United States) were reported first. Two years later, the same virus was widely found in the state of California in nemesia (Fig. 119).

The virus has isometric particles that are approximately 30 nm in diameter. Particles were observed in leaf dips and virion preparations from both angelonia and *Nicotiana benthamiana* and in thin sections of angelonia flower tissue by electron microscopy. Virion preparations were used to produce virus-specific polyclonal antisera in both Israel and the United States. The antisera did not react with *Pelargonium flower break virus* (PFBV), *Carnation mottle virus* (CarMV), or *Saguaro cactus virus* (SgCV) by either enzyme-linked immunosorbent assay (ELISA) or immunoblotting.

A complete nucleotide sequence has been obtained for an isolate of AnFBV from Florida (United States). The RNA genome is composed of 3,964 nucleotides, with four open reading frames arranged similarly to those of other carmoviruses. The gene sequence for an AnFBV coat protein has been determined for Israeli and Maryland (United States) isolates; it is closely related to that of PFBV. The AnFBV replicase, however, is most closely related to that of PFBV, CarMV, and SgCV.

Selected References

Adkins, S., Hammond, J., Gera, A., Maroon-Lango, C. J., Sobolev, I., Harness, A., Zeidan, M., and Spiegel, S. 2006. Biological and molecular characterization of a novel carmovirus isolated from *Angelonia*. Phytopathology 96:460-467.

Assis Filho, F. M., Harness, A., Tiffany, M., Gera, A., Spiegel, S., and Adkins, S. 2006. Natural infection of verbena and phlox by a recently described member of the *Carmovirus* genus. Plant Dis. 90:1115.

Mathews, D. M., and Dodds, J. A. 2008. First report of Angelonia flower break virus in *Nemesia* spp. and other ornamental plants in California. Plant Dis. 92:651.

Winter, S., Hamacher, A., Engelmann, J., and Lesemann, D.-E. 2006. Angelonia flower mottle, a new disease of *Angelonia angustifolia* caused by a hitherto unknown carmovirus. Plant Pathol. 55:820.

(Prepared by A. R. Chase and M. L. Daughtrey)

Bean Yellow Mosaic

Eustoma grandiflorum (lisianthus) infected with *Bean yellow mosaic virus* (BYMV), a potyvirus, shows symptoms of mosaic, leaf curling, chlorotic spotting, and flower color breaking. If *Tobacco mosaic virus* (TMV) or *Cucumber mosaic virus* (CMV) is present, as well, plants will also be stunted. When older plants are inoculated, milder symptoms are produced.

BYMV has flexuous rods, 700–760 nm in length, that contain single-stranded RNA. Inclusion bodies can be observed within the epidermis and are helpful for diagnosis. Cytoplasmic inclusions are granular or crystalline.

Individual viruses in the genus *Potyvirus* have relatively restricted host ranges. Some of the members are seed transmitted.

Potyviruses are transmitted by sap inoculation and by aphids. More than 20 aphid vectors are known to transmit BYMV in a nonpersistent manner, including the melon or cotton aphid (*Aphis gossypii*). Excluding aphid vectors and using virus-free seed are important for keeping BYMV out of the greenhouse. Aphid control and roguing out symptomatic plants will reduce virus spread within the greenhouse.

Selected References

Gera, A., and Cohen, J. 1990. The natural occurrence of bean yellow mosaic, cucumber mosaic and tobacco mosaic viruses in lisianthus in Israel. Plant Pathol. 39:561-564.

Lisa, V., and Dellavalle, G. 1987. Bean yellow mosaic virus in *Lisianthus russelianus*. Plant Pathol. 36:214-215.

(Prepared by A. R. Chase and M. L. Daughtrey)

Bidens Mottle

The natural occurrence of bidens mottle on *Rudbeckia hirta* (black-eyed Susan), *Ageratum houstonianum* 'Blue Mink,' and *Zinnia elegans* was reported from Florida (United States) in 1984. The original host, *Bidens pilosa*, is a common weed throughout Florida, and the virus disease was originally described on it in 1968. Many other hosts of *Bidens mottle virus* (BMoV) have since been identified, including vegetable and bedding plant crops. Stunting, mottling, leaf distortion, vein necrosis, leaf tip necrosis, flower color breaking, and bud abortion characterized infection on one or more of the bedding plants tested (Table 17).

BMoV has flexuous rods approximately 725 nm long. Pinwheel and aggregate inclusions can be observed using light microscopy. Purification from a *Nicotiana* hybrid was accomplished using chloroform carbon tetrachloride precipitation with 8% polyethylene glycol 6000, followed by differential centrifugation and separation in cesium chloride. This procedure gave approximately 4 mg of virus per 100 g of the tobacco host material.

The majority of BMoV hosts are in the family Asteraceae, and among them are many common weeds found throughout the southeastern United States. Control of BMoV has been based on excluding weed hosts and aphid vectors from the growing area and roguing out symptomatic plants when found. The disease has also been found in Florida and New York, where it caused symptoms in lettuce and endive.

Selected References

Christie, S. R., Edwardson, J. R., and Zettler, F. W. 1968. Characterization and electron microscopy of a virus isolated from *Bidens* and *Lepidium*. Plant Dis. Rep. 52:763-768.

Logan, A. E., Zettler, F. W., and Christie, S. R. 1984. Susceptibility of *Rudbeckia, Zinnia, Ageratum,* and other bedding plants to Bidens mottle virus. Plant Dis. 68:260-262.

(Prepared by A. R. Chase and M. L. Daughtrey)

TABLE 17. Symptoms Caused by *Bidens mottle virus* (BMoV) Infection in Some Bedding Plant Hosts[a]

Host	Family	Symptom(s)
Ageratum houstonianum	Asteraceae	Stunting; leaf distortion; flower bud abortion
Calendula officinalis	Asteraceae	Stunting; leaf mottling, distortion, and vein necrosis
Callistephus chinensis (China aster)	Asteraceae	Leaf distortion
Dimorphotheca pluvialis (rain daisy)	Asteraceae	Stunting; leaf distortion
Gaillardia × *grandiflora*	Asteraceae	Latent infection—no symptoms
Helianthus annuus (sunflower)	Asteraceae	Stunting; leaf mottling and distortion; flower bud abortion or deformation
Helichrysum bracteatum (strawflower)	Asteraceae	Stunting; leaf mottling and distortion; leaf tip necrosis
Petunia hybrida	Solanaceae	Stunting; leaf mottling and distortion; flower color breaking
Rudbeckia hirta (black-eyed Susan)	Asteraceae	Stunting; leaf mottling and distortion; flower bud abortion; flower color breaking
Stokesia laevis (Stokes' aster)	Asteraceae	Leaf mottling
Verbena hybrida	Verbenaceae	Stunting; leaf mottling and distortion
Zinnia elegans	Asteraceae	Stunting; leaf mottling and distortion; flower color breaking

[a] Courtesy A. R. Chase and M. L. Daughtrey—© APS.

Broad Bean Wilt

Broad bean wilt virus (BBWV) has been seen in only a few bedding plants, but it has the potential to be very destructive. This virus has been split into two species, BBWV-1 and BBWV-2, both of which have known ornamental hosts.

BBWV-1

BBWV-1 affects many hosts, including *Begonia* and *Petunia* spp. In *B. semperflorens* (wax begonia), it causes symptoms that are clearly different from those caused by the more common *Tobacco ringspot virus* (TRSV). Symptoms in wax begonia infected with BBWV-1 include leaf mottling, development of faint ring spots, and stunting—all of which are more pronounced during the winter months. In pink- and red-flowered varieties, flower color breaking occurs, as well. Ring spot symptoms caused by BBWV-1 infection have also been found in petunia. In addition, BBWV-1 is likely to cause disease in ornamental peppers; in addition to *Capsicum* spp., this virus has many other solanaceous hosts.

BBWV-1 is a member of the genus *Fabavirus*. Fabaviruses possess a bipartite genome of single-stranded RNA packaged in isometric particles that sediment as three separate components. Fabaviruses are transmitted in a nonpersistent manner by aphids and may be sap transmitted.

BBWV-1 has many natural hosts in 41 families, including the Asteraceae, Brassicaceae, Fabaceae, and Solanaceae. Virions are 25 nm in diameter. Depending on the virus strain, various types of inclusion bodies may be observed with a light microscope.

BBWV-1 is transmitted by the green peach aphid (*Myzus persicae*). Control of the green peach aphid will limit spread of the virus and reduce the chance that the disease will be introduced to wax begonia or another susceptible crop from weeds in the vicinity of the greenhouse or garden.

BBWV-2

BBWV-2 has been reported in *Eustoma grandiflorum* (lisianthus) and *Catharanthus roseus* (vinca). BBWV-2 is also likely to cause disease on ornamental peppers; in addition to *Capsicum* spp., this virus has many other solanaceous hosts.

BBWV-2 is a member of the genus *Fabavirus*. Fabaviruses possess a bipartite genome of single-stranded RNA packaged in isometric particles that sediment as three separate components. Fabaviruses are transmitted in a nonpersistent manner by aphids and may be sap transmitted.

BBWV-2 has many natural hosts in 41 families, including the Asteraceae, Brassicaceae, Fabaceae, and Solanaceae. Virions are 25 nm in diameter. Depending on the virus strain, various types of inclusion bodies may be observed with a light microscope.

BBWV-2 is transmitted by the green peach aphid (*Myzus persicae*). Control of the green peach aphid will limit spread of the virus and reduce the chance that the disease will be introduced to lisianthus, vinca, or another susceptible crop from weeds in the vicinity of the greenhouse or garden.

Selected References

Iwaki, M., Maria, E. R. A., Hanada, K., Onogi, S., and Zenbayashi, R. 1985. Three viruses occurred in lisianthus plants. Ann. Phytopathol. Soc. Jpn. 52:355.

Kobayashi, Y. O., Nakano, M., Kashiwazaki, S., Naito, T., Mikoshiba, Y., Shiota, A., Kameya-Iwaki, M., and Honda, Y. 1999. Sequence analysis of RNA-2 of different isolates of broad bean wilt virus confirms the existence of two distinct species. Arch. Virol. 144:1429-1438.

Lockhart, B. E. L., and Betzold, J. A. 1982. Broad bean wilt virus in begonia in Minnesota. Plant Dis. 66:72-73.

Uyemoto, J. K., and Provvidenti, R. 1974. Isolation and identification of two serotypes of broad bean wilt virus. Phytopathology 64:1547-1548.

(Prepared by M. L. Daughtrey, R. L. Wick, and J. L. Peterson; revised by A. R. Chase and M. L. Daughtrey)

Calibrachoa Mottle

In 2002, a virus with spherical particles was isolated from *Calibrachoa* hybrids with symptoms of leaf mottling, chlorotic blotching, and interveinal yellowing (Fig. 120). Depending on

Fig. 120. Interveinal yellowing (mottling) of calibrachoa leaves, symptomatic of infection by *Calibrachoa mottle virus*. (Courtesy A. R. Chase—© APS)

the cultivar and growing conditions, symptoms can range from severe mottling to no symptoms at all. *Calibrachoa mottle virus* (CbMV) commonly occurs in calibrachoa even though plants appear healthy. If plants are co-infected with CbMV and *Tobacco mosaic virus* (TMV), they will show severe stunting and interveinal yellowing.

Based on enzyme-linked immunosorbent assays (ELISA), CbMV appears to be unrelated to a number of known spherical viruses, including *Arabis mosaic virus, Broad bean wilt virus, Carnation mottle virus, Cucumber mosaic virus, Prunus necrotic ringspot virus, Tobacco ringspot virus, Tobacco streak virus, Tomato aspermy virus, Tomato bushy stunt virus,* and *Tomato ringspot virus.* The name *Calibrachoa mottle virus* (CbMV) was established in 2003, and the virus was fully described and characterized in 2010.

CbMV is not transmitted experimentally by any of the five common insect vectors of viruses but is easily mechanically transmitted. Commercial production of calibrachoa depends on vegetative propagation, and CbMV can be readily propagated with the host. To eliminate CbMV from calibrachoa, virus-free mother stock plants must be obtained and maintained using the best sanitation practices.

Selected References

Gulati-Sakhuja, A., and Liu, H.-Y. 2010. Complete nucleotide sequence and genome organization of *Calibrachoa mottle virus* (CbMV)—A new species in the genus *Carmovirus* of the family *Tombusviridae.* Virus Res. 147:216-223.

Liu, H.-Y., Sears, J. L., and Morrison, R. H. 2003. Isolation and characterization of a carmo-like virus from *Calibrachoa* plants. Plant Dis. 87:167-171.

Liu, H.-Y., Sears, J. L., Bandla, M., Harness, A. M., and Kulemeka, B. 2003. First report of Calibrachoa mottle virus infecting petunia. Plant Dis. 87:1538.

(Prepared by A. R. Chase and M. L. Daughtrey)

Cucumber Mosaic

Cucumber mosaic virus (CMV) is the type member of the genus *Cucumovirus* in the family *Bromoviridae.* CMV is very commonly encountered because of its wide host range (more than 1,200 species), worldwide distribution, and large number of aphid vectors. It has been found in *Eustoma grandiflorum* (lisianthus) and *Delphinium* spp. (larkspur) in Taiwan and in *Aubrieta* sp. (purple rock cress), *Dahlia pinnata, Lavandula* spp. (lavender), and *Petunia hybrida* from New Zealand. Other hosts include *Anemone* spp., *Iberis sempervirens* (candytuft), *Aquilegia* spp. (columbine), *Pelargonium* spp. (geranium), *Tagetes* spp. (marigold), petunia, *Phlox* spp., *Viola cornuta,* and *Zinnia elegans,* as well as many bulb crops.

Symptoms

Although some CMV infections are symptomless, a number of flower crops show dramatic systemic symptoms when infected (Table 18). Typically, cucumber mosaic disease causes a leaf mosaic, but ring spots, flower color breaking, and stunting are also seen. In some cases, CMV alone causes symptoms in plants, but it may also contribute to the expression of symptoms in mixed infections with other viruses. There are many strains of CMV, and they produce different symptoms.

Causal Agent

Cucumoviruses (including CMV) have tripartite genomes with three single-stranded, positive-sense RNA-containing isometric particles (each approximately 28 nm), all of which are necessary for infection. The particular virus strain or the presence of satellite RNA affects the host response.

TABLE 18. Symptoms of *Cucumber mosaic virus* (CMV) Infection in Some Bedding Plants[a]

Host	Symptom(s)
Begonia spp.	Stunting; ring spots on petals and leaves
Calendula officinalis	Chlorotic leaf mottling and deformation
Campanula carpatica	Flower streaking; plant distortion
Catharanthus roseus (annual vinca)	Flower streaking; plant distortion
Cyclamen spp.	Flower streaking; plant distortion
Dahlia spp.	Interveinal leaf mosaic
Eustoma grandiflorum (lisianthus)	Leaf mosaic, necrosis, and distortion; severe stunting; flower color breaking; flower distortion
Gerbera jamesonii (gerber daisy)	Flower color breaking; flower distortion; stunting; and leaf mottling
Impatiens hawkeri (New Guinea impatiens)	Leaf strapping and curling
Impatiens walleriana (garden impatiens)	Leaf curling
Lobelia erinus	Stunting; mild foliar mosaic
Pelargonium spp. (geranium)	Leaf breaking and mosaic
Primula spp. (primrose)	Necrosis; stunting; leaf mottling, vein clearing, and mosaic; green islands left within chlorotic leaf tissue; altered flower color

[a] Courtesy A. R. Chase and M. L. Daughtrey—© APS.

Host Range and Epidemiology

CMV has been sap transmitted experimentally to plants in at least 100 families, both monocots and dicots. It is transmitted in a nonpersistent manner by more than 80 species of aphids, including two very common greenhouse pests: the green peach aphid (*Myzus persicae*) and the melon or cotton aphid (*Aphis gossypii*). An aphid can acquire CMV within 5–10 sec and typically transmits the virus most efficiently for the first 2 min after acquisition (and generally for no longer than 2 h). Aphid transmission of CMV to geranium might occur via the foxglove aphid (*Aulacorthum solani*).

Seed transmission of CMV is known for 19 host species in the families Amaranthaceae, Brassicaceae, Caryophyllaceae, Fabaceae, and Labiatae. Handling plants is unlikely to spread CMV.

Management

For management of CMV, symptomatic plants should be rogued promptly. Weeds, which may harbor the virus, should be removed from inside the greenhouse and from the surrounding area, and aphid populations should be kept under control.

Virus indexing for CMV should be carried out when virus-free stock is created of any ornamental that is a potential host.

Selected References

Flasinski, S., Scott, S. W., Barnett, O. W., and Sun, C. 1995. Diseases of *Peperomia, Impatiens,* and *Hibbertia* caused by cucumber mosaic virus. Plant Dis. 79:843-848.

Francki, R. I. B., Mossop, D. W., and Hatta, T. 1979. Cucumber mosaic virus. No. 213. Online. Descriptions of Plant Viruses. Association of Applied Biologists.

Gera, A., and Cohen, J. 1990. The natural occurrence of bean yellow mosaic, cucumber mosaic and tobacco mosaic viruses in lisianthus in Israel. Plant Pathol. 39:561-564.

Milosevic, D., Ignjatov, M., Nikolic, Z., Gvozdnovic-Varga, J., Petrovic, G., Stankovic, I., and Krstic, B. 2015. First report of *Cucumber mosaic virus* causing chlorotic mottle on pot marigolds (*Calendula officinalis*) in Serbia. Plant Dis. 99:736.

Nameth, S. G. P., and Fisher, J. R. 2001. First report of a *Cucumber mosaic virus*-associated satellite RNA in *Lobelia erinus.* Plant Dis. 85:802.

Verma, N., Singh, A. K., Singh, L., Kulshreshtha, S., Raikhy, G., Hallan, V., Ram, R., and Zaidi, A. A. 2004. Occurrence of *Cucumber mosaic virus* in *Gerbera jamesonii* in India. Plant Dis. 88:1161.

Zitter, T. A., and Murphy, J. F. 2009. Cucumber mosaic virus. Plant Disease Lessons: Viruses and Viroids. Online. The Plant Health Instructor. doi:10.1094/PHI-I-2009-0518-01

(Prepared by M. L. Daughtrey, R. L. Wick, and J. L. Peterson; revised by A. R. Chase and M. L. Daughtrey)

Dahlia Mosaic

Viruslike symptoms have been observed on *Dahlia* spp. since the early 1900s, and the disease dahlia mosaic is now found worldwide.

Symptoms

On many virus-infected hybrid *Dahlia* cultivars, chlorotic bands develop down the midribs and along the larger veins of the leaves; this banding can be followed by development of mosaic and leaf distortion. Dahlias may also show systemic chlorosis, vein banding, and flower color breaking, and tubers may be short and thick. The most sensitive cultivars become stunted and develop a bushy habit. Some cultivars, however, are nearly symptomless.

Causal Agents

Two *Caulimovirus* spp. and one endogenous caulimovirus-like sequence are associated with dahlia mosaic. *Dahlia mosaic virus* (DMV [syn. DMV-Portland]) and Dahlia common mosaic virus (DCMV) have isometric particles that contain genomes of 7–8 kb; the sequence, in contrast, is integrated into the plant genome and referred to as "DvEPRS" (*Dahlia variabilis* endogenous plant pararetroviral sequence [syn. DMV-D10]). The symptoms particular to DMV, DCMV, and DvEPRS have not been clearly differentiated, but each has been found alone in dahlias with viruslike symptoms. Only DvEPRS has been found in wild dahlias, and in cultivated dahlias, it is the most prevalent of the three agents. In a 2007–2008 survey, 94% of the dahlias tested in the United States were positive for DvEPRS, whereas DMV and DCMV were less prevalent (in 48.5 and 23% of the samples, respectively).

DMV has spherical, DNA-containing particles that are 50 nm in diameter; DCMV particles are likely to be similar in size and shape. With a DMV infection, spherical or ellipsoidal inclusion bodies may be seen using light microscopy, especially in epidermal, palisade, and spongy parenchyma cells. The genome of DvEPRS is integrated into the chromosome of its host, and no unbound form has been detected.

Host Range and Epidemiology

DMV has been experimentally sap transmitted to other composites and to certain members of the Solanaceae, Chenopodiaceae, and Amaranthaceae. However, only *Dahlia* spp. have been found naturally infected with DMV, DCMV, and DvEPRS.

DMV and DCMV are both vectored in a noncirculative manner by the green peach aphid (*Myzus persicae*). Twelve other aphid species have also been shown to transmit DMV. In experiments, both DMV and DCMV were easily transmitted by the green peach aphid; attempts to transmit DvEPRS by aphids have not been successful.

DvEPRS is spread by vegetative propagation and is transmitted through seed to progeny with a very high rate of success.

Management

Infected dahlia stock should be discarded. Aphid exclusion and control are important for preventing the spread of disease from symptomless cultivars and from weed hosts (yet to be identified).

Because DvEPRS is integrated into the host genome, it is impossible to eliminate using heat treatment followed by meristem tip cultures. Whether dahlias are propagated by seed or by cuttings, there is no reduction from one generation to the next in widespread contamination with DvEPRS. It would be helpful to learn what, if any, environmental factors might trigger symptom development of DvEPRS in dahlias; those factors could then be avoided.

Selected References

Eid, S., and Pappu, H. R. 2014. Biological studies of three caulimoviruses associated with dahlia (*Dahlia variabilis*). Can. J. Plant Pathol. 36:110-115.

Eid, S., Druffel, K. L., Saar, D. E., and Pappu, H. 2009. Species of caulimoviruses in dahlia (*Dahlia variabilis*). HortScience 44:1498-1500.

Pahalawatta, V., Druffel, K., and Pappu, H. 2008. A new and distinct species in the genus *Caulimovirus* exists as an endogenous plant pararetroviral sequence in its host. Virology 376:253-257.

Pahalawatta, V., Miglino, R., Druffel, K. L., Jodlowska, A., van Schadewijk, A. R., and Pappu, H. R. 2007. Incidence and relative distribution of distinct caulimoviruses (genus *Caulimovirus*, family *Caulimoviridae*) associated with dahlia mosaic in *Dahlia variabilis*. Plant Dis. 91:1194-1197.

Pappu, H. F., Druffel, K. L., Miglino, R., and van Schadewijk, A. R. 2008. Nucleotide sequence and genome organization of a member of a new and distinct *Caulimovirus* species from dahlia. Arch. Virol. 153:2145-2148.

Pappu, H. R., Wyatt, S. D., and Druffel, K. L. 2005. Dahlia mosaic virus: Molecular detection and distribution in dahlia in the United States. HortScience 40:697-699.

(Prepared by A. R. Chase and M. L. Daughtrey)

Lisianthus Necrosis

Lisianthus necrosis virus (LNV), which is identified as a tentative species in the family *Tombusviridae,* was first reported in Japan in 1987, where it caused systemic necrosis in *Eustoma grandiflorum* (lisianthus). Symptoms appeared sporadically on field-grown lisianthus and included stunting; development of numerous tiny, necrotic spots or rings on both leaves and stems; and tip necrosis. Flower color breaking was observed in purple flowers. Infected plants contained icosahedral particles (30 nm in diameter), which had single-stranded RNA. LNV was not transmitted experimentally by the green peach aphid (*Myzus persicae*) but was found to be soilborne. Healthy *Nicotiana clevelandii* plants were infected with LNV when the soil was inoculated with a vector fungus (*Olpidium* sp.) along with crude sap from infected plants. LNV infected 21 species in 10 plant families by experimental inoculation. Only lisianthus, *N. clevelandii,* and *Gomphrena globosa* (common globe amaranth) were infected systemically. Sixteen other species developed local lesions after inoculation, whereas *Petunia hybrida* and *Dianthus superbus* were symptomlessly infected.

A different strain of LNV was reported from Taiwan in 2000. The virus was believed to have been imported from Europe on infected seedlings. Virus particles were 32–33 nm in diameter. This strain of LNV exhibited a different experimental host range from the virus described from Japan. Lisianthus that were naturally infected in Taiwan showed severe necrotic spotting, as well as color breaking and deformities. Tiny, bright-yellow, chlorotic leaf spots became necrotic and coalesced to kill leaves and eventually entire plants. Streaking that progressed to necrosis on stems was also noted. Transmission by aphids was attempted but not successful; however, lisianthus were infected by dipping wounded roots into a virus preparation and by planting them into soil in which diseased plants had grown. The same strain of LNV was later found in imported *D. caryophyllus* (carnation) and *Zantedeschia* spp. (calla lily) in Taiwan; both showed systemic necrosis.

At least nine other viruses, including *Broad bean wilt virus* (BBWV) and *Cucumber mosaic virus* (CMV), have been reported on lisianthus. Plants determined to have LNV should be destroyed.

Selected References

Chen, C. C., and Hsu, H. T. 2002. Occurrence of a severe strain of *Lisianthus necrosis virus* in imported carnation seedlings in Taiwan. Plant Dis. 86:444.

Chen, C. C., Chen, Y. K., and Hsu, H. T. 2000. Characterization of a virus infecting lisianthus. Plant Dis. 84:506-509.

Chen, Y. K., Jan, F. J., Chen, C. C., and Hsu, H. T. 2006. A new natural host of *Lisianthus necrosis virus* in Taiwan. Plant Dis. 90:1112.

Iwaki, M., Hanada, K., Maria, E. R. A., and Onogi, S. 1987. Lisianthus necrosis virus, a new necrovirus from *Eustoma russellianum*. Phytopathology 77:867-870.

(Prepared by A. R. Chase and M. L. Daughtrey)

Nemesia Ring Necrosis

Nemesia ring necrosis virus (NeRNV) is a member of the plant virus family *Tymoviridae*. It was first reported from Germany in 2000 and has since been reported from the United Kingdom and the United States (California and Florida). In *Nemesia* and *Verbena* spp., NeRNV causes chlorotic flecks in leaves, which later become necrotic. Virus particles are isometric and approximately 30 nm in diameter.

In addition to *Nemesia* and *Verbena* spp., NeRNV has been found in species of *Diascia* (twinspur) and *Alonsoa* (mask flower) and in *Chaenostoma cordatum* (bacopa). Thus, the virus is potentially problematic for crops in the Plantaginaceae, Scrophulariaceae, and Verbenaceae.

NeRNV has been completely sequenced. The antiserum is known to cross-react with that for another tymovirus that also infects various ornamentals: *Scrophularia mottle virus* (ScrMV). Primers for RT-PCR can differentiate NeRNV from ScrMV. NeRNV does not cause symptoms in *Chenopodium quinoa* (quinoa), nor does it systemically infect *Datura stramonium* (jimsonweed); these host range traits distinguish NeRNV from ScrMV.

NeRNV is most likely spread by vegetative propagation of infected stock plants. The virus is mechanically transmissible.

There are no practical cures for infection by NeRNV, so preventing losses is the responsibility of the specialist propagator. Cuttings should be supplied from stock plants that have been carefully indexed for NeRNV and other relevant viruses. Virus-indexed stock plants should be maintained by propagators in insect-free greenhouses. The tools used to make the cuttings should be disinfested between plants, and stock should be periodically tested for viruses of concern.

Selected References

Koenig, R., Barends, S., Gultyaev, A. P., Lesemann, D. E., Vetten, H. J., Loss, S., and Pleij, C. W. A. 2005. Nemesia ring necrosis virus: A new tymovirus with a genomic RNA having a histidylatable tobamovirus-like 3' end. J. Gen. Virol. 86:1827-1833.

Mathews, D. M., and Dodds, J. A. 2006. First report of *Nemesia ring necrosis virus* in North America in ornamental plants from California. Plant Dis. 90:1263.

Mumford, R. A., Jarvis, B., Harju, V., Elmore, J., and Skelton, A. 2005. The first identification of two viruses infecting trailing verbena in the UK. Plant Pathol. 54:568.

Skelton, A. L., Jarvis, B., Koenig, R., Lesemann, D. E., and Mumford, R. A. 2004. Isolation and identification of a novel tymovirus from nemesia in the UK. Plant Pathol. 53:798.

(Prepared by A. R. Chase and M. L. Daughtrey)

Pelargonium Flower Break

Pelargonium flower break virus (PFBV) was originally discovered in England in *Pelargonium domesticum* (regal or Martha Washington geranium). PFBV occurs naturally only in *Pelargonium* spp. (geranium) and is one of the most common viruses in *P.* × *hortorum* (florist's geranium) in the United States, Europe, and most recently, China.

Symptoms

Symptoms of PFBV usually appear during the spring. Florist's geranium may exhibit a line pattern or ring spots; *P. peltatum* (ivy geranium) may also show ring spots.

Infection of the very PFBV-sensitive florist's geranium cultivar Springtime Irene produces enlarging, chlorotic spots (2–5 mm in diameter) on primarily the younger leaves, as well as flower color breaking (white streaking and rugosity of petals). Chlorotic vein clearing and leaf deformities are also observed. These symptoms are enhanced when there is a simultaneous infection with *Tomato ringspot virus* (ToRSV). Infection with the two viruses increases the degree of leaf deformity and intensifies the colors of chlorotic streaks and ring patterns.

Some geraniums infected with PFBV are symptomless.

Causal Agent

PFBV is in the genus *Carmovirus* of the family *Tombusviridae*. PFBV has isometric particles that are 27–34 nm in diameter and contain single-stranded RNA. The nucleotide sequence of the genomic RNA of PFBV has been determined. Inclusion bodies have not been described.

Host Range and Epidemiology

PFBV occurs naturally only in regal or Martha Washington geranium, florist's geranium, and ivy geranium, but it has been transmitted by sap inoculation to 15 plant species. The virus can be transmitted from geranium to geranium with a contaminated cutting knife, via surface-contaminated pollen, and by thrips in the presence of infested pollen. PFBV may also be spread through recirculating nutrient solution.

The virus occurs at high concentrations in some cultivars but may be present at very low concentrations in others. In the Netherlands, detection with enzyme-linked immunosorbent assay (ELISA) was found to be reliable year-round in the sensitive cultivar Springtime Irene. Because the virus is unevenly distributed within plants, samples that include a mixture of root, leaf, and flower tissues are desirable for diagnosis by ELISA. Immunosorbent electron microscopy has also been used to detect PFBV in geraniums throughout the year.

Symptom development after inoculation may take longer than 1 year.

Management

PFBV can be eliminated from geraniums by using heat treatment. However, the plants are severely stressed by the treatment, and it may be difficult to obtain viable cuttings at the end of the 4-week, 35°C treatment necessary to accomplish the therapy. PFBV has been reported to be insensitive to virus-inactivating treatments, including 2% formaldehyde, trisodium orthophosphate, ultraviolet light, and ultrasound.

If virus-free material is obtained, it can be propagated in a clean stock program. In recirculating nutrient solution, slow-sand filtration will slow and reduce but not eliminate the spread of PFBV. To remove the virus from drain water, it may be necessary to employ ultrafiltration or heat treatment (greater than 90°C). Because PFBV can be spread among geraniums during plant production, virus-free geraniums should never be grown in the same greenhouse with PFBV-infected geraniums.

Selected References

Bouwen, I., and Maat, D. Z. 1992. Pelargonium flower-break and pelargonium line pattern viruses in the Netherlands; purification, antiserum preparation, serological identification, and detection in pelargonium by ELISA. Neth. J. Plant Pathol. 98:141-156.

Krczal, G., Albouy, J., Damy, I., Kusiak, C., Moreau, J. P., Berkelmann, B., and Wohanka, W. 1995. Transmission of Pelargonium flower break virus (PFBV) in irrigation systems and by thrips. Plant Dis. 79:163-166.

Paludan, N. 1976. Virus diseases in *Pelargonium hortorum,* especially concerning tomato ringspot virus. Acta Hortic. 59:119-130.

Paludan, N., and Begtrup, J. 1987. *Pelargonium* × *hortorum.* Pelargonium flower break virus and tomato ringspot virus: Infection trials, symptomatology and diagnosis. Dan. J. Plant Soil Sci. 91:183-194.

Stone, O. M., and Hollings, M. 1973. Some properties of pelargonium flower-break virus. Ann. Appl. Biol. 75:15-23.

Wei, M. S., Li, G. F., Ma, J., and Kong, J. 2015. First report of *Pelargonium flower break virus* infecting *Pelargonium* plants in China. Plant Dis. 99:735.

(Prepared by M. L. Daughtrey, R. L. Wick, and J. L. Peterson; revised by A. R. Chase and M. L. Daughtrey)

Petunia Vein Clearing

Petunia vein clearing virus (PVCV) is a pararetrovirus that affects *Petunia hybrida.* It was first described in 1973 from Germany and now occurs worldwide.

Symptoms

The major leaf veins of a petunia plant infected with PVCV become yellowed, presenting a symptom known as "vein clearing." The veins may be the only part of the leaf that becomes yellowed, the chlorosis may include parts of the leaf blades next to the veins, or chlorotic spots may develop along the veins. In addition, flower coloring may be abnormal, the shoots may be stunted, and the leaves may be malformed and show epinasty.

The range and strength of displayed symptoms can vary in cultivars of petunia.

Causal Agent

PVCV has a double-stranded DNA genome and isometric particles 43–46 nm in diameter; it is the type virus of the genus *Petuvirus* in the family *Caulimoviridae.* Dark inclusion bodies form in cells infected with PVCV, and virus particles may be seen in the cytoplasm adjacent to the inclusion bodies. Complete viral sequences (estimated 50–100 copies) are also integrated into the genome of *P. hybrida.* No vector has been identified for horizontal (plant-to-plant) transmission of PVCV. Therefore, disease outbreaks are believed to originate from the activation of integrated viral copies that can be transcribed into full-length viral genomes.

Endogenous PVCV sequences have also been seen in *P. integrifolia* and *P. axillaris,* the parents of hybrid petunia. Whether the PVCV integrants also trigger disease in wild species of *Petunia* has not been investigated.

DNA from episomal (free) virions of PVCV is easily detected by PCR, but the integrated DNA copies of the virus (although less numerous) may also serve as templates in PCR, producing faint bands. Extra controls must therefore be included to indicate the presence of genomic sequences. Electron microscopy is needed to visualize the episomal particles that cause the vein-clearing symptoms.

The presence of PVCV in the petunia genome makes this virus unavoidable, other than by management of the conditions that lead to the production of episomal virions (which in turn lead to the development of virus symptoms).

Host Range and Epidemiology

Only plants in the family Solanaceae are hosts of PVCV, and petunia is the only natural host. Wounding the petunia host appears to induce the production of infectious virus particles from endogenous sequences. Other stresses that commonly occur in plant production, such as water stress and high temperatures, may also cause disease expression.

Co-infections with other viruses—including *Potato virus Y* (PVY), *Tobacco mosaic virus* (TMV), and *Cucumber mosaic virus* (CMV)—have been seen in petunia. No vectors for PVCV are known; the virus is seed and graft transmissible but not mechanically transmissible.

Management

Virologists have discouraged the use of tissue culture methods in petunia production because they involve wounding, which triggers the formation of episomal particles and disease expression. Environmental stresses should likewise be avoided at the finishing stage of petunia culture.

P. hybrida cultivars seem to vary in ability to control PVCV activation and thus show different levels of susceptibility to virus induction after stress exposure. Choosing appropriate cultivars may reduce the frequency of symptom development.

Given the unique nature of this virus, additional research is required to devise a practical means of management.

Selected References

Harper, G., Richert-Pöggeler, K. R., Hohn T., and Hull, R. 2003. Detection of petunia vein-clearing virus: Model for the detection of DNA viruses in plants with homologous endogenous pararetrovirus sequences. J. Virol. Methods 107:177-184.

Lockhart, B. E. L., and Lesemann, D.-E. 1998. Occurrence of petunia vein-clearing virus in the U.S.A. Plant Dis. 82:262.

Noreen, F., Akbergenov, R., Hohn, T., and Richert-Pöggeler, K. R., 2007. Distinct expression of endogenous Petunia vein clearing virus and the DNA transposon dTph1 in two *Petunia hybrida* lines is correlated with differences in histone modification and siRNA production. Plant J. 50:219-229.

Richert-Pöggeler, K. R., and Lesemann, D.-E. 2007. Petunia vein-clearing virus. No. 417. Online. Descriptions of Plant Viruses. Association of Applied Biologists.

Richert-Pöggeler, K. R., Noreen, F., Schwarzacher, T., Harper, G., and Hohn, T. 2003. Induction of infectious petunia vein-clearing (pararetro) virus from endogenous provirus in petunia. EMBO J. 22:4836-4845.

(Prepared by A. R. Chase and M. L. Daughtrey)

Tobacco Mosaic

Tobacco mosaic virus (TMV) is a tobamovirus. It is distributed worldwide and affects a wide range of ornamentals, as well as many other plants. TMV is highly infectious because it persists in the form of unusually stable virus particles that are easily transmitted when plant materials are handled.

Symptoms

The symptoms of TMV vary according to the virus isolate and host cultivar; they are also influenced by environmental factors such as temperature and light intensity. Double infection with a second virus is commonly encountered, and this may intensify symptoms.

Typical symptoms of TMV in *Petunia hybrida* include mosaic and distortion, flower color breaking, chlorotic mottling, and stunting (Figs. 121, 122, and 123). *Cyclamen persicum* infected with both TMV and *Potato virus X* (PVX) shows curled, deformed petals with necrotic streaks. Cyclamen infected with TMV alone shows only mild symptoms of stunting and chlorosis, whereas cyclamen infected with both TMV and *Tomato aspermy virus* (TAV) also shows deformed flowers. In *Eustoma grandiflorum* (lisianthus), TMV produces a mosaic, whereas a mixed infection of TMV and either *Cucumber mosaic virus* (CMV) or *Bean yellow mosaic virus* (BYMV) causes more severe symptoms, including necrotic spotting and stunting. Infection of *Gerbera jamesonii* (gerber daisy) has been reported from Turkey, and infection of *Pelargonium* spp. (geranium) has been reported from Belgium. In *Impatiens hawkeri* (New Guinea impatiens), TMV causes stunting and leaf distortion. In *Calibrachoa* hybrids, necrotic lesions or systemic mosaic develop, depending on the virus isolate (Fig. 124).

Fig. 121. Mosaic and distortion of petunia leaves, caused by *Tobacco mosaic virus.* (Courtesy A. R. Chase—© APS)

Fig. 122. Color breaking of a petunia flower, caused by *Tobacco mosaic virus.* (Courtesy M. L. Daughtrey—© APS)

Plants can also be infected but remain symptomless. In one study, wild-type TMV and four other *Tobamovirus* spp. infected four petunia cultivars without producing obvious viral symptoms. One species of *Tobamovirus, Tomato mosaic virus* (ToMV), was shown to cause either symptomless infections or necrotic local lesions when inoculated onto petunia cultivars Surfinia, Sanguna, and Supertunia. In contrast, an isolate of TMV collected from petunia in Ohio (TMV-pet9) that was highly similar to typical TMV isolates produced symptoms on all but one of the tested petunia cultivars. Another TMV isolate (TMV-petRF) was collected from petunia during a 2014 outbreak, and its coat protein sequence was found to most closely resemble that of the TMV-pet9 isolate.

Causal Agent

TMV, the type member of the genus *Tobamovirus,* has rod-shaped particles that measure 300 × 18 nm. TMV is a very stable virus and has been well studied since the first report on *Nicotiana tabacum* in 1886.

Many closely related isolates are found around the world. A TMV isolate causing an outbreak on trailing petunia in Europe in the mid-1990s was found to be genetically quite similar to isolates reported from Japan and Korea.

Host Range and Epidemiology

TMV is widely distributed worldwide and has a wide host range of hundreds of species. Although TMV is easily transmitted through handling of plant materials, it is not commonly vectored by arthropods. Plants may become infected from root contact with virus particles or bits of infected plant tissue. Even transmission of TMV by splashing during normal overhead watering has been shown, making sanitation of production areas essential. Tobamoviruses are sometimes distributed via contamination of the seed coat.

TMV infects cyclamen, gerber daisy, New Guinea impatiens, lisianthus, geranium, and petunia, as well as other greenhouse flower crops. Among bedding plants, symptoms have been seen most commonly on petunia. Many vegetable crops, particularly those in the Solanaceae (e.g., tomato, eggplant, and pepper), are

Fig. 123. Petunia plant infected by *Tobacco mosaic virus* (left) versus a healthy petunia plant (right). (Courtesy A. R. Chase—© APS)

Fig. 124. Necrotic spots on calibrachoa leaves, associated with *Tobacco mosaic virus* in a mixed infection with *Calibrachoa mottle virus.* (Courtesy M. L. Daughtrey—© APS)

also susceptible, although resistant lines are typically available for food crops.

The sap of infected plants may be packed with virus particles. TMV transmission has been shown to occur up to the twentieth petunia plant cut with the same blade following a single cut made on a TMV-infected plant.

Management

Workers should wash their hands after handling tobacco products, and smoking should not be allowed in greenhouses and other production areas. TMV is very stable outside plants and easily transmitted when plants are handled, pruned, or transplanted. Plant debris and surfaces of benches and tools can harbor TMV. Virus particles can be transferred to doorknobs, forklift steering wheels, and other objects, where they can be picked up by workers and transferred to crops.

Treatment of TMV-contaminated tools with a 20% (weight/volume) solution of nonfat dry milk (NFDM) plus 0.1% Tween 20 or a 1:10 dilution of household bleach (0.6% sodium hypochlorite) has been shown to completely eliminate TMV transmission to petunia. Treatment of contaminated tools with 1% (weight/volume) Virkon S or 20% NFDM has also been shown to significantly reduce the incidence of infected petunias.

Infected living plants, especially those without symptoms, are another source of TMV. The purchase of virus-free plants and regular testing of production areas for TMV can help prevent introduction of the virus. Any plants found infected with TMV should be removed and steamed to inactivate the virus; contaminated growing areas should be sanitized with one of the treatments identified in the previous paragraph. All surfaces that have come into contact with TMV-infected plants—including flats, benches, and other containers—should be discarded or thoroughly disinfested prior to use.

Selected References

Allen, W. R. 1981. Dissemination of tobacco mosaic virus from soil to plant leaves under greenhouse conditions. Can. J. Plant Pathol. 3:163-168.

Cohen, J., Sikron, N., Shuval, S., and Gera, A. 1999. Susceptibility of vegetatively propagated petunia to tobamovirus infection and its possible control. HortScience 34:292-293.

Gera, A., and Cohen, J. 1990. The natural occurrence of bean yellow mosaic, cucumber mosaic and tobacco mosaic viruses in lisianthus in Israel. Plant Pathol. 39:561-564.

Kim, B.-S., Ruhl, G., Creswell, T., and Loesch-Fries, L. S. 2014. Molecular identification of a *Tobacco mosaic virus* isolate from imported petunias. Online. Plant Health Progress. doi:10.1094/PHP-BR-14-0018

Lewandowski, D. J., Hayes, A. J., and Adkins, S. 2010. Surprising results from a search for effective disinfectants for *Tobacco mosaic virus*-contaminated tools. Plant Dis. 94:542-550.

Li, R., Baysal-Gurel, F., Abdo, Z., Miller, S. A., and Ling, K.-S. 2015. Evaluation of disinfectants to prevent mechanical transmission of viruses and a viroid in greenhouse tomato production. Virol. J. 12:5.

Spence, N. J., Sealy, I., Mills, P. R., and Foster, G. D. 2001. Characterization of a tobamovirus from trailing petunia. Eur. J. Plant Pathol. 107:633-638.

Yorganici, U., and Karaca, I. 1974. Tobacco mosaic virus on *Gerbera jamesonii* in Turkey. J. Turk. Phytopathol. 3:116-123.

(Prepared by M. L. Daughtrey, R. L. Wick, and J. L. Peterson; revised by A. R. Chase and M. L. Daughtrey)

Tobacco Ringspot

Tobacco ringspot virus (TRSV) is an important nepovirus that affects a wide range of annual and perennial crops in North America, along with greenhouse ornamentals on occasion. The name Pelargonium ringspot virus has been used to refer to both TRSV and *Tomato ringspot virus* (ToRSV) on geranium (*Pelargonium* spp.).

Symptoms

TRSV-infected garden impatiens (*Impatiens walleriana*) are stunted and develop mosaic and chlorotic ring spots and line patterns. On *Begonia* spp., TRSV causes begonia yellow spot disease. Fibrous-rooted begonias are stunted and develop pale-yellow or white ring spots or a pale mottle. The symptoms of TRSV in begonia are similar to genetic variegations. *Broad bean wilt virus* (BBWV) causes similar symptoms in begonia. TRSV has also been identified in *Dahlia* spp.

Florist's geraniums (*P.* × *hortorum*) inoculated with TRSV as young seedlings develop chlorotic rings and line patterns that fade as the plants mature. Older geraniums are not susceptible to mechanical inoculation with TRSV. Symptomless infected geranium plants carry a low titer of the virus.

Causal Agent

TRSV is a member of the genus *Nepovirus*, which includes viruses that have three isometric particles and a bipartite genome of single-stranded RNA. Nepoviruses cause ring spots and mottling on a wide range of plants. Transmission occurs naturally via nematodes or pollen and experimentally via sap inoculation. Seed transmission may occur with nepoviruses, and a high percentage of seeds may carry the virus. Latent infections of seedlings grown from virus-infected seed have been noted.

TRSV has three types of isometric particles—all about 28 nm in diameter. Two of the particle types contain the RNA genome; the third contains solely protein. Satellite RNA is associated with some isolates of TRSV. Individual virus particles are hard to distinguish when viewed with an electron microscope because they resemble ribosomes, but the tubular or crystalline inclusions facilitate virus recognition. Serological tests and molecular sequencing can distinguish TRSV from ToRSV and other nepoviruses.

Host Range and Epidemiology

TRSV is endemic in eastern and central North America and has been reported occasionally from many other parts of the world, presumably resulting from the movement of infected nursery stock. Natural hosts of TRSV include both woody and herbaceous species.

Nematodes in the genus *Xiphinema* (*X. americanum* and others) are the best-known vectors of TRSV. Nematodes acquire the virus within 24 h of feeding and transmit it as larvae or adults with high efficiency. Other vectors may be involved, as well, including the onion thrips (*Thrips tabaci*), mites (*Tetranychus* spp.), grasshoppers (*Melanoplus* spp.), the tobacco flea beetle (*Epitrix hirtipennis*), and aphid species. Each may serve as a vector in specific circumstances.

Seed transmission is believed to be possible in all hosts and occurs in up to 100% of seeds when the virus is associated with the embryo. Seed transmission has not been demonstrated for florist's geranium.

Management

TRSV is not commonly found in the greenhouse industry. Virus indexing has been effective in minimizing its occurrence in florist's geranium.

In the landscape, nematodes may reside in the soil and vector the virus to newly planted annuals; nonhost species of bedding plants should be used when this problem has been observed. New flower beds without a history of TRSV may need to be established for susceptible hosts. Growers should be careful to avoid moving soil from contaminated beds to areas free of TRSV-bearing nematodes.

Selected References

Abo El-Nil, E., Hildebrandt, A. C., and Evert, R. F. 1976. Symptoms induced on virus-free geranium seedlings by tobacco ringspot and cucumber mosaic viruses. Phyton 34:61-64.

Lockhart, B. E., and Betzhold, J. A. 1979. Begonia yellow spot: A disease caused by tobacco ringspot virus infection. Plant Dis. Rep. 63:1046-1047.

Lockhart, B. E., and Pfleger, F. L. 1979. Identification of a strain of tobacco ringspot virus causing a disease of impatiens in commercial greenhouses. Plant Dis. Rep. 63:258-261.

Walsh, D. M., Horst, R. K, and Smith, S. H. 1974. Factors influencing symptom expression and recovery of tobacco ringspot virus from geranium. Phytopathology 64:588.

(Prepared by M. L. Daughtrey, R. L. Wick, and J. L. Peterson; revised by A. R. Chase and M. L. Daughtrey)

Tobacco Streak

Tobacco streak virus (TSV) is a member of the genus *Ilarvirus* in the family *Bromoviridae.* It is likely distributed worldwide.

Symptoms

In *Dahlia* hybrids, infection by TSV may be asymptomatic or cause mottling. A strain of TSV has been noted to infect *Impatiens walleriana* (garden impatiens), causing stunting of the whole plant and twisting and deformation of leaves; flowers were reduced in number but not deformed.

TSV-infected *Eustoma grandiflorum* (lisianthus) in Brazil showed irregularly shaped, necrotic ring spots on leaves, and flowers were smaller and had a shorter than usual lifespan. An ilarvirus disease reported from Italy in 1994 that caused necrotic line patterns on lisianthus was tentatively named Lisianthus line pattern virus (LLPV). It was serologically related to TSV but only distantly; it was most closely related to *Parietaria mottle virus* (ParMV). LLPV is mechanically transmissible, but there is no information on how it is spread naturally.

Causal Organism

TSV is a member of the genus *Ilarvirus,* whose members have a three-part RNA genome in three separate isometric particles. TSV has three quasi-isometric particles (27, 30, and 35 nm in diameter) and four single-stranded RNA species. Characteristics of inclusion bodies have not been defined.

Host Range and Epidemiology

TSV has a broad host range, including species in more than 30 families of monocots and dicots. Thrips, seed, and pollen transmission are known to occur for ilarviruses. Both the western flower thrips (*Frankliniella occidentalis*) and the onion thrips (*Thrips tabaci*) have been shown to transmit TSV. Pollen transmission may occur to the pollinated plant and to the resulting seed. In one instance, thrips were reported to facilitate spread of TSV by feeding on leaves of field crops that were dusted with windborne pollen from adjacent virus-infected weeds.

Management

Control of weed hosts and thrips vectors is critical, particularly during hybrid seed production. Because of the virus's wide host range, TSV may become a significant threat to bedding plant production in cases in which it is introduced on a crop plant and there is a significant population of thrips in the greenhouse.

Selected References

de Freitas, J. C., Kitajima, E. W., and Rezende, J. A. M. 1996. First report of tobacco streak virus on lisianthus in Brazil. Plant Dis. 80:1080.

Lockhart, B. E., and Betzold, J. A. 1980. Leaf-curl of impatiens caused by tobacco streak virus infection. Plant Dis. 64:289-290.

Scott, S. W. 2001. Tobacco streak virus. No. 381. Online. Descriptions of Plant Viruses. Association of Applied Biologists.

(Prepared by M. L. Daughtrey, R. L. Wick, and J. L. Peterson; revised by A. R. Chase and M. L. Daughtrey)

Tomato Spotted Wilt, Impatiens Necrotic Spot, and Other Tospovirus Diseases

Tomato spotted wilt and impatiens necrotic spot are two similar diseases caused by distinct tospoviruses. The genus *Tospovirus* is composed of 28 viruses, of which 11 are recognized and 17 are proposed species. Until the 1980s, many of these diverse viruses were considered to be strains of (or closely related to) *Tomato spotted wilt virus* (TSWV), which was first reported in tomato in Australia in 1915. TSWV has long been considered unique among plant viruses. It is the only one placed within the family *Bunyaviridae,* which is otherwise made up of viruses that cause diseases of mammals and insects.

During the mid-1980s, TSWV became more widely distributed in the United States as the range of its vector, the western flower thrips (*Frankliniella occidentalis*), extended east across the country. At the same time, the western flower thrips began to be a serious and common greenhouse pest, and what was thought to be TSWV was encountered much more frequently in greenhouses across the United States and Canada. Because of the difficulty of controlling the vector and the frequent exchange of plants and cuttings across long distances, both the thrips and the virus quickly became distributed throughout the North American greenhouse industry. The European greenhouse industry also began experiencing outbreaks of TSWV during the late 1980s, coinciding with importation of the western flower thrips.

In 1989, a serologically distinct strain of TSWV was reported from *Impatiens* spp. and designated TSWV-I to distinguish it from the common or lettuce strain, TSWV-L. The use of antibodies differentiating these two strains allowed data to be collected on their relative incidences. In a Pennsylvania (United States) survey of ornamentals from late 1989 to early 1991, 95% of the tospovirus-infected bedding plants were infected with TSWV-I, rather than TSWV-L. In 1991, TSWV-I was more thoroughly characterized and determined to be a distinct virus, at which time it was named *Impatiens necrotic spot virus* (INSV).

INSV is a serious threat to a wide range of North American bedding plants, especially *I. walleriana* (garden impatiens), *I. hawkeri* (New Guinea impatiens), *Lobelia erinus, Begonia* × *hiemalis* (hiemalis or Rieger begonia), *Antirrhinum majus* (snapdragon), *Eustoma grandiflorum* (lisianthus), and *Cyclamen persicum.* INSV has continued to be noted more often in greenhouse crops, whereas TSWV is seen occasionally as the cause of greenhouse epiphytotics and more commonly in ornamentals, vegetables, and field crops outdoors. In Europe, TSWV has been reported to be the predominant tospovirus in greenhouse-grown *Solanum lycopersicum* (tomato), *Capsicum annuum* (pepper), *Gerbera jamesonii* (gerber daisy), *Chrysanthemum morifolium,* and *Dieffenbachia seguine* (dumb cane). INSV has been noted in *Begonia* spp. in the Netherlands and in *Anemone coronaria* in Germany, as well as in garden impatiens across Europe and in Australia.

New tospoviruses are continually being discovered. Some of them may become important to bedding plant crops, especially if they are vectored by the western flower thrips.

Symptoms

TSWV and INSV cause an extraordinarily broad range of symptoms on many herbaceous plants, and the diseases they cause are visually similar. Symptoms include stunting, necrotic spotting, and chlorotic spotting, as well as areas of black or brown stem necrosis, ring spots, mosaic, mottling, line patterns, vein necrosis, and flower color breaking (Figs. 125–131). Plants infected while young are the most severely affected and may die. Any of the plant hosts may exhibit several symptom types, and symptoms will vary from cultivar to cultivar and from plant to plant. Latent (symptomless) infections are common in certain cultivars of susceptible crops and in some weed species.

In *B. semperflorens* (wax begonia) infected with a tospovirus, the petioles may turn brown and collapse or an area of

Fig. 125. Necrotic spots on leaves of annual vinca, caused by *Tomato spotted wilt virus.* (Courtesy M. L. Daughtrey—© APS)

Fig. 126. Stem necrosis in ranunculus, characteristic of infection by *Tomato spotted wilt virus.* (Courtesy A. R. Chase—© APS)

Fig. 127. Ring spots on a garden impatiens leaf, caused by *Impatiens necrotic spot virus.* (Courtesy A. R. Chase—© APS)

Fig. 128. Ring spots on a lisianthus leaf, caused by *Impatiens necrotic spot virus.* (Courtesy A. R. Chase—© APS)

Fig. 129. Mottling on a begonia leaf, caused by *Impatiens necrotic spot virus.* (Courtesy M. L. Daughtrey—© APS)

Fig. 130. Necrotic line patterns on coleus leaves, caused by *Impatiens necrotic spot virus.* (Courtesy M. L. Daughtrey—© APS)

the leaf blade at the petiole end may become necrotic; zigzag line patterns may appear in the leaves. *Catharanthus roseus* (annual vinca) is used as a diagnostic species for TSWV. Local, black spots form 10–14 days after inoculation, and leaves may yellow and abscise. Eventually, mosaic and leaf deformation develop. Both of these symptoms are also caused by an INSV infection.

Pericallis hybrida (cineraria) and *Calceolaria crenatiflora* (slipperwort or calceolaria) infected with INSV or TSWV often develop dramatic necrotic or chlorotic leaf spots or ring spots, and plants may be stunted. Cyclamen infected with INSV often show round, brown, necrotic leaf spots. Brown or yellow ring spots may appear on leaves, or browning may develop at the petiole ends of leaves. Brown, zigzag line patterns have also been observed on cyclamen, as has flower distortion and color breaking. For cyclamen, symptoms of a tospovirus infection may develop months after the initial infection; other hosts, such as impatiens, may show symptoms soon after infection.

Dahlia hybrids produced from tubers are frequently infected by TSWV. Yellow line patterns in the leaves are typical, as are yellow ring spots or blotches. Dahlias produced from seed show light chlorotic spots or line patterns when infected with INSV.

On New Guinea impatiens, symptoms of INSV infection include plant stunting; brown, black, or purple leaf spots; ring spots on petals; black, diffuse spotting on leaves; black stem sections; and leaf stunting, distortion, and chlorotic mottling. Plants may wilt and collapse. Variations in symptoms are extreme from cultivar to cultivar, and many cultivars are symptomless. Symptoms appear during the winter and may be masked during the summer.

Symptoms of INSV infection on garden impatiens include leaf and plant stunting and yellow mosaic, as well as the development of small, scattered, black or brown spots or flecks; brown or black ring spots, often preceding leaf yellowing and abscission; and black stem sections. Double-flowered impatiens (which are most often vegetatively propagated) are highly attractive to the thrips vector and often show dramatic symptoms of INSV infection. TSWV infection is less common. The symptoms of infection are similar for the two viruses.

INSV-infected *Ranunculus asiaticus* may show brown to black petiole, stem, and foliar necrosis. Flower buds may become necrotic, and the entire plant may be stunted.

Symptomless infections are known to occur on *Pelargonium* × *hortorum* (florist's geranium), but this host does not generally show symptoms. TSWV can cause yellow rings on *P. peltatum* (ivy geranium). Infection of *Euphorbia pulcherrima* (poinsettia) or rose (*Rosa hybrida*) with INSV or TSWV has not been noted.

Environmental conditions can affect symptom expression. For example, previously symptomless garden impatiens and New Guinea hybrids may show symptoms during the winter months.

Because the symptoms caused by INSV and TSWV are visually indistinguishable in most cases, enzyme-linked immunosorbent assay (ELISA) is typically used to identify whether the symptoms are caused by a tospovirus and if so, which one. Growers without access to a diagnostic laboratory may apply this technology in the form of an easy-to-use test strip or lateral-flow device (Fig. 132).

Causal Agents

One of the criteria for a distinct virus species is that there be less than a 90% similarity in the amino acid sequence of the nucleocapsid (N) proteins. Using this criterion and serological relatedness of the N protein, close to 30 viruses are grouped within the genus *Tospovirus*.

TSWV and INSV virions are pleomorphic, and particles are 70–120 nm in diameter. The genetic information of the viruses is contained in three single-stranded RNA genome segments, designated as large (L), medium (M), and small (S) RNAs. The L-, M-, and S-RNAs are approximately 9, 5, and 3 kb, respectively, in size and encode structural and nonstructural proteins. The ambisense S-RNA codes for a structural (N) and nonstructural (NSs) protein. M-RNA codes for the glycoprotein precursor, which is cleaved post-translationally into two mature glycoproteins (G_N and G_C) and the movement (NSm) protein in an ambisense manner. The negative-sense L-RNA codes for the RNA-dependent RNA polymerase. The L-, M-, and S-RNAs are encapsidated separately by the N protein and packaged within a host-derived, lipid envelope membrane. The two glycoproteins (G_N and G_C) are embedded into the envelope membrane and appear as spikelike projections on the surfaces of virions.

Although TSWV and INSV are distinguished by serologically distinct N proteins, they cause similar symptoms and have overlapping host ranges. TSWV and INSV also differ in the nature of the inclusion bodies formed in the cytoplasms of diseased tissues. INSV typically forms viroplasms that contain protein filaments in a paracrystalline array, whereas TSWV isolates form loose bundles of filaments. There are significant variations among isolates of both viruses in the abundance of inclusion bodies and virions.

New virus forms may be detected in the future that are not currently detectable with available antisera. The USDA Florist and Nursery Crops Laboratory reported in 1991 that an INSV isolate from a *Gloxinia* sp. that had been maintained at a high temperature in a *Nicotiana* sp. no longer reacted with INSV antisera because of the lack of the nucleoprotein accumulations in viroplasms typical of INSV. Bioassays of inoculated plants such as *N. benthamiana* are valuable for diagnoses of variants

Fig. 131. *Impatiens necrotic spot virus* infection in annual vinca, showing stunting and strapping (leaf narrowing) of terminal leaves. (Courtesy A. R. Chase—© APS)

Fig. 132. Leaf spots on snapdragon and positive ImmunoStrip results, indicating infection by *Impatiens necrotic spot virus.* (Courtesy A. R. Chase—© APS)

from the TSWV and INSV types currently known (those that do not test positive by ELISA).

New species of *Tospovirus* may also be encountered. *Tomato chlorotic spot virus* (TCSV) was most recently identified in annual vinca. This tospovirus was described from vinca in Florida (United States) that developed dark-brown stem lesions, chlorotic leaves, brown line patterns, ring spots, and blackened petioles. Additional ornamental hosts include lisianthus, *Hoya wayetii,* and *Schlumbergera truncata* (Christmas cactus). TCSV is cross-reactive with TSWV ImmunoStrips and was previously reported from vegetables, including tomato and pepper, from Florida and Puerto Rico, as well as from *Datura stramonium* (jimsonweed).

Several tospoviruses have been reported on lisianthus in Japan, including Lisianthus necrotic ringspot virus, *Iris yellow spot virus* (IYSV), and Chrysanthemum stem necrosis virus (CSNV), in addition to TSWV and INSV. IYSV has also been reported on lisianthus in the United Kingdom and Israel, as well as on ornamentals in the family Liliaceae. CSNV has been found on chrysanthemum in Brazil and Europe, on tomato in Brazil, and on gerber daisy in Slovenia.

Host Range and Epidemiology

TSWV is considered common in subtropical and tropical regions of the world, and together with INSV, it has become more important in the Northern Hemisphere because of the increased range of one of its vectors: the western flower thrips. Collectively, tospoviruses have an extremely broad experimental host range, encompassing more than 1,000 species (including monocots and dicots). Both TSWV and INSV have wide host ranges, including monocots as well as dicots. TSWV is important for ornamentals and for vegetable and field crops, and INSV is a concern primarily for ornamental crops.

Many flower crops are hosts of both TSWV and INSV, as are many vegetables and weeds. In addition to the hosts mentioned earlier in this section, the following species are among the other flowering potted plant hosts for INSV and/or TSWV: *Anemone, Aquilegia, Browallia, Calceolaria, Campanula, Capsicum, Clerodendrum, Diascia, Fuchsia, Gardenia, Gerbera, Kalanchoe, Lantana, Lobelia, Mimulus, Plectranthus, Ranunculus, Saintpaulia, Schlumbergera, Solanum, Schizanthus,* and *Streptocarpus.* Double infections of the two viruses are occasionally detected.

TSWV is vectored only by certain species of thrips: *F. fusca* (tobacco thrips); *F. intonsa* (flower thrips)*; F. occidentalis* (western flower thrips)*; F. schultzei* (common blossom thrips); *Scirtothrips dorsalis* (chilli thrips); *Thrips palmi* (melon thrips); *T. setosus* (Japanese flower thrips); and *T. tabaci* (onion thrips). Many of these thrips vectors have wide host ranges, facilitating spread of the virus. TSWV is acquired by larval thrips that feed on infected plant tissue; acquisition occurs during 15–30 min of feeding. The virus is transmitted by feeding of the adult (winged) thrips, and for TSWV, this requires as little as 5 min. Thrips that have acquired TSWV or INSV during larval feeding on infected plants are able to efficiently spread the virus as adults, creating epiphytotics on other susceptible crops within a greenhouse. Although TSWV multiplies in the thrips (providing circulative, persistent transmission), there is no evidence of transovarian transmission. INSV is vectored similarly, but the western flower thrips is the only vector known for this virus.

Three species of thrips have been shown to vector TCSV: western flower thrips, common blossom thrips, and flower thrips.

A tospovirus and a thrips vector may be introduced into the greenhouse together on a thrips-infested, virus-infected plant, or they may arrive in separate plant shipments. Weeds in the greenhouse and plants held over from one year to the next may be reservoirs of inoculum for a thrips population that is introduced separately. Infected flowering potted plant crops are important sources of viral inoculum for infection of vegetable and flower bedding plant crops grown from seed.

Seed transmission of TSWV has been reported for *Pericallis* × *hybrida* (cineraria) and tomato. (The contamination is believed to be carried on the seed coat, rather than in the embryo.) However, seed transmission of neither TSWV nor INSV has been observed in greenhouse production. Movement of the viruses between greenhouses is caused largely by the movement of infected cuttings and plants, whereas distribution within a single greenhouse is caused largely by the western flower thrips, which develops to adulthood after eggs are laid on infected plants.

Management

Management of TSWV and INSV requires eliminating infected plants and controlling the thrips vectors. Specialist propagators of susceptible crops should work closely with reputable diagnostic laboratories to develop and maintain clean stock. Exclusion is an important management strategy for tospoviruses, and growers should insist on using propagation materials that are free of TSWV, INSV, and thrips.

Growers should also familiarize themselves with the symptoms of tospoviruses on their crops. Diseased plants should be rogued out immediately. In the event of an outbreak, it may be helpful to halt production of highly susceptible cultivars or crop species until a successful virus and thrips management program has been established. Having overlapping crops of different ages that are highly susceptible makes it very difficult to eliminate tospoviruses from a greenhouse; a virus may be retained within adult western flower thrips, infected weeds, or infected crop plants. Crop materials should not be carried over from season to season.

Significant losses caused by tospoviruses are generally correlated with high populations of thrips, and thrips management is challenging. (See guidelines for thrips management in Part III, the section "Western Flower Thrips.") The western flower thrips feeds deep within vegetative and flower buds and on pollen at the backs of deep-throated flowers, so it is difficult to contact with insecticides. In addition, there are three stages in the life cycle of the thrips that present no target for chemical control: the eggs, which are inserted into host plant tissue, and the prepupa and pupa, which are nonfeeding stages typically located in the soil of the greenhouse floor or in the growing medium. Also, resistance to many pesticide groups (including pyrethroids, organophosphates, and carbamates) has been documented for the western flower thrips.

Monitoring for thrips allows a management program to get under way promptly by letting the grower know when the thrips population has risen above an experience-based, self-determined action threshold. An insecticide treatment threshold of 10–20 thrips per yellow sticky card per week has been adopted by some North American growers who finish crops for consumers; specialist propagators have a much lower tolerance for thrips. Petunias have been used as indicators of virus-carrying thrips; cultivars that develop local lesions, rather than systemic infection, are needed for this purpose. Indicator petunias may be made more attractive to thrips by adding nonsticky yellow or blue cards. Brown to black rings will develop around the whitish thrips feeding scars if a tospovirus has been transmitted. Monitoring is particularly critical within groups of plants that have been newly introduced from another greenhouse operation.

When thrips are detected at a level above the action threshold, an effective insecticide should be applied without delay and then repeated after an appropriate interval. Rotation of chemicals in different Insecticide Resistance Action Committee (IRAC) groups is important to slow the development of resistance. All the plants in the greenhouse should be treated. The next cycle of treatment should be initiated when card counts indicate that another population increase has occurred.

The use of biological control agents is being explored for the western flower thrips. The predators *Neoseiulus* (syn. *Amblyseius*) *cucumeris, N.* (syn. *A.*) *barkeri,* and *Orius tristicolor* have shown some promise for control of thrips in greenhouse cucumber. The predatory nematode *Steinernema feltiae* is commonly used against prepupal and pupal stages of thrips in potting mixes. The mite *Stratiolaelaps scimitus* (syn. *Hypoaspis miles*) and entomophagous fungi such as *Verticillium lecanii* are being utilized for their ability to control various stages of the thrips life cycle.

Paying careful attention to which crops are grown together within the same greenhouse will reduce crop losses. Seed crops should be separated from cutting crops, and susceptible vegetables should not be grown with vegetatively propagated ornamentals. Breaking the replication cycle of viruliferous thrips may be effective—for example, with a crop-free, weed-free period in midsummer or with production of a nonhost, such as poinsettia, during the fall. In southern and western areas of the United States, where both the virus and thrips are endemic, it may be necessary to screen the growing area to reduce thrips incursions from outdoors. Screening may also be used to isolate one house for the production of highly susceptible cultivars, crops, or propagation blocks. Positive air pressure has been used to counteract movement of thrips into the greenhouse. Areas adjacent to growing structures should be kept free of weeds.

Plant breeding and genetic engineering efforts are being directed toward making crop improvements that will reduce the impact of diseases caused by TSWV and INSV.

Selected References

Allen, W. R., and Matteoni, J. A. 1988. Cyclamen ringspot: Epidemics in Ontario greenhouses caused by the tomato spotted wilt virus. Can. J. Plant Pathol. 10:41-46.

Allen, W. R., and Matteoni, J. A. 1991. Petunia as an indicator plant for use by growers to monitor for thrips carrying the tomato spotted wilt virus in greenhouses. Plant Dis. 75:78-82.

Cho, J. J., Mau, R. F. L., Mitchell, W. C., Gonsalves, D., and Yudin, L. S. 1987. Host list of plants susceptible to tomato spotted wilt virus (TSWV). Res. Ext. Ser. Publ. RES-078. College of Tropical Agriculture and Human Resources, University of Hawaii, Honolulu.

de Ávila, A. C., de Haan, P., Kitajima, E. W., Kormelink, R., Resende, R. de O., Goldbach, R., and Peters, D. 1992. Characterization of a distinct isolate of tomato spotted wilt virus (TSWV) from *Impatiens* sp. in the Netherlands. Neth. J. Plant Pathol. 134:133-151.

de Ávila, A. C., de Haan, P., Smeets, M. L. L., Resende, R. de O., Kormelink, R., Kitajima, E. W., Goldbach, R. W., and Peters, D. 1993. Distinct levels of relationships between tospovirus isolates. Arch. Virol. 128:211-227.

German, T., Ullmann, D., and Moyer, J. W. 1992. *Tospoviruses:* Diagnosis, molecular biology, phylogeny, and vector relationships. Annu. Rev. Phytopathol. 30:315-348.

Gilbertson, R. L., Batuman, O., Webster, C. G., and Adkins, S. 2015. Role of the insect supervectors *Bemisia tabaci* and *Frankliniella occidentalis* in the emergence and global spread of plant viruses. Annu. Rev. Virol. 2:67-93.

Hausbeck, M. K., Welliver, R. A., Derr, M. A., and Gildow, F. E. 1992. Tomato spotted wilt virus survey among greenhouse ornamentals in Pennsylvania. Plant Dis. 76:795-800.

Hsu, H. T., and Lawson, R. H., eds. 1991. Virus–Thrips–Plant Interactions of Tomato Spotted Wilt Virus. Proc. U.S. Dep. Agric. Workshop Publ. ARS-87. U.S. Department of Agriculture, Agricultural Research Service, Washington, DC.

Law, M. D., and Moyer, J. W. 1990. A tomato spotted wilt-like virus with a serologically distinct N protein. J. Gen. Virol. 71:933-938.

Law, M. D., Speck, J., and Moyer, J. W. 1992. The M RNA of impatiens necrotic spot *Tospovirus* (Bunyaviridae) has an ambisense genomic organization. Virology 188:732-741.

Sether, D. M., and DeAngelis, J. D. 1992. Tomato Spotted Wilt Virus Host List and Bibliography. Ore. State Univ. Agric. Exp. Stn. Spec. Rep. 888.

Shimomoto, Y., Kobayashi, K., and Okuda, M. 2014. Identification and characterization of Lisianthus necrotic ringspot virus, a novel distinct tospovirus species causing necrotic disease of lisianthus (*Eustoma grandiflorum*). J. Gen. Plant Pathol. 80:169-175.

Urban, L. A., Huang, P.-Y., and Moyer, J. W. 1991. Cytoplasmic inclusions in cells infected with isolates of L and I serogroups of tomato spotted wilt virus. Phytopathology 81:525-529.

Urban, L. A., Speck, J., Moyer, J. W., and Daub, M. E. 1992. Transformation of chrysanthemum with the nucleocapsid gene of tomato spotted wilt virus. Phytopathology 82:1147.

Warfield, C. Y., and Clemens, K. 2015. First report of *Tomato chlorotic spot virus* on annual vinca (*Catharanthus roseus*) in the United States. Plant Dis. 99:895.

(Prepared by M. L. Daughtrey, R. L. Wick, and J. L. Peterson; revised by A. R. Chase and M. L. Daughtrey)

Verbena virus Y

Analysis of a yellow leaf mottle in *Verbena* cultivar Taylortown Red led to the discovery of a new potyvirus in 2010. The virus showed only 63% nuclear identity to isolates of *Potato virus Y* (PVY) and *Pepper mottle virus* (PeMoV), the most similar potyviruses.

The new virus, *Verbena virus Y* (VVY), belongs to the PVY subgroup within the genus *Potyvirus.* Polyclonal antibodies to PVY were found to cross-react against the new potyvirus, which suggests that the virus might have been misidentified as PVY in the past. Two other viruses were present in the mottled verbena containing the new potyvirus: *Broad bean wilt virus-1* (BBWV-1) and *Coleus vein necrosis virus* (CVNV). The originally observed symptoms were likely caused by one or both of these accompanying viruses (BBWV-1 and CVNV), because symptoms in plants inoculated with the new potyvirus remained latent for 2 years.

The green peach aphid (*Myzus persicae*) has been shown to transmit VVY, and additional vectors may be found in the future. No other natural hosts are known for VVY, but the virus has been transferred experimentally to some solanaceous plants.

Selected Reference

Kraus, J., Cleveland, S., Putnam, M. L., Keller, K. E., Martin, R. R., and Tzanetakis, I. E. 2010. A new *Potyvirus* sp. infects verbena exhibiting leaf mottling symptoms. Plant Dis. 94:1132-1136.

(Prepared by A. R. Chase and M. L. Daughtrey)

Part II. Abiotic Diseases and Disorders

All bedding plants are subject to diseases and disorders caused by abiotic agents, both during production and after being established in the landscape. Abiotic diseases and disorders have nonliving causes and can be thought of as stress-induced conditions (Table 19). Whereas abiotic diseases and disorders often affect many different species of plants growing near one another, biotic (contagious) diseases are often host specific. The overall pattern of symptom expression can indicate which kind of disease or disorder is involved: A biotic disease is usually randomly distributed and affects plants of one species or closely related species to various degrees, whereas an abiotic disease or disorder tends to be distributed more uniformly and can at times affect all the plants in a greenhouse or all the plants in a garden bed (regardless of plant species), depending on what kind of stress created the problem.

The symptoms caused by abiotic factors vary and may appear identical to those caused by biotic agents. Leaf chlorosis, for example, can be caused by such diverse abiotic agents as insufficient fertilizer, pesticide phytotoxicity, and cold soil temperature, just as it might be caused by a pathogen that produces root rot, which interferes with proper nutrient uptake. Factors such as air pollution, pesticide application, and too high or too low levels of light, nutrients, and water may all cause abiotic diseases and disorders of bedding plants at times.

Part II describes some of the abiotic factors that commonly cause symptoms in bedding plants. Many additional sources of information are available on this topic; a partial list is provided in the following "Selected References" section.

Selected References

Blanpied, G. D. 1985. Ethylene in post harvest biology and technology of horticultural crops. HortScience 20:39-60.

Bridgen, M. P. 1984. Be aware of gases in your greenhouse. Conn. Greenhouse Newsl. 123:1-2.

Costello, L. R., Perry, E. J., Matheny, N. P., Henry, M. J., and Geisel, P. M. 2003. Abiotic Disorders of Landscape Plants: A Diagnostic Guide. Publ. 3420, Division of Agriculture and Natural Resources, University of California, Oakland.

Gibson, J. L., Pitchay, D. S., Williams-Rhodes, A. L., Whipker, B. E., Nelson, P. V., and Dole, J. M. 2007. Nutrient Deficiencies in Bedding Plants: A Pictorial Guide for Identification and Correction. Ball Publishing, Chicago.

Hanan, J. J. 1973. Ethylene pollution from combustion in greenhouses. HortScience 8:23-24.

Howe, T. K., and Woltz, S. S. 1981. Symptomology and relative susceptibility of various ornamental plants to acute airborne sulfur dioxide exposure. Proc. Fla. State Hortic. Soc. 94:121-123.

Mastalerz, J. 1977. The Greenhouse Environment. John Wiley & Sons, New York.

Nau, J., ed. 2011. Ball Redbook, 18th ed. Vol. 2: Crop Production. Ball Publishing, Chicago.

White, J. W., ed. 1993. Geraniums IV. Ball Publishing, Chicago.

Williams, K. A., Craver, J. K., Miller, C. T., Rud, N., and Kirkham, M. B. 2015. Differences between the physiological disorders of intumescences and edemata. Acta Hortic. 1104:401-406.

Williams, K. A., Miller, C. T., and Craver, J. K. 2016. Light quality effects on intumescence (oedema) on plant leaves. Pages 275-286 in: LED Lighting for Urban Agriculture. T. Kozai, F. Kazuhiro, and E. Runkle, eds. Springer, Singapore.

(Prepared by A. R. Chase and M. L. Daughtrey)

Air Pollution

Although annuals planted in landscapes may, from time to time, show symptoms from ozone and other outdoor pollutants, symptoms of air pollution are more likely encountered when bedding plants are produced in a greenhouse. The most common and troublesome air pollutant is ethylene, which is produced by both internal combustion engines and horticultural products. Ethylene damage may develop during the winter in greenhouses and other enclosed structures when heaters are not properly vented outside. Ethylene damage may also occur at any time of year when bedding plants are shipped or stored with horticultural products that produce ethylene (e.g., apples) and when ethylene is applied to enhance the ripening of fruits or vegetables.

Ethylene exposure can cause symptoms such as chlorosis and water soaking of leaves, abortion of shoot meristems, changes in leaf angles relative to the stems (epinasty), and dramatic leaf dropping. Flowers may be deformed or abscise, and plant growth may be stunted because of the shortening of internodes. Symptoms of ethylene damage increase in severity as the level and duration of exposure to the gas increase. Prolonged exposure to as little as 10 parts per billion of ethylene can produce symptoms in plants.

Some plants create enough ethylene during shipping to damage themselves, particularly when they are wounded (e.g., cuttings for propagation) or shipped at a stressfully high temperature. Ethylene problems occur most commonly, however, when the greenhouse environment is polluted.

To minimize the opportunity for ethylene damage, growers should provide adequate ventilation for space heaters and conduct regular maintenance of all heating units. The oxygen supplies to burners, furnaces, and boilers must be adequate for complete combustion or ethylene will be released. Chimney height should be sufficient to ensure that combustion gases are carried away from the greenhouse and not drawn back in through vents.

Additional sources of information are provided in the "Selected References" at the beginning of Part II.

(Prepared by A. R. Chase and M. L. Daughtrey)

TABLE 19. Symptoms of Abiotic Diseases and Disorders in Some Bedding Plants and Possible Causes[a]

Symptom(s)	Cause(s)
Chlorosis	
Entire plant chlorotic	Poor nutrition
	High light intensity
Young leaves chlorotic	Inadequate micronutrient levels
	Poorly aerated soil mix
Older leaves chlorotic	Low nitrogen or potassium
	High soluble salts
	Overwatering
	Poorly aerated soil mix
Marginal chlorosis	Low magnesium or potassium
	High soluble salts
Interveinal chlorosis	Iron or manganese deficiency
	Sulfur dioxide air pollution injury
Necrosis	
Necrotic leaf margins or tips	Potassium deficiency
	Boron or fluoride toxicity
	High soluble salts
	Temperature extremes
	Desiccation injury
Necrotic leaf spots or sections	Cold-water injury
	Fertilizer toxicity
Marginal and internal leaf necrosis	Sun scorch
	Cold injury
	Air pollution injury
	Fertilizer toxicity
Growth Abnormalities: Stems	
Stem tips stunted	Boron, calcium, or copper deficiency
Fasciated stems	Genetic variation
	Herbicide injury
Stems browned at soil line	High soluble salts
	Slow-release fertilizer against stems
	Excessive irrigation
	Poor soil aeration
Stem lesions or cankers	Sunscald
	Mechanical injury
Wilting	High soluble salts
	High leaf temperature and cool soils
	High temperature
	Insufficient soil moisture
	Low humidity
	Poor root system
Tip dieback or blight	Calcium, copper, or boron deficiency
	Desiccation
Epinasty	Low light
	Ethylene toxicity
Stems thin and weak with elongated internodes	Fertilizer extremes
	Low light
	Close plant spacing
	High temperature
Stems thick with shortened internodes	High light
	Wide plant spacing
Growth Abnormalities: Leaves	
Leaves abnormally small	Fertilizer deficiency
	Copper deficiency
	High soluble salts
	High light
	Low light
	Low humidity
	Root-bound plants
Elongated petioles	Low light
Shortened petioles	High light
Leaves very thin	Excess nitrogen
	Low light
Leaves very thick	High light
Leaf margins split	Mechanical injury
	Fluctuating moisture supply
Leaves cupped	Nutritional disorder
	Air pollution
Leaf margins notched	Nutritional disorder
	Mechanical injury
	High temperature
Holes in leaves	Mechanical injury
Defoliation	High soluble salts
	Abrupt light reduction
	Prolonged shipping
	Chilling injury
	Desiccation
	Poor soil aeration
	Ethylene toxicity
	Low soil moisture
Growth Abnormalities: Roots	
Shallow roots	Excessive bottom heat
	Excessive soil moisture
	Poor soil aeration
	High soil compaction
Slow root development	High soil salinity
	Soil temperature extremes
	Plant potted too deeply
Roots browned	High soil salinity
	High soil moisture
	Poor soil aeration
	High soil compaction
Plant stunted	Fertilizer extremes
	Soil pH extremes
	Low light
	Temperature extremes
	High soil moisture
	Poor soil aeration
	Pot-bound root system

[a] Courtesy A. R. Chase and M. L. Daughtrey—© APS.

Excess or Insufficient Light

Because light is required for photosynthesis, excesses and insufficiencies affect plant growth profoundly. Little research has been published concerning abiotic diseases and disorders of bedding plants caused by inappropriate light levels. Even so, light is key for successful crop production and for choosing appropriate planting sites in the landscape.

There are also important interactions among the basic inputs that plants need for growth. For example, plants produced under high light levels use more fertilizer than those grown under low light levels. Thus, moving a plant from high light conditions to low light conditions can result in fertilizer toxicity, and moving a plant from a shadier to a brighter location can result in chlorosis and underfertilization.

Rapid and extreme changes in light level frequently cause leaf abscission, necrosis, and bleaching (Fig. 133). When a crop is exposed to a precipitous increase in light, the foliage may

Fig. 133. Coleus with bleached leaves, caused by excessive light. (Courtesy A. R. Chase—© APS)

develop necrotic areas. Plants being produced in a shaded area will burn if they are suddenly exposed to full sun, even briefly; this is true even of plants that can ordinarily tolerate full-sun conditions. Insufficient light may also cause damage to bedding plants—particularly delays in flowering and reductions in number of flowers.

One of the key considerations when choosing appropriate bedding plants for the landscape is whether the site is sunny or shaded. Plants with tolerance for the light level or levels available will perform best at that site.

Certain bedding plants, such as *Cuphea ignea* (cigar flower) and *Ipomoea batatas* (sweetpotato vine), are prone to a somewhat oedema-like disorder on foliage and stems termed "intumescence," in which the epidermal or parenchymal cells form outgrowths. Intumescence is particularly likely to develop during winter greenhouse production, and it is ameliorated by an increase in blue wavelengths or exposure to ultraviolet B light. The oedema that is common in *Pelargonium peltatum* (ivy geranium), however, is not regulated by ultraviolet light.

Additional sources of information are provided in the "Selected References" at the beginning of Part II.

(Prepared by A. R. Chase and M. L. Daughtrey)

Nutritional Imbalances

During bedding plant production, nutritional imbalances from excesses and deficiencies of macro- and micronutrients are among the most frequent causes of symptoms of abiotic diseases and disorders. High quantities of macronutrients are needed for plant growth, yet some of the most dramatic symptoms are caused by deficiencies or toxicities of micronutrients, such as boron, iron, and fluoride.

The pH of the growing medium determines the availability of nutrients. Bedding plants are generally grown best at a pH of 5.6–6.2. Micronutrient deficiency (too little iron, manganese, boron, copper, or zinc) will arise if the growing mix has a pH of 6.5 or greater. Some macronutrients (particularly calcium and magnesium) are less available to plants when the pH is less than 5.4.

Regular monitoring of nutrients in the foliage and the growing medium is important for ensuring plant health during bedding plant production. The following sections describe the generic symptoms caused by various nutrient imbalances.

Additional sources of information are provided in the "Selected References" at the beginning of Part II.

(Prepared by A. R. Chase and M. L. Daughtrey)

Nutritional Deficiencies

Nitrogen (N)

Because N is a very mobile element, symptoms of deficiency appear first on the older leaves and progress upward. Nitrogen-deficient leaves are chlorotic and may be reduced in size and have shortened internodes. Eventually, the lower leaves may turn necrotic and abscise. Leaf reddening is seen in some plants, including *Begonia* spp., *Tagetes* spp. (marigolds), and *Viola* × *wittrockiana* (pansy). If the deficiency is not treated, a general loss in plant vigor and growth will occur.

Phosphorous (P)

The initial indication of P deficiency is stunted growth, including small leaves and shortened internodes. Foliage is at first a deep green; older leaves may become dull and eventually chlorotic and then necrotic. Because of the loss of green pigment (chlorophyll) in leaves, red and blue pigments become evident, especially on the undersurfaces of leaves near the main veins. Dead patches develop along the edges of leaves in extreme cases.

Phosphorous uptake by *Pelargonium* spp. (geraniums) and other plants is reduced at cold growing temperatures (less than 13°C). The result is a reddening of foliage (Fig. 134), which can be confused with the symptoms of Pythium root rot.

Potassium (K)

Leaves and stems are stunted when plants are deficient in K. In some species, leaf color remains normal, while in others, leaves turn chlorotic and necrotic. Chlorosis and necrosis first develop on the margins of older leaves, as K is a mobile element. Eventually, older leaves turn brown and dry.

Magnesium (Mg)

Plants with low levels of Mg have less vigor and smaller leaves than normal. Interveinal chlorosis is characteristic on older leaves, beginning at the leaf margins and extending between the veins. Veins maintain their green color.

Calcium (Ca)

Calcium deficiency may appear occasionally in bedding plant production. Unlike the more mobile elements, for which deficiency is seen in the lower leaves, a deficiency of Ca is expressed in the upper part of the plant. Young leaves may be chlorotic, stunted, and distorted and sometimes become strap-like or crinkled. Scorching develops along the edges of leaves, shoot growth ceases, and pedicels may collapse.

Allowing the soil to dry somewhat between waterings will facilitate plants' ability to take up calcium.

Iron (Fe)

Because Fe is relatively immobile in plants, interveinal chlorosis first develops on young leaves. Yellowing, tip scorching, and stunting of new leaves can also occur.

Certain bedding plants are especially likely to show signs of Fe deficiency when they are grown at pH levels that are too high (Fig. 135)—among them, *Antirrhinum majus* (snapdragon), *Begonia* spp., *Calibrachoa* hybrids, *Chaenostoma cordatum* (bacopa), *Diascia* hybrids, *Nemesia fruticans,* pansy, *Petunia hybrida, Verbena hybrida,* and *Zinnia* spp. Because the soil pH strongly influences the availability of Fe to the plant, it should be monitored regularly, especially in these crops.

Fig. 134. Petunia plant with reddened foliage, symptomatic of a phosphorous deficiency caused by cold growing temperatures. (Courtesy A. R. Chase—© APS)

Fig. 135. Calibrachoa with chlorosis on new leaves, symptomatic of an iron deficiency caused by high pH. (Courtesy A. R. Chase—© APS)

Poor root health resulting from inadequate soil aeration in a water-logged growing mix will also reduce the ability of roots to take up Fe.

(Prepared by A. R. Chase and M. L. Daughtrey)

Excessive Levels of Soluble Salts

Toxicity sometimes results from excessive applications of fertilizer. Even when fertilizer is applied at the appropriate rate, continuous irrigation without leaching may result in the buildup of salts that are toxic to plants in general.

Salvia spp. and *Impatiens hawkeri* (New Guinea impatiens) are particularly sensitive to high levels of soluble salts. Symptoms usually include development of marginal leaf burn or leaf tip chlorosis and necrosis, but at times, the main indication is chlorosis of the lower leaves. Plants may die as a result of overfertilization. Root systems on plants that are overfertilized are reduced and may appear to be infected with a root pathogen.

Under high light conditions, applying fertilizer at every irrigation (known as "constant feed") can result in phytotoxicity on some plugs. Growers must be extra careful when using soluble fertilizer at each irrigation if a slow-release fertilizer is already in the growing medium. If the level of soluble salts becomes excessive, leaching will reduce the irrigation-applied fertilizer but result in additional release from the slow-release fertilizer.

(Prepared by A. R. Chase and M. L. Daughtrey)

Micronutrient Toxicities

Micronutrients may also cause toxicity. This is most frequently seen on *Pelargonium* × *hortorum* (florist's geranium), *Impatiens hawkeri* (New Guinea impatiens), and other bedding plants that overaccumulate iron and manganese at a low growing pH, at which these elements are more available. For florist's geranium, maintaining a growing pH of 5.8–6.3 is recommended to avoid this problem. Toxicity caused by iron or manganese causes brown speckling and some chlorosis on the outer edges of lower leaves of florist's geranium, which can progress and develop into more extensive necrosis. Although in the same genus, *P. peltatum* (ivy geranium) is not prone to this problem and is best suited to a slightly lower growing pH (pH 5.5–6.0). Other iron- and manganese-sensitive plants grown at too low a pH may also show necrosis (Fig. 136).

Fig. 136. Injury on New Guinea impatiens caused by iron–manganese toxicity. (Courtesy M. L. Daughtrey—© APS)

Symptoms caused by growing plants at too low a pH are often similar to those of plants improperly treated with an iron chelate. Iron toxicity can result unless the treatment is directed at the growing medium or rinsed off the foliage immediately after application.

(Prepared by A. R. Chase and M. L. Daughtrey)

Pesticide Toxicity (Phytotoxicity)

Pesticides are routinely used by many bedding plant producers to control pests. Using these products is generally safe on the crops for which they are labeled and helps producers meet the standards for aesthetic quality of ornamentals. Even so, phytotoxicity (plant injury) from pesticide applications sometimes occurs. The symptoms of phytotoxicity reduce plant quality and can make plants unsalable. When attempting to diagnosis symptoms on a bedding plant crop, growers should check recent application records. Consideration of pesticide injury as the cause of symptoms is most appropriate when there is a fairly uniform distribution of symptoms across a cultivar or crop and when the pattern of damage in the crop coincides with the pattern of spray application.

Pesticides labeled for use on large and diverse plant groups—such as *Salvia* spp., *Viola* × *wittrockiana* (pansy), and *Impatiens* spp.—should not be assumed to be equally safe on all species and cultivars. Periodically, individual cultivars show sensitivity to chemical applications. Growers should always check product labels for listed precautions; manufacturers are careful to provide information on plants with known sensitivity to their products. Labels also instruct growers to treat a small number of plants as a test for plant safety before making widespread use of a product. Even with normal, routine spray or drench applications, a shift in environmental conditions may occasionally result in phytotoxicity on a crop that is not usually sensitive to a treatment.

Choosing the wrong pesticide active ingredient, plant protection product (spreader–sticker), or rate of application can cause plant injury. Certain bedding plants (notably, pansy and impatiens) tend to be more pesticide sensitive than others and

Fig. 137. Necrotic spotting on dusty miller plants, caused by a phytotoxic spray application. (Courtesy A. R. Chase—© APS)

Fig. 138. White necrosis on salvia, caused by a phytotoxic spray application. (Courtesy A. R. Chase—© APS)

are thus more frequently injured. Under some conditions, even a water spray can cause tissue damage, especially on open blooms. The symptoms of phytotoxicity will vary with environmental conditions, with plant species and cultivar, and even with plant age or tissue age.

Plant vigor at the time of pesticide application can also affect the margin of safety. In some cases, stressed plants are more susceptible to pesticide phytotoxicity than unstressed plants. Stresses may involve a wide range of causes, including extremes of fertilizer, water, temperature, and light. Maintaining plants under optimal conditions will lessen the chances of phytotoxicity.

Although the practice of tank mixing pesticides, fertilizers, and/or spreader–sticker products saves labor, it can injure plants. It is also risky to apply pesticides to water-stressed plants or under adverse environmental conditions, such as when the plant surface will remain wet for a long time after application. In addition, extremes in temperature can affect the safety of applying pesticides. Applications should be made at an air temperature between 15 and 32°C using water at or near the air temperature.

Chemicals used for purposes other than pest management, such as algaecides, growth regulators, and disinfestants, can also cause phytotoxicity. Even if documenting product use is not required by government regulations, producers should maintain a record of every product applied on or near crops and note the general environmental conditions that prevailed during and soon after an application.

Phytotoxicity can be expressed in a variety of ways, some obvious and some subtle. Symptoms range from minor leaf speckling to plant death. One of the most common symptoms of phytotoxicity is chlorosis; this is also a symptom of nutrient deficiency, so diagnosing the cause can be confusing. Sometimes, the chlorosis is slight and barely noticeable, while at other times, the tissue turns completely white within a few weeks of pesticide application. The leaves may also remain green while the petioles or stems become chlorotic. Chlorosis caused by pesticide application is frequently confined to the newest leaves, which is also symptomatic of iron deficiency.

Another commonly recognized symptom of phytotoxicity is necrosis, or "burning." Necrosis can occur on leaf tips, margins, or interveinal tissues or be scattered across the leaf or petal surface (Figs. 137 and 138). Burning is the most obvious symptom of phytotoxicity short of plant death. It usually appears within 1 week of pesticide application but may take as long as 6–8 weeks if multiple applications are required and the pesticide is applied only once a month. Burning can be caused in several ways. Contact burns are characterized by general speckling of the leaf surface; spots appear wherever droplets of spray landed. Some burns occur at places on the leaf where spray has puddled or residue has accumulated and caused localized damage. One aid to diagnosing this kind of injury is to look for a shading effect from other leaves, in which portions of leaf surfaces are uninjured because other leaves overlapped and protected them during the spray application. Not all necrosis is from direct contact injury. In some cases, the pesticide is absorbed through the leaves or roots and redistributed, so that injury typically appears at the leaf margins, where the pesticide accumulates. Some burn symptoms are limited to young plant tissue, because the plant's physical defenses were not yet fully developed at the time of exposure.

Additional sources of information are provided in the "Selected References" at the beginning of Part II.

(Prepared by A. R. Chase and M. L. Daughtrey)

Water Imbalances

Imbalances in water result in classic abiotic stress symptoms in many bedding plants. All plants are more prone to wilt, leaf necrosis, and leaf abscission when irrigation is withheld, and injuries can be compounded by the phytotoxic effects of pesticides. In some crops, water deficits are obvious only when the foliage loses its normal color; in some cases, foliage becomes grayish.

Excess water and restricted transpiration have traditionally been blamed for the blisterlike swellings referred to as "oedema" on plants such as *Pelargonium peltatum* (ivy geranium), *Begonia* spp., *Cleome houtteana* (spider flower), *Cuphea ignea* (cigar flower), *Ipomoea batatas* (sweetpotato vine), and *Solanum lycopersicum* (tomato). However, research has shown that in some of these species, other environmental factors are important in the development of oedema (see the section "Excess or Insufficient Light"). The term "intumescence" is used to distinguish an oedema-like injury associated with the absence of ultraviolet light, such as that seen in tomato and sweetpotato vine. Oedema is characterized by small, blister-like areas primarily on the lower surfaces of affected leaves (although sometimes the upper surfaces) (Fig. 139); these areas develop when swollen parenchymal cells burst through the epidermis. When young lesions are viewed by holding them up to the light, they appear lighter in color and may be concentrated around the leaf margins. Studies on *P.* × *hortorum* (florist's geranium), ivy geranium, and geranium hybrids have indicated that several factors influence development of oedema on *Pelargonium* spp.; high light intensity, low nutritional levels, high soil moisture, and high humidity are all conducive to development.

Additional sources of information are provided in the "Selected References" at the beginning of Part II.

(Prepared by A. R. Chase and M. L. Daughtrey)

Fig. 139. Oedema, a physiological problem characterized by formation of warty tissue on leaf undersides, on ivy geranium. (Courtesy A. R. Chase—© APS)

Part III. Arthropod Pests

Overview

Annual bedding plant season is always a busy time of year, as greenhouse producers concentrate on growing and selling multiple crops. Given this focus, pest management is generally not a top priority. However, by identifying the major arthropod (insect and mite) pests that occur during all phases of the production cycle and beyond and by incorporating pest management into the daily schedule, greenhouse producers will be able to implement the appropriate management strategies to avoid extensive outbreaks of arthropod pests.

A variety of arthropod pests may be encountered during the production of annual bedding plants and after they have been incorporated into the landscape, although the pests present may vary depending on the geographic location and time of year (Table 20 and Appendix III). Furthermore, the damage caused by arthropod pests may be influenced by the types (cultivars/varieties) of annual bedding plants that are grown and the arthropod pests' feeding behaviors, which are associated with the damage symptoms that will be evident.

The primary arthropod pests of annual bedding plants are aphids, beetles, caterpillars, fungus gnats, leafhoppers, leafminers, mealybugs, mites, plant bugs, shore flies, thrips, and whiteflies. Although slugs and snails are not arthropods, they are pests of many annual bedding plants. The arthropod pests with piercing–sucking mouthparts (xylem or phloem feeders) include aphids, mealybugs, leafhoppers, and whiteflies, and those with chewing mouthparts include beetles, caterpillars, and fungus gnat larvae. Mites and thrips have very discrete feeding behaviors. Mites feed within leaf cells and on chloroplasts, whereas thrips feed on the mesophyll and epidermal cells of leaf tissues. Thus, the damage symptoms are very characteristic of each pest. Nearly all the major arthropod pests feed on the aboveground portions of plants, with the exception of fungus gnat larvae, which are located in the growing medium and feed on plants' roots or at the bases of cuttings, thus inhibiting the ability of annual bedding plants to take up water and nutrients.

The following sections outline the conditions and major arthropod pests of bedding plants associated with propagation and production and following installation into landscapes.

(Prepared by R. A. Cloyd)

Seed and Cutting Propagation

The major arthropod pests associated with seed and cutting propagation of annual bedding plants are fungus gnats and shore flies (Fig. 140). Both insect pests thrive under moist conditions, which provide an environment conducive for development and population growth. The number of fungus gnat larvae that can cause damage to seedlings is lower than the number required to damage larger plants. Therefore, greenhouse producers must maintain fungus gnat populations at low levels to alleviate potential damage to seedlings and cuttings. In fact, only one fungus gnat larva is needed to kill a seedling or unrooted cutting.

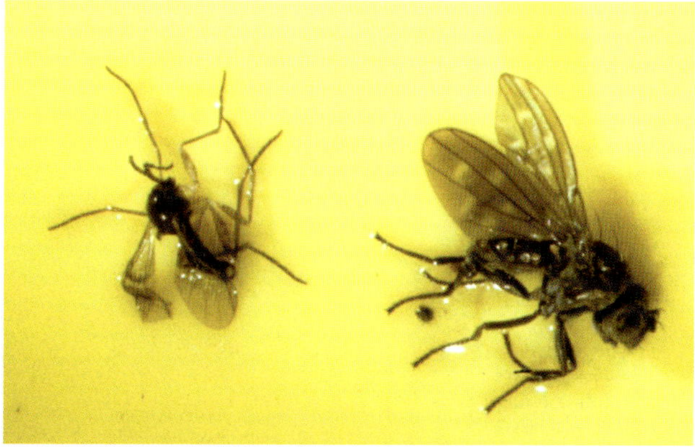

Fig. 140. Adult fungus gnat (*Bradysia* sp.) (left) and shore fly (*Scatella* sp.) (right) on a yellow sticky card. (Courtesy R. A. Cloyd—© APS)

(Prepared by R. A. Cloyd)

Production of Annual Bedding Plants

A range of arthropod pests can be problematic during the production of annual bedding plants, including the American serpentine leafminer, several species of aphids, the broad mite, several species of caterpillars, the citrus mealybug, the cyclamen mite, fungus gnats, the greenhouse whitefly, the sweetpotato whitefly, the twospotted spider mite, and the western flower thrips (Table 21). A number of operational procedures that occur during production may contribute to pest problems:

1. Annual bedding plants are typically irrigated and well fertilized, providing an attractive food source for arthropod pests, and the succulent growth is easy for arthropod pests to feed on. Furthermore, the abundant levels of nutrients (in the form of amino acids) in plant tissues may result in increased growth, development, and reproduction of some arthropod pests.
2. Plant production is continuous, without any fallow periods, which means a food source is always available.
3. Environmental conditions such as temperature, light, and relative humidity that are maintained to promote plant growth and development are also conducive to arthropod pest development.

TABLE 20. Annual Bedding Plants and Associated Arthropod (Insect and Mite) and Mollusk Pests[a,b]

Scientific Name (Common Name)	Aphids	Beetles	Caterpillars	Fungus Gnats	Leafhoppers	Leafminers	Mealybugs	Mites	Plant Bugs	Slugs and Snails	Thrips	Whiteflies
Ageratum (ageratum)	X		X					X			X	X
Antirrhinum (snapdragon)	X		X					X	X		X	
Begonia (wax begonia and others)				X			X	X		X	X	X
Brassica (ornamental kale)	X		X							X		
Browallia (amethyst flower)												
Calendula (pot marigold)	X	X			X				X		X	X
Calibrachoa (calibrachoa)	X										X	
Campanula (bellflower)				X	X							
Capsicum (ornamental pepper)	X							X			X	
Catharanthus roseus (annual vinca)	X			X						X		
Celosia (cockscomb)											X	X
Chrysanthemum (chrysanthemum)	X	X	X	X	X	X	X	X	X	X	X	X
Coreopsis (tickseed)		X							X			
Cosmos (cosmos)		X			X				X			
Dahlia (dahlia)	X	X	X		X	X		X	X	X	X	X
Dianthus (Sweet William, pinks)										X	X	
Eustoma (lisianthus)											X	
Gaillardia (blanket flower)									X			
Gazania (gazania)	X	X									X	
Gerbera (gerber daisy)	X	X			X	X	X	X		X	X	X
Gomphrena (common globe amaranth)								X		X		
Helichrysum (strawflower)			X									
Impatiens (garden and New Guinea impatiens)								X	X	X	X	
Lobelia (garden lobelia)				X						X		
Lobularia maritima (sweet alyssum)			X									
Matthiola (stock)								X				
Nemesia (nemesia)	X									X		
Nicotiana (flowering tobacco)	X		X									X
Ocimum basilicum (basil)	X											
Osteospermum (trailing African daisy)	X	X							X	X	X	
Pelargonium (ivy and zonal geraniums)	X		X	X				X		X	X	X
Petunia (petunia)	X		X		X	X				X	X	X
Phlox (phlox)								X			X	
Plectranthus scutellarioides (coleus)				X			X			X	X	X
Portulaca (moss rose, purslane)											X	
Primula (primrose)							X	X		X		
Salvia (sage)	X				X			X		X	X	X
Scaevola (fan flower)								X				
Senecio (cineraria, dusty miller)	X									X		
Tagetes (marigold)	X	X			X	X	X	X	X	X	X	
Torenia (wishbone flower)								X				
Tropaeolum (nasturtium)	X				X	X				X	X	
Verbena (verbena)	X					X			X		X	X
Viola (pansy)	X		X		X	X		X		X		
Zinnia (zinnia)	X	X			X	X		X	X	X	X	

[a] Courtesy R. A. Cloyd—© APS.
[b] The host plants included in this table are those for which data are available on specific arthropod (insect and mite) and mollusk (snail and slug) pest damage.

4. Not enough natural enemies (biological control agents), such as parasitoids and predators, are able to migrate into the greenhouse to regulate existing arthropod pest populations because of the limited number of openings to the outdoors and lack of vegetation located immediately outside the greenhouse.
5. Greenhouse doors and vents are commonly left open, which allows flying insects—including thrips, whiteflies, leafminers, moths, beetles, plant bugs, and winged aphids—to enter easily.
6. Air circulation fans, which are useful in reducing problems with foliar diseases such as Botrytis blight (gray mold), can easily disperse arthropod pests such as thrips throughout the greenhouse; this is particularly the case with hanging baskets.
7. The movement of personnel along benches, especially personnel wearing yellow- or blue-colored clothing, can transport and distribute insects within and between greenhouses.

(Prepared by R. A. Cloyd)

TABLE 21. Arthropod (Insect and Mite) and Mollusk (Slug and Snail) Pest Damage on Annual Bedding Plants[a]

Common Name (Scientific Name)	Damage
Aphids	
Cotton/Melon aphid (*Aphis gossypii*)	Feeding causes leaf yellowing, distorted leaf growth, and stunted plant growth.
Foxglove aphid (*Aulacorthum solani*)	Feeding causes leaf yellowing, distorted leaf growth, and stunted plant growth.
Green peach aphid (*Myzus persicae*)	Feeding causes leaf yellowing, distorted leaf growth, and stunted plant growth.
Beetles	
Japanese beetle (*Popillia japonica*)	Adults chew holes in flower petals; feeding on leaves causes a lacelike, skeletonized appearance.
Spotted cucumber beetle (*Diabrotica undecimpunctata howardi*)	Adults chew irregularly shaped holes in leaves and flowers.
Caterpillars	
Beet armyworm (*Spodoptera exigua*)	Caterpillars consume plant leaves, stems, and flowers; fecal deposits may be present.
Diamondback moth (*Plutella xylostella*)	Caterpillars initially mine plant leaves and then feed on leaves; fecal deposits may be present.
Imported cabbageworm (*Pieris rapae*)	Caterpillars consume plant leaves, stems, and flowers; fecal deposits may be present.
Pansyworm/Variegated fritillary (*Euptoieta claudia*)	Caterpillars consume plant leaves, stems, and flowers; fecal deposits may be present.
Saltmarsh caterpillar (*Estigmene acrea*)	Caterpillars consume plant leaves, stems, and flowers; fecal deposits may be present.
Tobacco budworm (*Heliothis virescens*)	Caterpillars consume plant leaves, stems, and flowers; fecal deposits may be present.
Yellowstriped armyworm (*Spodoptera ornithogalli*)	Caterpillars consume plant leaves, stems, and flowers; fecal deposits may be present.
Fungus Gnats	
Darkwinged fungus gnats (*Bradysia coprophila*, *Bradysia impatiens*)	Larvae feed on plant roots and crowns, causing leaf yellowing and plant stunting and wilting.
Leafhoppers	
Aster leafhopper (*Macrosteles quadrilineatus*)	Feeding causes spotting of plant leaves; can transmit the aster yellows phytoplasma.
Potato leafhopper (*Empoasca fabae*)	Feeding causes stippling of plant leaves; also leaf browning, or necrosis ("hopperburn").
Leafminer	
American serpentine leafminer (*Liriomyza trifolii*)	Larvae create meandering or serpentine mines beneath the leaf cuticle.
Mealybug	
Citrus mealybug (*Planococcus citri*)	Feeding causes leaf yellowing and plant stunting and wilting.
Mites	
Broad mite (*Polyphagotarsonemus latus*)	Feeding causes leaf curling and distortion; leaves appear darker green than normal.
Cyclamen mite (*Phytonemus pallidus*)	Feeding causes leaf curling and distortion; flower buds may abort.
Twospotted spider mite (*Tetranychus urticae*)	Feeding causes leaves to appear bleached and have white to silver-gray speckles; may also cause leaf yellowing and bronzing.
Plant Bugs	
Fourlined plant bug (*Poecilocapsus lineatus*)	Feeding causes leaf distortion and stippling; tan- to brown-colored spots coalesce, forming brown blotches.
Tarnished plant bug (*Lygus lineolaris*)	Feeding by adults and nymphs causes leaf yellowing and twisting or distortion of terminal growth; leaves may appear ragged or discolored.
Shore Fly	
Shore fly (*Scatella stagnalis*)	Larvae do not cause direct plant damage; adults may leave black fecal deposits.
Slugs and Snails	
Black slug (*Arion ater*)	Feeding causes large, irregularly shaped, ragged holes on plant leaves.
Brown garden snail (*Cornu aspersum*)	Feeding causes large, irregularly shaped, ragged holes on plant leaves.
Giant gardenslug (*Limax maximus*)	Feeding causes large, irregularly shaped, ragged holes on plant leaves.
Gray fieldslug (*Deroceras reticulatum*)	Feeding causes large, irregularly shaped, ragged holes on plant leaves.
Greenhouse slug (*Milax gagates*)	Feeding causes large, irregularly shaped, ragged holes on plant leaves.
Orange-banded arion/White-soled slug (*Arion fasciatus*)	Feeding causes large, irregularly shaped, ragged holes on plant leaves.
Three-band gardenslug (*Lehmannia valentiana*)	Feeding causes large, irregularly shaped, ragged holes on plant leaves.
Thrips	
Chilli thrips (*Scirtothrips dorsalis*)	Feeding causes leaf curling, thickened new leaf growth, and deformed leaves.
Greenhouse thrips (*Heliothrips haemorrhoidalis*)	Feeding may cause damaged leaves to appear silvery or bleached and distorted; black fecal spots may be present.
Western flower thrips (*Frankliniella occidentalis*)	Feeding causes scarring, necrotic spotting, and distorted growth of leaves and flower deformation; damaged leaves and flowers may have a white or silvery appearance; black fecal deposits may be present; can transmit *Impatiens necrotic spot virus* and *Tomato spotted wilt virus*.
Whiteflies	
Bandedwinged whitefly (*Trialeurodes abutiloneus*)	Feeding by nymphs and adults causes leaf yellowing, leaf distortion, and plant wilting.
Greenhouse whitefly (*Trialeurodes vaporariorum*)	Feeding by nymphs and adults causes leaf yellowing, leaf distortion, and plant wilting.
Sweetpotato whitefly (*Bemisia tabaci*)	Feeding by nymphs and adults causes leaf yellowing, leaf distortion, and plant wilting.

[a] Courtesy R. A. Cloyd—© APS.

After Installation into Landscapes

Once annual bedding plants have been placed into the landscape, the arthropod pests that feed on them are subject to environmental conditions (e.g., rainfall, high relative humidity, and sunlight) and natural enemies that can cause mortality, thus reducing pest populations. Moreover, the tolerance for arthropod pests in the landscape is higher than that in the greenhouse, because there is less scrutiny for plant damage by homeowners and landscapers than by greenhouse producers.

A number of arthropod pests may occur on annual bedding plants once the plants leave the greenhouse and are incorporated into the landscape: namely, aster leafhopper, fourlined plant bug, Japanese beetle, potato leafhopper, spotted cucumber beetle, tarnished plant bug, and twospotted spider mite. In addition, slugs and snails (which are not arthropod pests) feed on many annual bedding plants in landscapes.

(Prepared by R. A. Cloyd)

Management Strategies

A variety of management and plant protection strategies can be implemented to reduce problems with arthropod pests of annual bedding plants both during production and after planting in the landscape, including cultural controls, scouting, pesticide applications, and biological control agents.

Cultural Controls

Cultural control strategies should always be the primary means of managing arthropod pests. Proper fertilization may lead to fewer problems with arthropod pests. For example, overfertilizing plants (particularly with water-soluble, nitrogen-based fertilizers) may result in the production of succulent growth, which can increase plants' susceptibility to phloem-feeding arthropods, such as aphids and whiteflies, along with spider mites and thrips. The high leaf nitrogen content and high levels of amino acids (both primary food sources for arthropod pests) can increase female reproduction. Furthermore, too much fertilizer may result in excessive plant growth; consequently, the leaf cuticle may be thinner and increase the ease by which arthropod pests can penetrate the leaf with their mouthparts.

Reducing excess moisture content of the growing medium by watering only when needed and repairing pipe leaks when they occur will minimize problems with fungus gnats and shore flies. Both insect pests thrive under moist conditions. Moreover, excess moisture can lead to the growth of algae, which provides an ideal breeding environment for populations of fungus gnats and shore flies. Growing media in which the upper 25–50 mm can be easily remoistened if allowed to dry down provide an unfavorable environment for fungus gnat females to lay eggs. Eggs that are laid in such media tend to have high mortality levels, thus reducing the number of potential damaging larvae.

Weed management both inside and outside the greenhouse will prevent insects from moving from weeds into the greenhouse or migrating from outside onto annual bedding plants. Many weeds provide refuge and are alternative hosts for insect and mite pests, and some weeds serve as potential inoculum sources for diseases (e.g., viruses) that are vectored by insects such as aphids and thrips.

Scouting

Scouting, whether by using yellow or blue sticky cards or visually inspecting plants, is helpful in determining the status or population dynamics of arthropod pests (Table 22). Early detection through scouting will minimize later efforts to address large arthropod populations if pest management strategies are implemented immediately.

Incoming plant material should be inspected, and if feasible, plants infested with pests should be quarantined in an isolated greenhouse or area so the plants can be treated before they are introduced into the main production area. The entire greenhouse does not have to be scouted; rather, selected areas near vents and doors and along the edges of benches should be the focus of scouting, because these are the areas in which initial arthropod pest infestations are most likely to start. Plants on the floor, on benches, and in hanging baskets should also be inspected.

Growers should take advantage of previous experience by focusing scouting efforts on plant species that tend to be susceptible to arthropod pests. These plants can be used as indicator plants when scouting. Scouting may provide information on the effectiveness of pest management programs and track pest population trends during the growing season.

TABLE 22. Quick Guide to Diagnosis of Arthropod Pests by Leaf Symptoms[a]

Symptom(s)	Pest(s)
Meandering mines or tunnels beneath the epidermal layers of leaves	American serpentine leafminer
Leaves yellowed (chlorosis); leaves distorted and curled (upward or downward)	Aphids (cotton/melon, foxglove, green peach, potato) Bandedwinged whitefly Citrus mealybug Greenhouse whitefly Potato leafhopper Sweetpotato whitefly
Leaves spotted, speckled, or stippled	Aster leafhopper Potato leafhopper Twospotted spider mite
Leaves curled, distorted, cupped, or puckered; leaves harder and darker green than normal	Broad mite Cyclamen mite
Portions of or entire leaves missing	Caterpillars (beet armyworm, cabbage looper, diamondback moth, imported cabbageworm, pansyworm/variegated fritillary, saltmarsh caterpillar, tobacco budworm, yellow-striped armyworm)
Leaf undersides silvery with brown, necrotic areas; leaves curl upward and have thickened new growth	Chilli thrips
Leaves silvery or bleached; leaves distorted	Greenhouse thrips, western flower thrips
Leaves distorted and stippled and have small holes; develop tan to red-brown, rounded, depressed spots; later develop large, brown spots or blotches	Fourlined plant bug

[a] Courtesy R. A. Cloyd—© APS.

Pesticide Applications

Pesticides (insecticides and miticides) may be used to suppress arthropod pest populations on annual bedding plants grown in greenhouses and outdoor landscapes (Table 23). The types of pesticides that may be used in greenhouses and landscapes are as follows:

1. Contact pesticides kill insect and mite pests by direct contact or when pests walk or crawl over a treated surface. After an insect or mite pest walks across a treated surface, the pesticide residue enters the body and moves to sites of action.
2. Stomach poison pesticides act through the insect pest feeding on a treated surface, such as a plant leaf, and ingesting the pesticide residue, which is then absorbed through the pest's stomach lining. Insects stop feeding within 24–48 h, and they usually die within 2–4 days.
3. Translaminar pesticides penetrate the leaf tissue and form a reservoir of active ingredient within the leaf that provides residual activity against foliar-feeding insects and mites.
4. Systemic pesticides, when applied to the growing medium either as a drench or as granules, involve the active ingredient being taken up by the plant's root system and then translocated or distributed throughout the plant. However, the plant must have a well-established root system and be actively growing to take up the active ingredient. Systemic pesticides can also be applied directly to plants' leaves as a spray application. Systemic pesticides are used primarily to prevent infestations of phloem-feeding insects, such as aphids, leafhoppers, mealybugs, and whiteflies.

Growers should irrigate plants prior to applying any pesticide to maintain turgidity and thus avoid plant injury (phytotoxicity).

In addition, growers should read the label directions before using any pesticide and wear the appropriate personal protective equipment during application (Fig. 141). Effective suppression is best achieved when plants are small, which makes it easier to obtain sufficient coverage throughout the entire plant canopy, including the upper and lower leaf surfaces. However, obtaining sufficient suppression of arthropod pest populations may be problematic for several reasons, including poor coverage, improper timing, pH of the spray solution, and frequency of application.

Poor coverage, especially when using a short-residual contact pesticide, will often result in continued problems with arthropod pests. These products are most effective when the wet spray contacts the targeted arthropod pest. Improper timing of pesticide applications also generally results in insufficient suppression of arthropod pest populations. Effective suppression will be obtained when the most susceptible life stage of the targeted arthropod pest is present. In general, the egg and pupal stages are tolerant of most insecticides and miticides, whereas the larva, nymph, and adult stages are more susceptible. If the age structure of the population of the targeted arthropod pest is primarily in the less-susceptible pupal stage and an insecticide application is performed, the result will more than likely be inadequate suppression.

The pH of the spray solution can significantly influence the effectiveness of a pesticide. For a spray solution with a pH greater than 7, there is generally a breakdown of the active ingredient into molecules that have no insecticidal or miticidal properties; this breakdown is referred to as "alkaline hydrolysis." Many pesticides are susceptible to breakdown if the pH of the spray solution is not within an acceptable range. The acceptable range for most pesticides is a pH of 5.5–6.8. However, growers should examine the product label to determine the appropriate pH for a specific pesticide. The water pH can be adjusted, although the process must be performed carefully. The use of pH paper or litmus paper is not a particularly accurate means of monitoring pH (although using pH paper may be valid, because a water pH of 6–7 is generally acceptable for most insecticides and miticides). Acetic acid (vinegar) can be added to the spray solution in small increments to lower the pH; however, the grower must routinely check the pH to avoid adding too much vinegar and to maintain the spray solution pH at around 6.5. The pH can be increased by adding household ammonia. Growers should adjust the water pH prior to adding any pesticide to the spray container.

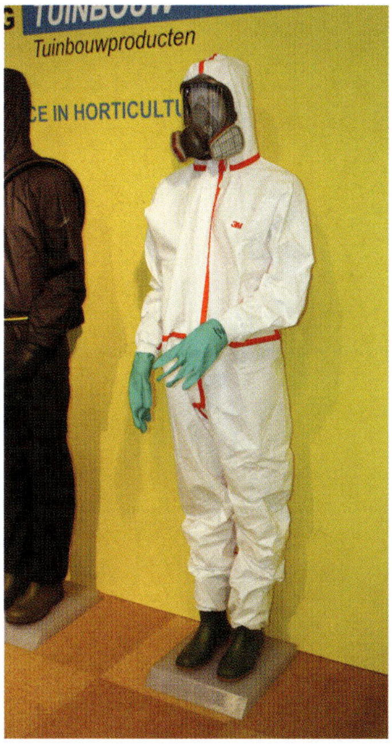

Fig. 141. Personal protective equipment appropriate for pesticide application. (Courtesy R. A. Cloyd—© APS. Reproduced, by permission, from Linderman, R. G., and Benson, D. M., eds. 2014. Compendium of Rhododendron and Azalea Diseases and Pests, 2nd ed. American Phytopathological Society, St. Paul, MN.)

The water pH is not constant and can be influenced by seasonal variability, resulting in changes during the growing season. Therefore, growers must regularly measure or monitor the water pH and make any needed adjustments using buffers or water-conditioning agents: compounds that reduce the potential for alkaline hydrolysis and subsequently modify the spray solution pH so it can be maintained within a range of 5–7. Buffers and water-conditioning agents are safer to use in lowering the spray solution pH than compounds such as sulfuric acid. The buffer should be added to the spray tank before the pesticide, because certain pesticides may begin to degrade after coming into contact with an alkaline solution. Growers should follow these guidelines to avoid or prevent the water pH from reducing the effectiveness of a pesticide application:

- Read the manufacturer's label directions to determine the appropriate water pH.
- Regularly test the water pH, because it can change during the growing season. Water samples should be collected in clean, nonreactive containers (e.g., glass jars), and the samples should be representative of what would be used in the spray solution. The water should be allowed to flow long enough to flush out any that remains in the spray hose. The pH should be determined immediately after collection using pH paper.
- Apply the spray solution as soon as possible after mixing. A pesticide spray solution (or mixture) should always be used within 6 h to avoid any potential problems associated with pH.
- Adjust the water pH with buffers or water-conditioning agents, which are compounds designed to sustain buffering capacity and suppress the process of alkaline hydrolysis. These compounds modify the spray solution pH and maintain the pH within a desired range. Both compounds are much easier and safer to use than sulfuric acid in lowering spray solution pH. Acetic acid (vinegar) can also be used to acidify the water.

The length of the spray interval and the frequency of application may also influence the effectiveness of pesticide applications. If the interval between spray applications is too long (e.g., 10–14 days), the result may be insufficient suppression of the arthropod pest population, especially when overlapping generations are present. Arthropod pest populations will be at various life stages (egg, larva or nymph, pupa, and adult) simultaneously. Spray intervals must be shortened to kill the most susceptible life stages (larva, nymph, and adult) that were previously in the egg or pupal stage.

Biological Control Agents

The use of biological control agents is another option for managing arthropod pest populations but may not always be feasible. Annual bedding plants are typically sold or moved rapidly (within 4–6 weeks), which may not provide natural enemies (parasitoids, predators, and entomopathogenic nematodes) enough time to establish and provide adequate regulation. However, hanging baskets and larger combination and specialty annual planters may be in production for much longer than 4–6 weeks, which will allow sufficient time for natural enemies to become established. Natural enemies for most of the arthropod pests encountered during annual bedding plant

TABLE 23. Modes of Action, Common Names (Active Ingredients), and Activity Types of Pesticides Used Against Insect, Mite, and Mollusk Pests of Annual Bedding Plants[a,b]

Mode of Action: Common Name[c]	Activity Type[d]	Target Pest						
		Aphids	Beetles	Caterpillars	Fungus Gnats	Leafhoppers	Leafminers	Mealybugs
Acetylcholine esterase inhibitor:								
Acephate (1B)	C, S, T	X	X			X	X	X
Carbaryl (1A)	C	X	X	X			X	X
Chlorpyrifos (1B)	C	X	X	X	X	X	X	X
Methiocarb (1A)	C	X						
Prolong opening of sodium channels:								
Bifenthrin (3A)	C	X	X	X	X	X		X
Cyfluthrin (3A)	C	X		X	X	X		X
Fenpropathrin (3A)	C	X	X	X	X		X	X
Fluvalinate (3A)	C	X	X	X	X	X		
Lambda-cyhalothrin (3A)	C	X	X	X		X	X	X
Permethrin (3A)	C	X	X	X	X		X	
Pyrethrins (3A)	C	X	X	X	X	X		X
Nicotinic acetylcholine receptor modulator:								
Acetamiprid (4A)	C, S, T	X				X	X	X
Clothianidin (4A)	C, S, T	X	X				X	X
Dinotefuran (4A)	C, S, T	X	X		X	X	X	X
Flupyradifurone (4D)	C, S, T	X				X		X
Imidacloprid (4A)	C, S, T	X	X		X	X	X	X
Thiamethoxam (4A)	C, S, T	X	X		X	X	X	X
Nicotinic acetylcholine receptor disruptor/agonist and GABA chloride channel activator:								
Spinosad (5)	C, ST, T		X	X			X	
GABA chloride channel activator:								
Abamectin (6)	C, T	X						
Juvenile hormone inhibitor:								
Fenoxycarb (7B)	C							
Kinoprene (7A)	C				X			
Pyriproxyfen (7C)	C, T				X			X
Chitin synthesis inhibitor:								
Buprofezin (16)	C					X		X
Cyromazine (17)	C				X			
Diflubenzuron (15)	C			X	X			
Novaluron (15)	C, T			X			X	
Growth and embryogenesis inhibitor:								
Clofentezine (10A)	C							
Etoxazole (10B)	C, T							
Hexythiazox (10A)	C							
Chordotonal organ TRPV channel modulator:								
Pymetrozine (9B)	C, S, T	X						
Pyrifluquinazon (9B)	C, ST, T	X				X		X
Insect midgut membrane disruptor:								
Bacillus thuringiensis subsp. *israelensis* (11A)	ST				X			
Bacillus thuringiensis subsp. *kurstaki* (11A)	ST			X				
Oxidative phosphorylation uncoupler:								
Chlorfenapyr (13)	C, T			X	X			
Mitochondria electron transport inhibitor:								
Acequinocyl (20B)	C							
Bifenazate (20D)	C							
Cyflumetofen (25)	C							
Fenazaquin (21A)	C							
Fenpyroximate (21A)	C							X
Pyridaben (21A)	C							
Tolfenpyrad (21A)	C	X		X		X		
Desiccation or membrane disruptor:								
Clarified hydrophobic extract of neem oil	C	X						X
Mineral oil	C	X			X		X	X
Paraffinic oil	C	X					X	X
Petroleum oil	C	X	X				X	X
Potassium salts of fatty acids	C	X	X		X	X	X	X
Lipid biosynthesis inhibitor:								
Spiromesifen (23)	C, T							
Spirotetramat (23)	C, S, T	X				X		X

[a] Courtesy R. A. Cloyd—© APS.

[b] Some of the pesticides listed are only for greenhouse use during the production of bedding plants. Growers should be sure to read the label of each product to determine its proper use.

[c] Insecticide Resistance Action Committee (IRAC) mode of action designations are in parentheses.

[d] Activity type codes: C = contact; S = systemic; ST = stomach poison; T = translaminar.

[e] Additional materials for slugs and snails include iron phosphate and metaldehyde.

Mode of Action: Common Name[c]	Activity Type[d]	Target Pest					
		Mites	Plant Bugs	Shore Flies	Slugs and Snails[e]	Thrips	Whiteflies
Acetylcholine esterase inhibitor:							
Acephate (1B)	C, S, T		X			X	X
Carbaryl (1A)	C						
Chlorpyrifos (1B)	C			X		X	
Methiocarb (1A)	C	X			X	X	
Prolong opening of sodium channels:							
Bifenthrin (3A)	C	X	X			X	X
Cyfluthrin (3A)	C					X	X
Fenpropathrin (3A)	C	X				X	X
Fluvalinate (3A)	C	X	X				X
Lambda-cyhalothrin (3A)	C	X	X			X	X
Permethrin (3A)	C					X	X
Pyrethrins (3A)	C		X			X	X
Nicotinic acetylcholine receptor modulator:							
Acetamiprid (4A)	C, S, T		X			X	X
Clothianidin (4A)	C, S, T					X	X
Dinotefuran (4A)	C, S, T					X	X
Flupyradifurone (4D)	C, S, T		X				X
Imidacloprid (4A)	C, S, T					X	X
Thiamethoxam (4A)	C, S, T					X	X
Nicotinic acetylcholine receptor disruptor/agonist and GABA chloride channel activator:							
Spinosad (5)	C, ST, T					X	
GABA chloride channel activator:							
Abamectin (6)	C, T	X				X	X
Juvenile hormone inhibitor:							
Fenoxycarb (7B)	C						
Kinoprene (7A)	C					X	X
Pyriproxyfen (7C)	C, T			X		X	X
Chitin synthesis inhibitor:							
Buprofezin (16)	C						X
Cyromazine (17)	C			X			
Diflubenzuron (15)	C			X			X
Novaluron (15)	C, T	X					
Growth and embryogenesis inhibitor:						X	X
Clofentezine (10A)	C						
Etoxazole (10B)	C, T	X					
Hexythiazox (10A)	C	X					
Chordotonal organ TRPV channel modulator:							
Pymetrozine (9B)	C, S, T						X
Pyrifluquinazon (9B)	C, ST, T					X	X
Insect midgut membrane disruptor:							
Bacillus thuringiensis subsp. *israelensis* (11A)	ST						
Bacillus thuringiensis subsp. *kurstaki* (11A)	ST						
Oxidative phosphorylation uncoupler:							
Chlorfenapyr (13)	C, T	X				X	
Mitochondria electron transport inhibitor:							
Acequinocyl (20B)	C	X					
Bifenazate (20D)	C	X					
Cyflumetofen (25)	C	X					
Fenazaquin (21A)	C	X					X
Fenpyroximate (21A)	C	X					
Pyridaben (21A)	C	X					X
Tolfenpyrad (21A)	C					X	X
Desiccation or membrane disruptor:							
Clarified hydrophobic extract of neem oil	C	X				X	X
Mineral oil	C	X	X			X	X
Paraffinic oil	C	X				X	X
Petroleum oil	C	X				X	X
Potassium salts of fatty acids	C	X				X	X
Lipid biosynthesis inhibitor:							
Spiromesifen (23)	C, T	X					X
Spirotetramat (23)	C, S, T	X				X	X

(Continued on next page)

TABLE 23. *(Continued from previous page)*

Mode of Action: Common Name[c]	Activity Type[d]	Target Pest						
		Aphids	Beetles	Caterpillars	Fungus Gnats	Leafhoppers	Leafminers	Mealybugs
Ryanodine receptor modulator:								
Chlorantraniliprole (28)	C, S		X					
Cyantraniliprole (28)	C, S, T	X	X	X				
Ecdysone agonist:								
Methoxyfenozide (18)	C			X				
Chordotonal organ modulator:								
Flonicamid (29)	C, S, T	X				X		X
Nicotinic acetylcholine receptor disruptor/agonist, GABA chloride channel activator, and nicotinic acetylcholine receptor modulator:								
Spinetoram (5) + Sulfoxaflor (4C)	C, S, T	X	X	X				X
Unknown or uncertain:								
Azadirachtin	C, ST	X		X	X	X	X	X
Pyridalyl	C, ST, T			X				
Entomopathogenic fungi:								
Beauveria bassiana	C	X				X		X
Isaria fumosorosea	C	X	X		X	X	X	X
Metarhizium anisopliae	C							

production are commercially available from distributors and suppliers (Table 24).

A long-term strategy when using natural enemies, particularly against aphids, is to utilize so-called banker plants. These are plants such as barley (*Hordeum vulgare*), cereal rye (*Secale cereale*), and wheat (*Triticum* spp.) that are hosts to alternative prey, including the corn leaf aphid (*Rhopalosiphum maidis*) and bird cherry-oat aphid (*R. padi*) for certain aphid parasitoids (Fig. 142). These aphid species do not feed on annual bedding plants. The parasitoids may move from the banker plants to attack aphids on annual bedding plants and then disperse back onto the banker plants. This system can reduce the number of parasitoid releases required to regulate aphid populations during the production of annual bedding plants. The use of ornamental pepper (*Capsicum annuum*) cultivars such as Purple Flash, Black Pearl, and Explosive Ember as banker plants may be an option when establishing populations of predatory bugs (e.g., *Orius* spp.) for regulation of thrips populations on longer-term bedding plants.

Another option early in annual bedding plant production is to apply a microbial pesticide (fungus or bacterium) containing *Beauveria bassiana, Isaria fumosorosea, Metarhizium anisopliae,* or *Bacillus thuringiensis* subsp. *israelensis* as the active ingredient. *B. bassiana* can be effective on aphids, whiteflies, and thrips; *I. fumosorosea* can have activity on thrips, aphids, and whiteflies; and *M. anisopliae* can be effective against thrips, whiteflies, and mites. *B. thuringiensis* subsp. *israelensis* is active only on fungus gnat larvae. To maximize effectiveness, the grower should apply a microbial pesticide early in the production cycle, because an entomopathogenic fungus or bacterium will not provide quick knockdown or suppression of a large arthropod pest population.

Fig. 142. Banker plants (cereal rye) with adult parasitoids, used for management of aphids. (Courtesy R. A. Cloyd—© APS)

Selected References

Baker, J. R. 1994. Insect and Related Pests of Flowers and Foliage Plants. N.C. Coop. Ext. Serv. Publ. AG-136. North Carolina State University, Raleigh.

Berg Stack, L., Cloyd, R., Dill, J., McAvoy, R., Pundt, L., Raudales, R., Smith, C., and Smith, T. 2015–2016. New England Greenhouse Floriculture Guide: A Management Guide for Insects, Diseases, Weeds, and Growth Regulators. New England Floriculture and New England State Universities.

Cloyd, R. A. 2007. Plant Protection: Managing Greenhouse Insect and Mite Pests. Ball Publishing, Chicago.

Cloyd, R. A. 2011. Managing insect and mite pests. Pages 107-119 in: Ball RedBook, 18th ed. Vol. 2: Crop Production. J. Nau, ed. Ball Publishing, Chicago.

Cloyd, R. A. 2012. Insect and mite management in greenhouses. Pages 391-441 in: Greenhouse Operation and Management, 7th ed. P. V. Nelson, ed. Prentice-Hall, Saddle River, NJ.

Cloyd, R. A. 2015. Effect of Water and Spray Solution pH on Pesticide Activity. Kans. State Univ. Agric. Exp. Stn. Coop. Ext. Serv. Publ. MF-3272. Kansas State University, Manhattan.

Dreistadt, S. H. 2001. Integrated Pest Management for Floriculture and Nurseries. Univ. Calif. State. Integr. Pest Manage. Proj. Publ. 3402. Division of Agriculture and Natural Resources, University of California, Oakland.

Flint, M. L., and Dreistadt, S. H. 1998. Natural Enemies Handbook: The Illustrated Guide to Biological Pest Control. Univ. Calif. State. Integr. Pest Manage. Proj. Publ. 3386. Division of Agriculture and Natural Resources, University of California, Oakland.

Gill, S., and Sanderson, J. 1998. Ball Identification Guide to Greenhouse Pests and Beneficials. Ball Publishing, Chicago.

Mahr, S. E. R., Cloyd, R., Mahr, D. L., and Sadof, C. S. 2001. Biological Control of Insects and Other Pests of Greenhouse Crops. North Centr. Reg. Publ. 581. University of Wisconsin, Madison.

Powell, C. C., and Lindquist, R. K. 1997. Ball Pest and Disease Manual, 2nd ed. Ball Publishing, Chicago.

(Prepared by R. A. Cloyd)

Mode of Action: Common Name[c]	Activity Type[d]	Target Pest					
		Mites	Plant Bugs	Shore Flies	Slugs and Snails[e]	Thrips	Whiteflies
Ryanodine receptor modulator:							
Chlorantraniliprole (28)	C, S						
Cyantraniliprole (28)	C, S, T					X	X
Ecdysone agonist:							
Methoxyfenozide (18)	C						
Chordotonal organ modulator:							
Flonicamid (29)	C, S, T		X			X	X
Nicotinic acetylcholine receptor disruptor/agonist, GABA chloride channel activator, and nicotinic acetylcholine receptor modulator:							
Spinetoram (5) + Sulfoxaflor (4C)	C, S, T			X		X	X
Unknown or uncertain:							
Azadirachtin	C, ST			X		X	X
Pyridalyl	C, ST, T					X	
Entomopathogenic fungi:							
Beauveria bassiana	C		X			X	X
Isaria fumosorosea	C	X	X			X	X
Metarhizium anisopliae	C	X				X	X

Arthropod Pests of Bedding Plants

American Serpentine Leafminer

The American serpentine leafminer (*Liriomyza trifolii*) is the most important leafminer species of annual bedding plants, including basil, chrysanthemum, dahlia, gerber daisy, marigold, nasturtium, pansy, petunia, Sweet William, and zinnia.

Life Cycle

An American serpentine leafminer adult is 1.0–3.5 mm long and dark gray in color, with yellow bands or markings on the abdomen. A leafminer adult may resemble a common housefly in shape but not in size (Fig. 143). The female creates small feeding punctures in leaves with her egg-laying device (ovipositor). Both the male and female feed on the fluids that exude from these wounds. The female deposits eggs into leaf tissue and is capable of laying more than 100 eggs. The eggs hatch in 4–5 days, producing larvae.

The larva is white to yellow and has a dark head and mouth hook. Three larval stages occur within the leaf. The larva is approximately 2 mm long when full grown. Eventually, larvae cut holes in leaves and fall to the soil or growing medium to pupate. Adults emerge in about 10 days and live 3–4 weeks. The life cycle from egg to adult is completed in 15–24 days; however, development is dependent on temperature and host plant.

Damage

Damage from American serpentine leafminer larval feeding is primarily aesthetic, as plants are rarely killed from an infestation. The larvae create meandering or serpentine mines just beneath the epidermal layer of the leaf (Fig. 144). The mines typically occur near leaf margins.

Any damage may reduce the salability of annual bedding plants, although variability may occur depending on the extent of the damage. Adult egg-laying punctures and feeding, as well as larval tunneling, can reduce plant salability. Certain cultivars of annual bedding plants may be more susceptible than others to feeding damage.

Management

Problems with the American serpentine leafminer can be minimized by weed removal and implementation of proper cultural practices. For instance, it is important to avoid overfertilizing annual bedding plants, especially with nitrogen-based, water-soluble fertilizers, which may cause succulent growth and thus increase plants' attractiveness to adult females. Removing weed and plant debris, both inside greenhouses and in outdoor landscapes, will eliminate alternative hosts.

Fig. 143. American serpentine leafminer (*Liriomyza trifolii*) adults. (Courtesy R. A. Cloyd—© APS)

Fig. 144. Serpentine mine in a dahlia leaf, caused by an American serpentine leafminer (*Liriomyza trifolii*) larva. (Courtesy R. A. Cloyd—© APS)

TABLE 24. Commercially Available Biological Control Agents for Various Insect and Mite Pests of Annual Bedding Plants[a]

Pest	Biological Control Agent [Type]
Aphids	*Adalia bipunctata* [predator] *Aphelinus abdominalis* [parasitoid] *Aphidius colemani* [parasitoid] *Aphidius ervi* [parasitoid] *Aphidius matricariae* [parasitoid] *Aphidoletes aphidimyza* [predator] *Chrysoperla carnea* [predator]
Beetles (larvae)	*Heterorhabditis bacteriophora* [beneficial nematode] *Heterorhabditis megidis* [beneficial nematode] *Steinernema kraussei* [beneficial nematode]
Broad mite and cyclamen mite	*Amblyseius swirskii* [predator] *Neoseiulus californicus* [predator] *Neoseiulus* (syn. *Amblyseius*) *cucumeris* [predator]
Chilli thrips	*Amblyseius swirskii* [predator] *Neoseiulus* (syn. *Amblyseius*) *cucumeris* [predator] *Orius insidiosus* [predator]
Fungus gnats (larvae)	*Dalotia coriaria* [predator] *Gaeolaelaps aculeifer* [predator] *Steinernema feltiae* [beneficial nematode] *Stratiolaelaps scimitus* (syn. *Hypoaspis miles*) [predator]
Greenhouse whitefly	*Encarsia formosa* [parasitoid]
Leafminers	*Diglyphus isaea* [parasitoid]
Mealybugs	*Anagyrus pseudococci* [parasitoid] *Cryptolaemus montrouzieri* [predator]
Shore flies	*Dalotia coriaria* [predator] *Gaeolaelaps aculeifer* [predator] *Steinernema carpocapsae* [beneficial nematode] *Steinernema feltiae* [beneficial nematode] *Stratiolaelaps scimitus* (syn. *Hypoaspis miles*) [predator]
Sweetpotato whitefly	*Amblyseius swirskii* [predator] *Delphastus catalinae* [predator] *Eretmocerus eremicus* [parasitoid]
Twospotted spider mite	*Amblyseius andersoni* [predator] *Amblyseius californicus* [predator] *Amblyseius fallacis* [predator] *Feltiella acarisuga* [predator] *Galendromus occidentalis* [predator] *Phytoseiulus persimilis* [predator] *Stethorus punctillum* [predator]
Western flower thrips	*Amblyseius swirskii* [predator] *Dalotia coriaria* [predator] *Neoseiulus* (syn. *Amblyseius*) *cucumeris* [predator] *Orius insidiosus* [predator] *Steinernema feltiae* [beneficial nematode] *Stratiolaelaps scimitus* (syn. *Hypoaspis miles*) [predator]

[a] Courtesy R. A. Cloyd—© APS.

Insecticides may be used to suppress American serpentine leafminer populations; however, the larvae are well protected within the leaf tissue, allowing them to escape exposure from contact insecticides. Insecticides with translaminar activity, in which the material penetrates the leaf tissue and forms a reservoir of active ingredient within the leaf, are more likely to be effective in killing leafminer larvae. Drench applications of systemic insecticides may also be effective when applied during the early stages of larval feeding. To avoid selecting for resistance in American serpentine leafminer populations, the grower should always rotate insecticides with different modes of action.

Commercially available natural enemies include the parasitoid *Diglyphus isaea*. It should be released preventively to regulate low leafminer populations on longer-term bedding plants.

Selected References

Baker, J. R. 1994. Insect and Related Pests of Flowers and Foliage Plants. N.C. Coop. Ext. Serv. Publ. AG-136. North Carolina State University, Raleigh.

Cloyd, R. A. 2007. Plant Protection: Managing Greenhouse Insect and Mite Pests. Ball Publishing, Chicago.

Powell, C. C., and Lindquist, R. K. 1997. Ball Pest and Disease Manual, 2nd ed. Ball Publishing, Chicago.

(Prepared by R. A. Cloyd)

Aphids

A number of aphid species—such as the green peach aphid (*Myzus persicae*), the cotton or melon aphid (*Aphis gossypii*), and the foxglove aphid (*Aulacorthum solani*)—feed on a wide range of annual bedding plants, including the following: ageratum, calibrachoa, celosia, chrysanthemum, dahlia, geranium, gerber daisy, marigold, nasturtium, pansy, salvia, snapdragon, sweetpotato vine, verbena, and zinnia.

Green Peach Aphid: This aphid has long cornicles, which are tubes that project from the back of the abdomen (Fig. 145). The cornicles are approximately the length of the body, and the tips are black. A winged adult is 1.8–2.1 mm long and has a black head and thorax; the abdomen is yellow-green and has a dark spot at the top. A wingless adult is 1.7–2.0 mm long; it is typically pale green and has a black head. A nymph is initially green but eventually turns yellow. The head of the green peach aphid has a distinct indentation at the base of the antennae.

Cotton/Melon Aphid: This aphid has shorter cornicles than the green peach aphid, and the antennae are usually shorter than the body. Unlike the green peach aphid, the cotton/melon aphid does not have an indentation at the base of the antennae. The cotton/melon aphid can vary in color from pale yellow to dark green (appearing almost black). In addition, this aphid is typically covered with wax secretions, which give the body a dull sheen. A winged adult female is 1.2 mm long; it has a black head and thorax and light-green mottling on the body. A winged adult male has a dark head and thorax, and the abdomen is yellow-green. A wingless adult is 1–2 mm long and yellow to dark green with light-green mottling and black cornicles. A nymph is 0.5–1.0 mm long and similar to the adults in color.

Fig. 145. Green peach aphid (*Myzus persicae*) adult and nymphs. (Note the long cornicles on the end of the abdomen.) (Courtesy R. A. Cloyd—© APS)

Foxglove Aphid: This aphid has dark-green spots or darkened areas at the base of the cornicles and black markings on the leg joints and antennae. An adult winged female is 1.8–3.0 mm long, pale green to dark yellow to orange, and shiny in comparison with other aphids. A wingless aphid may have black markings at the tip of the abdomen.

Additional aphid species that feed on annual bedding plants in greenhouses or landscapes include the cabbage aphid (*Brevicoryne brassicae*), the chrysanthemum aphid (*Macrosiphoniella sanborni*), and the potato aphid (*Macrosiphum euphorbiae*).

Life Cycle

Aphids, in general, are soft-bodied insects that have cornicles on the ends of their hind ends or abdomens (Fig. 146). Aphids are 1.0–2.5 mm long and vary in color from green to black to yellow to pink. Because the aphid's color will vary depending on the host plant on which it fed, color should never be used to identify an aphid to species. In the greenhouse, all aphids are females (i.e., no egg stage), and each can give birth to 100 live female nymphs, which can themselves give birth to live offspring. In outdoor landscapes, aphid females produce eggs that hatch into both females and males.

Aphids' rapid reproductive ability may result in the development of an extensive outbreak within a short time, which is why aphids can spread quickly among annual bedding plants. Aphids will spread more rapidly if plants are spaced closely together, because aphids (especially winged individuals) can easily move among plants. Certain annual bedding plants (both species and cultivars) are more susceptible to aphids than others.

Damage

Aphids feed on new terminal growth and the undersides of leaves (Fig. 147), and they cause both direct and indirect damage to annual bedding plants. Direct damage is associated with aphids removing plant fluids with their piercing–sucking mouthparts. Aphids feeding on new growth may cause leaf yellowing and plant stunting, and leaves can appear distorted or curled (upward or downward). Furthermore, many aphid species act as vectors of certain viruses associated with annual bedding plants.

Indirect damage from aphid feeding involves the production of honeydew, which is a clear, sticky liquid that serves as a growing medium for black sooty mold (Fig. 148). The presence of black sooty mold may inhibit a plant's ability to produce food by means of photosynthesis. Honeydew may also attract ants that protect aphids from natural enemies and move them among plants. An excessive aphid population may result in the presence of white cast (i.e., molted) skins (Fig. 149), which can negatively affect the aesthetic quality of plants. Furthermore, these white cast exoskeletons may be mistakenly identified by growers as whiteflies.

Aphids are typically the first arthropod pests noticed in the greenhouse during the initial stages of annual bedding plant production.

Management

Growers should avoid overfertilizing annual bedding plants, especially with nitrogen-based, water-soluble fertilizers, because doing so leads to the production of soft, succulent growth,

Fig. 146. Oleander aphid (*Aphis nerii*), displaying cornicles on the end of the abdomen. (Courtesy R. A. Cloyd—© APS)

Fig. 148. Black, mycelial growth of sooty mold on a leaf, resulting from aphid honeydew production. (Courtesy R. A. Cloyd—© APS)

Fig. 147. Oleander aphids (*Aphis nerii*) feeding on terminal growth. (Courtesy R. A. Cloyd—© APS)

Fig. 149. White cast (molted) skins of aphids on a rose leaf. (Courtesy R. A. Cloyd—© APS)

which is easier for aphids to penetrate with their mouthparts. Growers should also remove all weeds from inside and around the greenhouse perimeter, because many weeds serve as reservoirs for aphids.

Contact insecticides may be used early in production to prevent aphid populations from building to damaging levels. However, thorough coverage of all plant parts is essential, and repeat applications may be required. Systemic insecticides may also be effective against aphids, especially if applied early in the crop production cycle and before aphid populations reach excessive levels. Systemic insecticides can be applied as a drench or in granule form to the growing medium or as a spray to plant leaves.

A number of natural enemies of aphids are commercially available, including the parasitoids *Aphidius colemani* and *Aphidius ervi;* when released in a timely manner, they may be effective in regulating aphid populations. The parasitoid *Aphelinus abdominalis* can be effective when a larger aphid species (e.g., the foxglove aphid) is present and when temperatures are high. The predatory midge *Aphidoletes aphidimyza* should be released near a localized aphid population and can be used in combination with certain parasitoids. The green lacewing (*Chrysoperla carnea*) can be released among a localized aphid infestation, but it may be less effective on plants with trichomes (hairs). The effectiveness of the ladybird beetles *Adalia bipunctata* and *Hippodamia convergens* may be limited; both prefer high aphid populations, and the adults may leave the greenhouse when they cannot find an aphid colony. Naturally occurring or resident populations of predators and parasitoids frequently regulate aphid populations outdoors. Some of these natural enemies include ladybird beetle larvae and adults, green lacewing larvae, hover fly larvae, midge larvae, minute pirate bug nymphs and adults, and an assortment of parasitoids.

The use of so-called banker plants may also be worth consideration in managing aphid populations. (See earlier in Part III, the section "Biological Control Agents.")

Selected References

Berg Stack, L., Cloyd, R., Dill, J., McAvoy, R., Pundt, L., Raudales, R., Smith, C., and Smith, T. 2015–2016. New England Greenhouse Floriculture Guide: A Management Guide for Insects, Diseases, Weeds, and Growth Regulators. New England Floriculture and New England State Universities.

Capinera, J. L. 2001. Handbook of Vegetable Pests. Academic Press, San Diego.

Cloyd, R. A. 2012. Insect and mite management in greenhouses. Pages 391-441 in: Greenhouse Operation and Management, 7th ed. P. V. Nelson, ed. Prentice-Hall, Saddle River, NJ.

Gill, S., Cloyd, R. A., Baker, J. R., Clement, D. L., and Dutky, E. 2006. Pests and Diseases of Herbaceous Perennials, 2nd ed. Ball Publishing, Chicago.

van Emden, H. F., and Harrington, R., eds. 2007. Aphids as Crop Pests. CABI, Wallingford, Oxfordshire, UK.

(Prepared by R. A. Cloyd)

Aster Leafhopper

The aster leafhopper (*Macrosteles quadrilineatus*) feeds on a wide range of annual bedding plants. More importantly, however, the aster leafhopper serves as a vector for the aster yellows phytoplasma (AYP), which causes aster yellows disease. Annual bedding plants susceptible to aster yellows include aster, calendula, chrysanthemum, cosmos, dahlia, marigold, petunia, and zinnia.

Life Cycle

An aster leafhopper adult is green-yellow and has six black markings on the head; the markings are a common feature for

Fig. 150. Aster leafhopper (*Macrosteles quadrilineatus*) adult. (Courtesy W. S. Cranshaw—© APS. Reproduced, by permission, from Schwartz, H. F., and Mohan, S. K., eds. 2008. Compendium of Onion and Garlic Diseases and Pests, 2nd ed. American Phytopathological Society, St. Paul, MN.)

identification (Fig. 150). The adult female lays eggs into plant stems. The eggs hatch into nymphs that resemble adults, but they cannot fly because they do not have fully developed wings. There are four nymphal instars, which take 4–6 weeks to complete. Adults can live for several months.

Aster leafhoppers typically migrate north from southern regions from May through June. There can be multiple generations throughout the growing season.

Damage

The aster leafhopper injects a toxin (along with its saliva) into plant tissue during feeding. The direct damage from feeding is minimal—specifically, spotting of plant leaves (referred to as "speckling" or "stippling"). However, the indirect damage associated with transmitting the AYP is the primary damage caused by the aster leafhopper. Young nymphs cannot transmit the disease, but older nymphs and adults can. The pathogen incubates within the body of the aster leafhopper adult for 10–20 days, and the adult can transmit the pathogen during its entire life span.

Plants infected with aster yellows exhibit symptoms such as leaf yellowing, stunted growth, distorted flower growth, shortened internodes, excessive branching, and greening of flower petals.

Management

Contact insecticides can be used to manage aster leafhopper populations. However, for these insecticides to be effective, thorough coverage of all plant parts and repeat applications are required.

Plants that exhibit symptoms of aster yellows should be removed immediately from the greenhouse or the landscape. Removing weeds from within the greenhouse, around the production area, and in the landscape will eliminate alternative hosts for the aster leafhopper and overwintering of the AYP.

Selected References

Cranshaw, W., and Shetlar, D. 2017. Garden Insects of North America, 2nd ed. Princeton University Press, Princeton, NJ.

Westcott, C. 1973. The Gardener's Bug Book, 4th ed. Doubleday, Garden City, NY.

(Prepared by R. A. Cloyd)

Bandedwinged Whitefly

The bandedwinged whitefly (*Trialeurodes abutiloneus*) may be noticed on certain annual bedding plants, including geranium and petunia. However, in most instances, this whitefly species is not considered a major insect pest in greenhouses or landscapes.

Life Cycle

The bandedwinged whitefly female is 1.9–2.8 mm long, and it lays eggs on the undersides of leaves. The eggs hatch into nymphs that are 0.37 mm long; each is translucent white and has yellow spots on both sides of the abdomen. The nymphs move around on plants, looking for feeding sites. The nymphs eventually pupate, and then the adults emerge.

An adult is yellow and has a green tinge on the thorax (midsection of body). The forewings are marked with several zigzagged, smoky-gray bands (Fig. 151). When the wings are folded over the body, the lines appear to be continuous from wing to wing. The hind wings lack these bands. Waxy filaments are found around the edge of the fourth instar nymph (pupa); a dark patch often shows on the center of the body.

The development time from egg to adult takes about 20–28 days, depending on the temperature. There may be several generations per year in greenhouses.

Damage

Both the adult and the nymphs have piercing–sucking mouthparts, which are used to remove fluids from the plant and thus cause plant stunting and wilting and leaf yellowing (chlorosis). The bandedwinged whitefly also produces honeydew, which serves as a growing medium for certain black sooty mold. The presence of black sooty mold can reduce the aesthetic quality of annual bedding plants.

Management

Weed removal will eliminate reservoirs of bandedwinged whitefly populations. Overfertilizing annual bedding plants should be avoided, because the succulent growth that results may enhance plants' susceptibility to females.

Contact insecticides may be effective in suppressing bandedwinged whitefly populations; to be effective, frequent applications must be made and all plant parts must be thoroughly covered. Regular scouting of annual bedding plants using yellow sticky cards will help time insecticide applications appropriately.

The use of the parasitoid *Eretmocerus eremicus* may also be an option in regulating populations of the bandedwinged whitefly.

Selected References

Berg Stack, L., Cloyd, R., Dill, J., McAvoy, R., Pundt, L., Raudales, R., Smith, C., and Smith, T. 2015–2016. New England Greenhouse Floriculture Guide: A Management Guide for Insects, Diseases, Weeds, and Growth Regulators. New England Floriculture and New England State Universities.

Dreistadt, S. H. 2001. Integrated Pest Management for Floriculture and Nurseries. Univ. Calif. State. Integr. Pest Manage. Proj. Publ. 3402. Division of Agriculture and Natural Resources, University of California, Oakland.

Gill, S., and Sanderson, J. 1998. Ball Identification Guide to Greenhouse Pests and Beneficials. Ball Publishing, Chicago.

(Prepared by R. A. Cloyd)

Broad Mite

Damage caused by the broad mite (*Polyphagotarsonemus latus*) may be encountered on a number of annual bedding plants, including begonia, chrysanthemum, dahlia, geranium, gerber daisy, New Guinea impatiens, marigold, snapdragon, verbena, torenia, and zinnia.

Life Cycle

The broad mite is difficult to see with the naked eye, because an adult is only 0.25 mm long—about the size of a pinhead. An adult mite is oval or elongated in shape and varies in color from white to amber to dark green (Fig. 152). A female may lay up to 40 eggs during her life span, depending on temperature and relative humidity.

The eggs are oval and white and covered with protrusions, or bumps. Six-legged larvae emerge from the eggs and later transition into eight-legged nymphs. The nymphs eventually develop into adults. The development time from egg to adult is 5–6 days at temperatures between 21 and 26ºC and 7–10 days at temperatures between 10 and 18ºC.

Broad mites tend to feed in groups, primarily on the undersides of young leaves and in flowers, where females lay eggs. Broad mites do not produce webbing like twospotted spider mites.

Damage

Broad mites feed on young, tender growth and cause leaves to become curled, distorted, cupped, or puckered (Fig. 153). Leaves also appear glossy, shiny, or silvery and are darker green and harder or more brittle than normal. Damage tends to

Fig. 151. Bandedwinged whitefly (*Trialeurodes abutiloneus*) adults. (Courtesy Ronald Smith, Auburn University, from Bugwood.org. Reproduced, by permission, according to terms of Creative Commons Attribution 3.0 License.)

Fig. 152. Broad mite (*Polyphagotarsonemus latus*) adult. (Courtesy K. K. Rane—© APS)

Fig. 153. Cupping of English ivy leaves caused by broad mite (*Polyphagotarsonemus latus*) feeding. (Courtesy R. A. Cloyd—© APS)

be most noticeable when buds and leaves are mature. A large infestation of broad mites may result in plant stunting.

Broad mite damage may be misdiagnosed as a symptom of disease (e.g., a virus) or as an herbicide injury, environmental issue (e.g., temperature), or nutritional problem.

Management

Once plants exhibit symptoms of broad mite damage, few options are available other than to dispose of infested plants. Sanitation practices, including cleaning the greenhouse prior to introducing new plants and disinfecting benches, will help alleviate problems with broad mites.

Broad mites are extremely difficult to manage with miticides, because these pests tend to be located in the meristematic tissues of plants. However, miticides with translaminar activity (i.e., after application, the material penetrates the leaf tissue and new terminal growth, forming a reservoir of active ingredient within the leaf or new growing point) may be more effective than those with contact activity.

Scouting should focus on annual bedding plants that are susceptible to broad mites, and any plants that exhibit damage should be discarded immediately. Plants within 15.2–30.4 cm of symptomatic plants should also be discarded, because even if they do not exhibit symptoms, they may be infested with broad mites. Inspecting incoming plant material and isolating plants for 1 week may help prevent introducing infested plants into the main crop.

The release of predatory mites such as *Amblyseius swirskii, Neoseiulus californicus,* and *N. cucumeris* may help to regulate broad mite populations and prevent them from establishing.

Selected References

Cloyd, R. A. 2010. Broad Mite and Cyclamen Mite: Management in Greenhouses and Nurseries. Kans. State Univ. Agric. Exp. Stn. Coop. Ext. Serv. Publ. MF-2938. Kansas State University, Manhattan.

Gerson, U. 1992. Biology and control of the broad mite, *Polyphagotarsonemus latus* (Banks) (Acari: Tarsonemidae). Exp. Appl. Acarol. 13:163-178.

Westcott, C. 1973. The Gardener's Bug Book, 4th ed. Doubleday, Garden City, NY.

(Prepared by R. A. Cloyd)

Fig. 154. Tobacco budworm (*Heliothis virescens*) larva feeding on a petunia flower. (Courtesy R. A. Cloyd—© APS)

Caterpillars

Caterpillars are the larval, or immature, stages of butterflies and moths. Many different types of caterpillars feed on a wide range of annual bedding plants grown in greenhouses and incorporated into landscapes, including the tobacco budworm (*Heliothis virescens*), the beet armyworm (*Spodoptera exigua*), the diamondback moth (*Plutella xylostella*), the saltmarsh caterpillar (*Estigmene acrea*), the pansyworm or variegated fritillary (*Euptoieta claudia*), the imported cabbageworm (*Pieris rapae;* syn. *Artogeia rapae*), and the yellowstriped armyworm (*Spodoptera ornithogalli*).

Tobacco Budworm: An adult is pale green to light brown, and the forewings are marked with four light, wavy bands. The wingspan is approximately 38 mm. The caterpillar is 38 mm long when full grown and varies in color depending on the host plants upon which it has fed; it may be black, pale brown, yellow, green, and/or red. It may also have stripes that extend the length of the body. In addition, the caterpillar may have small hairs, or setae, on localized sections of the body. Tobacco budworm caterpillars feed on a number of annual bedding plants, including ageratum, chrysanthemum, flowering tobacco, geranium, petunia, snapdragon, and strawflower (Fig. 154).

Beet Armyworm: An adult has forewings that are mottled gray to brown and have a pale spot in the middle of each front margin; the wingspan is 25–30 mm. Caterpillars are pale green to yellow-green in the first two instars, but in the third instar, they develop wavy bands or stripes that extend the length of the body. Caterpillars eventually become variable in appearance and may have a green, pink, yellow, or white body. Moreover, a series of dark spots may be present on the top of the body. Beet armyworm caterpillars usually do not have hairs, or setae, but they have a black spot on the side, just above the second pair of legs. These caterpillars feed on chrysanthemum and geranium.

Diamondback Moth: An adult is 6 mm long and gray-brown in color; it has a 12.7 mm wingspan. When the wings are folded over the body, a distinct three-diamond pattern can be observed along the middle of the wings. The caterpillar is about 12 mm long when full grown and light green in color (Fig. 155). The early larval stages create mines in leaves, and the later larval stages chew entire leaves. Each caterpillar has a pair of prolegs that protrude from the back of the abdomen. When disturbed, this caterpillar wiggles vigorously and descends on a silken thread. Diamondback moth caterpillars feed on ornamental cabbage and kale, sweet alyssum, and stock.

Saltmarsh Caterpillar: An adult has a 35–45 mm wingspan and white forewings that are marked with a number of black spots. It is primarily yellow and has a series of black spots on the upper side of the abdomen. The caterpillar is initially dark brown but later transitions into a variety of colors, including yellow, brown, and black. A young caterpillar is 10 mm long and as it matures develops brown, yellow, and white longitudinal stripes that extend the length of the body. A fully mature caterpillar is 50–60 mm long and usually dark colored (but may be yellow-brown); its body is covered with hairs that are cream, gray, yellow, or brown. Saltmarsh caterpillars feed on chrysanthemum.

Pansyworm/Variegated Fritillary: An adult is orange and has a wingspan of 45–80 mm. The wings are spotted or checkered and have thick, dark veins and markings. There are distinct black spots near the margins of the wings. The caterpillar is orange-red and approximately 31 mm long when full grown. It is covered with black spines arranged in six rows on both the top and side of the body. The caterpillar also has black stripes that extend the length of the body. These caterpillars feed on sweet alyssum, pansy, and violet.

Imported Cabbageworm: An adult has a wingspan of 32–44 mm; the wings are white with three or four dark spots. The female has two dark spots on the forewings, whereas the male has only one. The female lays more than 100 yellow, bullet-shaped eggs on the undersides of leaves. The eggs hatch into caterpillars that are velvety, smooth, and light green, with pale-yellow stripes extending the length of the body. The body is covered with short, thin hairs. Caterpillars are 25 mm in length when full grown (Fig. 156). Adults emerge from pupae within 1 week. There can be two or more generations per year, depending on the geographic location, with potentially up to eight generations occurring in the southern United States. Caterpillars feed primarily on ornamental cabbage and kale.

Yellowstriped Armyworm: An adult has a wingspan of 34–41 mm; the forewings are brown-gray and display a complex pattern of light and dark markings. White, diagonal bands occur near the center of the wings, and there may be noticeable white colorations near the wing margins. Caterpillars vary in color from black to brown to yellow; mature caterpillars have a brown band that extends the length of the body and a light, white line in the center. There are distinct black, triangular markings along both sides of the body, and a yellow or white line extends the length of the body (Fig. 157).

Additional caterpillars that may feed on annual bedding plants grown in greenhouses or landscapes are the cabbage looper (*Trichoplusia ni*), European corn borer (*Ostrinia nubilalis*), and European pepper moth (*Duponchelia fovealis*).

Fig. 155. Diamondback moth (*Plutella xylostella*) larva on a pac choi leaf (*Brassica rapa* subsp. *chinensis*). (Note the black fecal deposits.) (Courtesy R. A. Cloyd—© APS)

Life Cycle

The life cycle consists of an egg, caterpillar (or larva), pupa, and adult. Adult female moths are generally active at night, but they may be present during the day. Adults are highly attracted to outdoor lighting.

The female lays eggs on the leaf undersides of many annual bedding plants. The number of eggs laid varies depending on the species, but in general, a female can lay between 20 and 100 eggs during her lifetime. There may be one to several generations per year, depending on geographic location.

Damage

Caterpillars hatch from eggs and cause damage by consuming plant leaves, stems, and flowers. Depending on the species, a caterpillar may consume the entire leaf or leave the midvein (skeletonization). Black to brown fecal deposits, or frass, may be present on plant leaves and stems and are an indication of caterpillar feeding.

Management

Weed removal is always important in alleviating problems with caterpillars, because adults may use certain weeds as alternative hosts for laying eggs.

Fig. 156. Imported cabbageworm (*Pieris rapae*) larva. (Courtesy R. A. Cloyd—© APS)

Fig. 157. Yellowstriped armyworm (*Spodoptera ornithogalli*) larva. (Courtesy R. A. Cloyd—© APS)

Contact or stomach poison insecticides (e.g., *Bacillus thuringiensis* subsp. *kurstaki*) may be effective if applied early, when caterpillars are small and not feeding extensively. Horticultural oils, whether petroleum or mineral based, may kill the eggs. However, thorough coverage of all plant parts and repeat applications are required to ensure effective suppression of caterpillar populations. Hand-picking caterpillars and placing them into containers of soapy water may be effective when handling a minor infestation.

Certain species of *Trichogramma* parasitoids may be purchased and released against specific caterpillar pests.

Selected References

Capinera, J. L. 2001. Handbook of Vegetable Pests. Academic Press, San Diego.

Cranshaw, W., and Shetlar, D. 2017. Garden Insects of North America, 2nd ed. Princeton University Press, Princeton, NJ.

Dreistadt, S. H. 2001. Integrated Pest Management for Floriculture and Nurseries. Univ. Calif. State. Integr. Pest Manage. Proj. Publ. 3402. Division of Agriculture and Natural Resources, University of California, Oakland.

Gill, S., Cloyd, R. A., Baker, J. R., Clement, D. L., and Dutky, E. 2006. Pests and Diseases of Herbaceous Perennials, 2nd ed. Ball Publishing, Chicago.

Westcott, C. 1973. The Gardener's Bug Book, 4th ed. Doubleday, Garden City, NY.

(Prepared by R. A. Cloyd)

Chilli Thrips

The chilli thrips (*Scirtothrips dorsalis*) feeds primarily on plants grown in landscapes. Annual bedding plants that may be susceptible to feeding by chilli thrips include begonia, celosia, chrysanthemum, coleus, dahlia, geranium, gerber daisy, marigold, petunia, salvia, snapdragon, verbena, and zinnia. The chilli thrips is well established in the U.S. states of Florida and Texas; thus, greenhouse producers that receive plant material from either state may inadvertently receive infested material.

Life Cycle

An adult is gray-white and 0.5–1.2 mm long; it has short, dark wings (Fig. 158). The female reproduces parthenogenetically (i.e., no mating required) and can lay up to 200 eggs during her life span. The eggs are inserted into plant tissue, and larvae emerge after 6–8 days. Larvae resemble adults. There are two larval stages (first and second instar) and two pupal stages (prepupa and pupa), which are located primarily in the soil.

Fig. 158. Female chilli thrips (*Scirtothrips dorsalis*) adult. (Courtesy Andrew Derksen, USDA–APHIS, from Bugwood.org. Reproduced, by permission, according to terms of Creative Commons Attribution 3.0 License.)

The life cycle can be completed in 15–21 days, but development depends on temperature. One generation can be completed in 14–20 days, and multiple generations occur throughout the growing season.

Damage

Chilli thrips adults and larvae feed mainly on plant leaves (more so than flowers). The undersides of leaves will have a silvery sheen, and brown, necrotic areas will eventually develop. Direct feeding can result in leaves curling upward, as well as thickened new leaf growth and stunted plant growth. Young leaves may be deformed by chilli thrips feeding, resembling herbicide injury. Symptoms may vary depending on plant type.

The chilli thrips does not transmit any viruses to annual bedding plants.

Management

Contact insecticides may be used to suppress populations of the chilli thrips, but thorough coverage of plant leaves and repeat applications are required to obtain adequate suppression. Translaminar insecticides (i.e., after application, the material penetrates the leaf tissue and new terminal growth, forming a reservoir of active ingredient within the leaf or new growing point) may be more effective against chilli thrips populations than contact insecticides.

On certain plants, regulation of chilli thrips populations may be provided by the natural enemies *Amblyseius swirskii, Neoseiulus cucumeris,* and *Orius* spp.

Selected References

Arthurs, S., McKenzie, C. L., Chen, J., Dogramaci, M., Brenna, M., Houben, K., and Osborne, L. 2009. Evaluation of *Neoseiulus cucumeris* and *Amblyseius swirskii* (Acari: Phytoseiidae) as biological control agents of chilli thrips, *Scirtothrips dorsalis* (Thysanoptera: Thripidae) on pepper. Biol. Control 49:91-96.

National Pest Alert: *Scirtothrips dorsalis* Hood. 2007. National Integrated Pest Management Center, USDA–APHIS, National Plant Diagnostic Network, and USDA–CSREES.

(Prepared by R. A. Cloyd)

Citrus Mealybug

The mealybug species commonly encountered on annual bedding plants is the citrus mealybug (*Planococcus citri*). The annual bedding plants susceptible to the citrus mealybug include begonia, chrysanthemum, coleus, English primrose, gerber daisy, and marigold.

Life Cycle

When young, the citrus mealybug is 1–3 mm long and orange to pink in color. Mature adults appear as white flecks or cottony masses; an adult has a fringe of waxy filaments on the periphery of the body (Fig. 159). Citrus mealybugs congregate in large numbers at leaf junctures, on the undersides of leaves, and on stems (Fig. 160). The female is white, wingless, and 2–5 mm long when full grown. The male is usually smaller.

The life cycle includes an egg, three nymphal stages, and adult. The adult female lays eggs underneath her body cavity. The eggs hatch into nymphs, or crawlers, which move around looking for places to settle and feed. The nymphs are yellow to orange and eventually turn white after several successive molts. The male citrus mealybug becomes winged, mates with a female, and then dies after 2–3 days. The female continues development and dies after laying eggs. A female is capable of laying up to 600 eggs.

The life cycle from egg to adult takes approximately 60 days. Development is dependent on temperature and/or host plant.

Damage

Citrus mealybugs use their piercing–sucking mouthparts to feed on the fluids in plants' vascular tissues, including the phloem and the mesophyll or both. Their feeding causes leaf yellowing and plant stunting and wilting. Also, similar to aphids, citrus mealybugs excrete honeydew, which serves as a growing medium for black sooty mold. Infestation by a large citrus mealybug population can result in plant death.

Management

Weed removal, proper fertilization, and disposal of old plant material will help manage the citrus mealybug. Heavily infested plants (>50%) should be disposed of immediately. Plants that have been fertilized with high concentrations of nitrogen-based, water-soluble products tend to develop succulent growth, which is more susceptible to the citrus mealybug.

Contact insecticides may be used to suppress citrus mealybug populations. However, these products are most effective on the young nymphs, or crawlers, because adults form a white, waxy, protective covering that is impervious to insecticides. Frequent applications and thorough coverage of all plant parts are required to achieve sufficient suppression, particularly when there are overlapping generations. Because the crawler stage does not have a waxy covering, it is most susceptible to many different types of insecticides. Citrus mealybug populations are difficult to suppress with standard contact and systemic insecticides.

Fig. 159. Citrus mealybug (*Planococcus citri*) early and late instars. (Note the fringe of waxy filaments on the periphery of the body.) (Courtesy R. A. Cloyd—© APS)

Fig. 160. Citrus mealybug (*Planococcus citri*) later instar feeding at an umbrella tree petiole juncture. (Courtesy R. A. Cloyd—© APS)

Natural enemies of the citrus mealybug include the adult and larval stages of the predatory ladybird beetle (*Cryptolaemus montrouzieri*) and the parasitoids *Anagyrus pseudococci* and *Leptomastix dactylopii*. Both parasitoids are commercially available for regulation of citrus mealybug populations, but the availability of *L. dactylopii* is limited.

Selected References

Cloyd, R. A. 2011. Mealybug: Management in Greenhouses and Nurseries. Kans. State Univ. Agric. Exp. Stn. Coop. Ext. Serv. Publ. MF-3001. Kansas State University, Manhattan.

Dreistadt, S. H. 2001. Integrated Pest Management for Floriculture and Nurseries. Univ. Calif. State. Integr. Pest Manage. Proj. Publ. 3402. Division of Agriculture and Natural Resources, University of California, Oakland.

Franco, J. C., Zada, A., and Mendel, Z. 2009. Novel approaches for the management of mealybug pests. Pages 233-278 in: Biorational Control of Arthropod Pests: Application and Resistance Management. I. Ishaaya and A. R. Horowitz, eds. Springer, Dordrecht, the Netherlands.

(Prepared by R. A. Cloyd)

Cyclamen Mite

The cyclamen mite (*Phytonemus pallidus*) is very similar to the broad mite with respect to biology, life cycle, and feeding damage (see the earlier section "Broad Mite"). The cyclamen mite feeds on a wide range of annual bedding plants, including ageratum, begonia, chrysanthemum, dahlia, geranium, gerber daisy, impatiens, marigold, snapdragon, stock, and zinnia.

Life Cycle

An adult cyclamen mite is about 0.25 mm long and oval shaped; it is yellow to brown and appears transparent. In contrast to the eggs of the broad mite, the eggs of the cyclamen mite are oval and smooth, with no bumps or protrusions.

The life cycle from egg to adult is completed in 1–3 weeks, and development depends on temperature. A female cyclamen mite lays 1–3 eggs per day and up to 16 eggs during her 2-week life span. The eggs are deposited in clusters within terminal buds.

Damage

The cyclamen mite, like the broad mite, feeds within plant cells. Feeding causes leaf distortion, twisting or curling, and bronzing, and leaves typically appear wrinkled, brittle, and rough. Plants that are heavily infested with cyclamen mites will be stunted and have small leaves that eventually turn brown to silver. In addition, flowers may be distorted and flower buds may abort or fail to open (Fig. 161).

Management

Sanitation practices such as cleaning greenhouses prior to introducing new plants and disinfecting benches will help reduce problems with cyclamen mites, as with broad mites. Cyclamen mites are very difficult to manage with miticides, because these pests are located primarily in the meristematic tissues of plants. Miticides with translaminar activity (i.e., after application, the material penetrates leaf tissue and new terminal growth, forming a reservoir of active ingredient within the leaf or new growing point) may be more effective than those with contact activity.

Scouting should focus on those annual bedding plants that are susceptible to cyclamen mites. Any plants exhibiting damage symptoms should be discarded immediately. Plants within

Fig. 161. Flower bud distortion on cyclamen caused by feeding of cyclamen mites (*Phytonemus pallidus*). (Courtesy R. A. Cloyd—© APS)

15.2–30.4 cm of symptomatic plants should also be discarded, because even if these plants do not exhibit symptoms, they may be infested with cyclamen mites. Growers should always inspect incoming plant material and isolate plants for 1 week to help prevent introducing infested plants into the main crop.

The predatory mites *Amblyseius swirskii, Neoseiulus californicus,* and *N. cucumeris* can be released to help regulate cyclamen mite populations and prevent them from establishing.

Selected References

Cloyd, R. A. 2010. Broad Mite and Cyclamen Mite: Management in Greenhouses and Nurseries. Kans. State Univ. Agric. Exp. Stn. Coop. Ext. Serv. Publ. MF-2938. Kansas State University, Manhattan.

Westcott, C. 1973. The Gardener's Bug Book, 4th ed. Doubleday, Garden City, NY.

(Prepared by R. A. Cloyd)

Fourlined Plant Bug

The fourlined plant bug (*Poecilocapsus lineatus*) is widely distributed throughout the eastern United States. Adults and nymphs feed on a number of annual bedding plants, including chrysanthemum, dahlia, heliotrope, snapdragon, and zinnia.

Life Cycle

A fourlined plant bug adult is 6 mm long, slender and oval, and green or yellow, with four distinct black stripes that extend down the wings (Fig. 162). The female inserts eggs into slits near the terminal shoots that were created during early summer. The eggs hatch from late April through June, and the emergent nymphs are orange-red and have black stripes on the wing buds. A nymph normally completes development and transitions into an adult within 30 days.

The fourlined plant bug feeds for approximately 6 weeks; adults feed primarily on the upper leaf surfaces. Both adults and nymphs move rapidly when disturbed. There is typically one generation produced per year, although this varies depending on geographic location.

Damage

Both nymphs and adults have piercing–sucking mouthparts, which they use to feed on plant fluids while simultaneously injecting a toxic secretion that results in the loss of chlorophyll. Feeding by fourlined plant bugs causes leaf distortion and

Fig. 162. Fourlined plant bug (*Poecilocapsus lineatus*) adult on a leaf. (Courtesy R. A. Cloyd—© APS)

Fig. 163. Leaf spots caused by feeding of the fourlined plant bug (*Poecilocapsus lineatus*). (Courtesy R. A. Cloyd—© APS)

stippling. After several weeks, necrotic spots may fall out, leaving small holes. The most common type of plant damage caused by the feeding of nymphs and adults is the formation of round, depressed (1–3 mm), tan to red-brown spots (Fig. 163). These spots eventually coalesce, forming large, brown blotches that are often confused with symptoms of a leaf spot disease.

Management

Weed removal will help alleviate problems with the fourlined plant bug. Hand-picking nymphs and adults and placing them into a soapy water solution may be effective in managing a low infestation. Contact insecticides, including horticultural oils (both petroleum and mineral based), may be more effective against nymphs than adults when applied from May through June. Thorough coverage of all plant parts is important to obtain sufficient suppression.

Selected References

Cranshaw, W., and Shetlar, D. 2017. Garden Insects of North America, 2nd ed. Princeton University Press, Princeton, NJ.

Krischik, V., and Davidson, J., eds. 2004. IPM (Integrated Pest Management) of Midwest Landscapes. Cooperative Project of NCR-193, North Central Committee on Landscape IPM, Minnesota Agricultural Experiment Station SB-07645. Minneapolis.

Westcott, C. 1973. The Gardener's Bug Book, 4th ed. Doubleday, Garden City, NY.

(Prepared by R. A. Cloyd)

Fungus Gnats

A number of fungus gnat species may be problematic during seed and cutting propagation of annual bedding plants, including two darkwinged fungus gnats: *Bradysia coprophila* and *B. impatiens.* Although fungus gnats have a wide host range, annual bedding plants with succulent stems, such as coleus and geranium, are especially susceptible to fungus gnat larval feeding damage. Begonia, campanula, chrysanthemum, and lobelia are susceptible to fungus gnats, as well. Younger plants, in general, are more susceptible to damage than older plants.

Life Cycle

Fungus gnat larvae are white, transparent or slightly translucent, and legless. A larva is about 3.1 mm long when full grown. A diagnostic characteristic of fungus gnat larvae is the black head capsule, which is absent in shore fly larvae (Fig. 164) (see the later section "Shore Flies").

A fungus gnat adult is winged, black, and 3.1 mm long, with long legs and antennae (Fig. 165). Each wing has a distinct y-shaped vein. Adults tend to fly near the growing medium and live 7–10 days. Fungus gnat adults are not strong fliers compared with shore fly adults.

The life cycle from egg to adult is completed in 21–28 days, depending on temperature.

Fig. 164. Fungus gnat (*Bradysia* sp.) larvae. (Note the black head capsule.) (Courtesy R. A. Cloyd—© APS)

Fig. 165. Fungus gnat (*Bradysia* sp.) adult female. (Courtesy T. Meers—© APS)

Damage

Fungus gnats may be problematic, especially during propagation of annual bedding plants, for several reasons:

1. Excessive populations of flying adults may be present, which could distract workers.
2. The larval and adult stages are capable of disseminating and transmitting soilborne plant pathogens such as *Phytophthora, Pythium,* and *Thielaviopsis* spp.
3. The larvae feed on plant roots, causing injury and creating wounds that may allow for the entry of soilborne plant pathogens.
4. The larvae may tunnel into the plant crown, resulting in plant death.

Typical symptoms of larval feeding damage are plant stunting and wilting and yellowing of lower leaves. The extent of larval transmission of soilborne plant pathogens may be less likely under greenhouse conditions than laboratory conditions and may vary depending on the specific pathogen species.

Management

Water management and sanitation are important to avoid problems with fungus gnats. Growers should use a well-drained growing medium, because root decay and microbial activity associated with wet conditions are attractive to fungus gnats. In addition, removing weeds from underneath benches and around the greenhouse perimeter will help alleviate problems with fungus gnats, because low-growing weeds, such as creeping woodsorrel (*Oxalis corniculata*), maintain moist conditions, providing a prime habitat for fungus gnats to develop. Moreover, the use of composts and microbially active growing media may result in higher fungus gnat populations, because these types of growing media are favorable for pest development and reproduction.

Treating the growing medium with insecticides, including insect growth regulators and contact and microbial insecticides, may be effective in suppressing fungus gnat larval populations. In addition, larval populations may be effectively regulated with natural enemies such as the entomopathogenic nematode *Steinernema feltiae,* the rove beetle *Dalotia coriaria,* and several soil-residing predatory mites (*Stratiolaelaps scimitus* and *Gaeolaelaps aculeifer*). Also, adult hunter flies (*Coenosia attenuata*), which feed on fungus gnat (and shore fly) adults, may inadvertently enter the greenhouse from outside or on incoming plant material.

Selected References

Cloyd, R. A. 2008. Management of fungus gnats (*Bradysia* spp.) in greenhouses and nurseries. Floricult. Ornament. Biotech. 2(2):84-89.

Cloyd, R. A. 2010. Fungus Gnats: Management in Greenhouses and Nurseries. Kans. State Univ. Agric. Exp. Stn. Coop. Ext. Serv. Publ. MF-2937. Kansas State University, Manhattan.

Harris, M. A., Gardener, W. A., and Oetting, R. D. 1996. A review of the scientific literature on fungus gnats (Diptera: Sciaridae) in the genus *Bradysia.* J. Entomol. Sci. 31(3):252-276.

Ugine, T. A., Sensenbach, E. J., Sanderson, J. P., and Wraight, S. P. 2010. Biology and feeding requirements of larval hunter flies *Coenosia attenuata* (Diptera: Muscidae) reared on larvae of the fungus gnat *Bradysia impatiens* (Diptera: Sciaridae). J. Econ. Entomol. 103:1149-1158.

(Prepared by R. A. Cloyd)

Greenhouse Thrips

The greenhouse thrips (*Heliothrips haemorrhoidalis*) feeds on annual bedding plants, including begonia, chrysanthemum, dahlia, nasturtium, phlox, and Sweet William. However, which plants are affected depends on geographic location, because this insect is most prevalent in the U.S. states of California and Florida and in greenhouses in the nation's Northeast and Southeast.

Life Cycle

A greenhouse thrips adult is 1.3–1.8 mm long, appears dark brown to black with a silver sheen, and has yellow legs (Fig. 166). It also has a network of lines on both the head and the body. The greenhouse thrips female is parthenogenic (i.e., can reproduce without mating), and males are rarely present.

A female can lay 25–50 eggs into leaf tissue; they hatch within 2–3 weeks into larvae that are pale yellow and have distinct red eyes. There are two pupal stages (prepupa and pupa); they do not feed and are located in the soil or growing medium. Adults then emerge and can live up to 7 weeks in the greenhouse; adults do not fly very well, so they tend to remain in shaded areas on plants.

The development time from egg to adult occurs in 30–40 days, depending on temperature. Overlapping generations may be present in a greenhouse.

Damage

The greenhouse thrips feeds primarily on plant leaves, including both the lower and upper leaf surfaces. Feeding causes discolored areas to develop between the lateral veins, and damaged leaves may appear silvery or bleached. Feeding may also result in leaf distortion. Black spots, or excrement, may be visible on plant leaves.

A large population of greenhouse thrips can cause considerable plant damage.

Management

Removing weeds and plant debris will help manage the greenhouse thrips. Furthermore, disposing of old stock plants will reduce reservoirs for greenhouse thrips.

Contact insecticides may be effective against the greenhouse thrips, but frequent enough applications must be made and all plant parts must be thoroughly covered. In the greenhouse, scouting plants regularly by means of yellow sticky cards and conducting visual inspections will help to estimate the population dynamics of the greenhouse thrips throughout the growing season. Estimating the population dynamics will help in timing insecticide applications.

Natural enemies can be purchased from suppliers and released to regulate populations of the greenhouse thrips. The grower should contact a supplier directly to determine the types and availability of natural enemies for use against this pest.

Selected References

Cranshaw, W., and Shetlar, D. 2017. Garden Insects of North America, 2nd ed. Princeton University Press, Princeton, NJ.

Dreistadt, S. H. 2001. Integrated Pest Management for Floriculture and Nurseries. Univ. Calif. State. Integr. Pest Manage. Proj. Publ. 3402. Division of Agriculture and Natural Resources, University of California, Oakland.

Johnson, W. T., and Lyon, H. H. 1988. Insects That Feed on Trees and Shrubs, 2nd ed. Cornell University Press, Ithaca, NY.

(Prepared by R. A. Cloyd)

Greenhouse Whitefly

The greenhouse whitefly (*Trialeurodes vaporariorum*) feeds on a variety of annual bedding plants, including ageratum, begonia, calendula, chrysanthemum, coleus, fuchsia, geranium, gerber daisy, salvia, and verbena.

Life Cycle

A greenhouse whitefly adult is about 4.2 mm long, and its wings are covered with a white, waxy powder. An adult holds its wings flat over the body, parallel to the leaf surface (Fig. 167). The female adult lays eggs on the undersides of plant leaves: up to 20 eggs per day and up to 300 eggs during her 30- to 45-day life span. Eggs are pale green to purple and hatch within 5–10 days.

The newly emerged nymphs, or crawlers, search for feeding sites on the undersides of leaves. Nymphs are typically flat and transparent to yellow-brown in color. They do not move after establishing a feeding site, remaining immobile for 2–3 weeks, and then they transition into a pupal stage, or fourth instar. A greenhouse whitefly pupa has elongated, waxy filaments that encircle the outer periphery of the body; they are elevated in profile and have vertical (perpendicular) sides. The pupae appear raised on the leaf surface. Adults emerge from pupae after approximately 1 week.

The life cycle from egg to adult is completed in 3–4 weeks, depending on the temperature and the plant type upon which the insects have fed. The nymphal stages last approximately 1 month, and there can be multiple overlapping generations. All the life stages (eggs, nymphs, pupae, and adults) are located primarily on the undersides of leaves.

Damage

Greenhouse whitefly nymphs and adults have piercing–sucking mouthparts, which they use to withdraw plant fluids

Fig. 166. Greenhouse thrips (*Heliothrips haemorrhoidalis*) adult. (Courtesy Chazz Hesselein, Alabama Cooperative Extension System, from Bugwood.org. Reproduced, by permission, according to terms of Creative Commons Attribution 3.0 License.)

Fig 167. Greenhouse whitefly (*Trialeurodes vaporariorum*) adults feeding on the underside of a gerber daisy leaf. (Courtesy R. A. Cloyd—© APS)

from the phloem sieve tubes. Their feeding causes plant wilting and stunting and leaf yellowing and distortion.

Similar to aphids, whiteflies excrete honeydew, which accumulates on plant leaves and serves as a growing medium for black sooty mold.

Management

Scouting is important to determine the number of whiteflies present in the greenhouse. Furthermore, scouting will detect seasonal trends in whitefly populations associated with various life stages throughout the growing season. Placing yellow sticky cards just above the crop canopy is the primary means of scouting for whitefly adults. The individuals captured on the cards should be counted weekly and the numbers of whitefly adults recorded on data sheets. In addition, visual inspection of plants is required to detect eggs, nymphs, and pupae. Growers should look at the undersides of leaves, which is where all the life stages of the greenhouse whitefly are primarily located.

Proper watering and fertilization practices are effective in minimizing potential problems with greenhouse whiteflies. Greenhouse whiteflies are attracted to and lay more eggs on plants that receive too much fertilizer, because these plants provide sufficient nutrition for nymphs. Growers should remove plant debris and old stock plants from the greenhouse or place them into containers with tight-sealing lids. Growers should also remove weeds in and around the greenhouse to eliminate potential refuge sites for the greenhouse whitefly. Certain weeds—such as annual sowthistle (*Sonchus oleraceus*), common chickweed (*Stellaria media*), and dandelion (*Taraxacum officinale*)—may harbor populations of greenhouse whitefly.

Contact, translaminar, and systemic insecticides may be used to suppress populations of the greenhouse whitefly. When contact insecticides are used, multiple applications and thorough coverage of all plant parts are required for effectiveness. Contact insecticides are most effective early in the crop production cycle, because smaller plants make it easier for sprays to penetrate the crop canopy, subsequently ensuring sufficient coverage of leaf undersides. Translaminar insecticides (i.e., those in which the spray material penetrates the leaf surface tissue, forming a reservoir of active ingredient within the leaf) provide up to 14 days of residual activity, even after the spray residues dissipate, but residual activity (persistence) will vary depending on the specific insecticide. Systemic insecticides are also effective against the greenhouse whitefly, especially if applied early in the crop production cycle and before the whitefly population has built to an excessive level. Systemic insecticides can be applied as a drench or in granule form to the growing medium or as a spray on plant leaves.

Releases of the parasitic wasp (parasitoid) *Encarsia formosa* may regulate greenhouse whitefly populations. In addition, spray applications of entomopathogenic fungi such as *Beauveria bassiana* and *Isaria fumosorosea* may be effective in suppressing whitefly populations if the relative humidity is greater than 75%.

Selected References

Cloyd, R. A. 2012. Insect and mite management in greenhouses. Pages 391-441 in: Greenhouse Operation and Management, 7th ed. P. V. Nelson, ed. Prentice-Hall, Saddle River, NJ.

Dreistadt, S. H. 2001. Integrated Pest Management for Floriculture and Nurseries. Univ. Calif. State. Integr. Pest Manage. Proj. Publ. 3402. Division of Agriculture and Natural Resources, University of California, Oakland.

Westcott, C. 1973. The Gardener's Bug Book, 4th ed. Doubleday, Garden City, NY.

(Prepared by R. A. Cloyd)

Japanese Beetle

The Japanese beetle (*Popillia japonica*) was discovered in the United States in 1916 and has since spread throughout most of the eastern United States, from Maine to Georgia. Distribution has continued to spread west beyond the Mississippi River, and the beetle has become established in certain states throughout the Great Plains and in urban environments in the western United States. Japanese beetle adults feed on certain annual bedding plants grown outdoors, including dahlia, marigold, and zinnia.

Life Cycle

The Japanese beetle adult is 9.5–13.0 mm long and metallic green with coppery-brown wing covers (Fig. 168). Approximately 14 tufts of white hair are present along the median of the abdomen.

Adults emerge from the soil and are usually present from May through September, depending on the geographic location. An adult lives from 30 to 45 days. After mating, the female deposits up to 60 eggs in small clusters into the soil at depths of 13–100 mm. The eggs hatch in 2 weeks, and C-shaped, white larvae (grubs) emerge (Fig. 169).

Damage

Japanese beetle adults congregate in large numbers on annual bedding plants; they feed primarily on the flowers, although they will also feed on the leaves. Adults chew holes in open flower petals and can completely destroy a flower. On

Fig. 168. Japanese beetle (*Popillia japonica*) adults feeding on a leaf. (Courtesy R. A. Cloyd—© APS)

Fig. 169. Japanese beetle (*Popillia japonica*) larva (grub) in the soil. (Note the C-shape.) (Courtesy R. A. Cloyd—© APS)

leaves, they feed between the veins, which results in leaves having a lacelike, skeletonized appearance. The larvae (grubs) are located in the soil and may feed on the roots of annual bedding plants.

Management

Japanese beetle adults can be managed by hand-picking them from annual bedding plants or by placing fine netting over plants. Contact insecticides may be effective in suppressing populations of Japanese beetle adults; however, repeat applications are required.

The use of Japanese beetle traps should be avoided, as the traps tend to lure or attract more adult beetles into an area than would normally occur. Furthermore, adult beetles may feed on annual bedding plants before they reach the traps, which may increase the amount of feeding damage. If traps are used to actually "trap-out" Japanese beetle adults, they must be placed away (>30 feet [9 m]) from plantings.

Insecticides and natural enemies, including entomopathogenic nematodes (e.g., *Heterorhabditis bacteriophora*), can be applied to the turfgrass surrounding greenhouse facilities and landscapes to suppress populations of the Japanese beetle larval (grub) stage.

Selected References

Fleming, W. E. 1972. Biology of Japanese Beetle. U.S. Dep. Agric. Tech. Bull. 1449.

Horst, R. K., and Cloyd, R. A. 2007. Compendium of Rose Diseases and Pests, 2nd ed. American Phytopathological Society, St. Paul, MN.

Johnson, W. T., and Lyon, H. H. 1988. Insects That Feed on Trees and Shrubs, 2nd ed. Cornell University Press, Ithaca, NY.

Potter, D. A., and Held, D. W. 2002. Biology and management of the Japanese beetle. Annu. Rev. Entomol. 47:175-205.

Westcott, C. 1973. The Gardener's Bug Book, 4th ed. Doubleday, Garden City, NY.

(Prepared by R. A. Cloyd)

Potato Leafhopper

The potato leafhopper (*Empoasca fabae*) is the primary leafhopper species that may be encountered feeding on annual bedding plants. This leafhopper feeds on a wide range of annual bedding plants, including campanula, chrysanthemum, dahlia, marigold, and zinnia.

Life Cycle

A potato leafhopper adult is 3.0–3.5 mm long, slender, and wedge shaped with a tapered end. It is yellow to pale green and has white spots on the head and thorax (midsection of body). It holds its wings rooflike over its body (Fig. 170).

Nymphs and adults are active and typically move sideways when disturbed. Adults are winged and can fly; nymphs do not have wings and cannot fly.

The female deposits eggs into leaves, petioles, or stems. In approximately 10 days, the eggs hatch into nymphs, which are similar in appearance to adults but smaller and without wings (Fig. 171). The five nymphal stages develop in 2 weeks, after which the adult emerges. White cast skins, which are an indication of the nymphs having molted, may be present on the undersides of leaves.

An adult generally lives for 1 month but may live longer. The life cycle from egg to adult is completed in 28 days, and there may be multiple generations per year.

Damage

The potato leafhopper has piercing–sucking mouthparts, which it uses to withdraw plant fluids and to inject toxins into the plant. This feeding not only destroys plant cells but also disrupts phloem movement throughout the plant. Feeding may cause stippling of plant leaves (Fig. 172), which is similar to the feeding damage caused by the twospotted spider mite (see the later section "Twospotted Spider Mite"). Potato leafhopper feeding may also result in plant stunting and leaf distortion, chlorosis or yellowing, curling, and browning, or necrosis (referred to as "hopperburn").

Management

Removing weeds will help manage populations of potato leafhoppers, because certain weeds serve as reservoirs for this pest. Contact insecticides may also be used to reduce potato leafhopper populations. Achieving effective management using insecticides is difficult, however; both nymphs and adults are very mobile, and new adults may enter a treated area after the spray residue has dried. Frequent applications are required

Fig. 170. Potato leafhopper (*Empoasca fabae*) adult. (Courtesy P. J. Jentsch—© APS. Reproduced, by permission, from Sutton, T. B., Aldwinckle, H. S., Agnello, A. M., and Walgenbach, J. F., eds. 2014. Compendium of Apple and Pear Diseases and Pests, 2nd ed. American Phytopathological Society, St. Paul, MN.)

Fig. 171. Potato leafhopper (*Empoasca fabae*) nymph. (Courtesy Frank Peairs, Colorado State University, from Bugwood.org. Reproduced, by permission, according to terms of Creative Commons Attribution 3.0 License.)

Fig. 172. Stippling of salvia leaves caused by feeding of the potato leafhopper (*Empoasca fabae*). (Courtesy R. A. Cloyd—© APS)

to effectively protect plants from potato leafhopper feeding damage.

Selected References

Baker, J. R. 1994. Insect and Related Pests of Flowers and Foliage Plants. N.C. Coop. Ext. Serv. Publ. AG-136. North Carolina State University, Raleigh.

Cranshaw, W., and Shetlar, D. 2017. Garden Insects of North America, 2nd ed. Princeton University Press, Princeton, NJ.

Westcott, C. 1973. The Gardener's Bug Book, 4th ed. Doubleday, Garden City, NY.

(Prepared by R. A. Cloyd)

Shore Flies

Shore flies (including the species *Scatella stagnalis*) are primarily considered nuisance insect pests during seed and cutting propagation of annual bedding plants. However, shore fly adults are often misidentified as fungus gnat adults, even though they look distinctly different. Adult shore flies may also be mistaken for adult hunter flies (*Coenosia attenuata*), which are predators, despite differences in appearance; hunter flies are somewhat larger, are dark gray, and do not have white spots on the wings (as shore flies do).

Life Cycle

A shore fly adult is 3.1 mm long and black, and its wings have five or six white or light-colored spots (Fig. 173). Fungus gnat and hunter fly adults do not have spots on their wings. In addition, a shore fly adult has short antennae and legs. Shore fly adults are stronger fliers than fungus gnat adults and are often seen resting on plant leaves; shore fly adults fly up when disturbed.

A shore fly larva is 6.3 mm long and opaque yellow-brown; it does not have a black head capsule like a fungus gnat larva. A pupa is a dark-brown, seedlike structure and may be seen around the edges of algal mats.

The life cycle of the shore fly, from egg to adult, is completed in 21–28 days, depending on temperature, similar to the life cycle of the fungus gnat (see the previous section "Fungus Gnats").

Damage

Shore fly larvae feed primarily on algae located on the surface of the growing medium or in other areas of the greenhouse where conditions are conducive for alga growth. The larvae generally do not directly feed on plant roots.

Shore flies may act as vectors of certain soilborne plant pathogens, such as *Pythium* spp. The extent or likelihood of shore flies transmitting disease may be negligible under greenhouse conditions, as is also the case with fungus gnats.

The presence of shore fly adults is a greater concern during propagation, because they are very noticeable flying around. They are generally considered a nuisance insect pest; however, they can leave black fecal deposits on plant leaves (Fig. 174), which may affect the aesthetic quality of annual bedding plants. For this reason, an infestation of a large population of shore fly adults may result in a shipment of plants being rejected by another grower.

Management

Water management and sanitation are important in controlling populations of shore flies. Growers should always use a growing medium that drains well to prevent the buildup of algae, which serves as a breeding habitat for shore flies. The placement of yellow or hopper-finder sticky tape in rows among plants may be effective in trapping large numbers of shore fly adults and possibly capturing adult females before they lay eggs.

Insecticides such as insect growth regulators may be effective in suppressing shore fly larval populations if timed accordingly and applied frequently enough. Entomopathogenic nematodes are commercially available, such as *Steinernema feltiae* and *S. carpocapsae;* when applied frequently and at high rates, they may be effective in regulating shore fly larval populations.

The rove beetle *Dalotia coriaria,* which is a generalist predator, may feed on shore fly larvae in the growing medium. In addition, adult rove beetles are winged and can readily disperse throughout a greenhouse. The hunter fly adult and the small

Fig. 173. Shore fly (*Scatella* sp.) adults. (Courtesy R. A. Cloyd—© APS)

Fig. 174. Shore fly (*Scatella* sp.) adult and black fecal deposits on a leaf. (Courtesy R. A. Cloyd—© APS)

parasitoid *Hexacola neoscatellae* may be present in the greenhouse and can help regulate shore fly larval populations.

Some natural enemies and soil-dwelling predatory mites, such as *Gaeolaelaps aculeifer* and *Stratiolaelaps scimitus*, may not be effective in sufficiently regulating shore fly populations. The reason is that shore fly larvae can survive under extremely moist conditions, which are not conducive for the survival of many natural enemies.

Selected References

Cloyd, R. A. 2012. Insect and mite management in greenhouses. Pages 391-441 in: Greenhouse Operation and Management, 7th ed. P. V. Nelson, ed. Prentice-Hall, Saddle River, NJ.

Hyder, N., Coffey, M. D., and Stanghellini, M. E. 2009. Viability of oomycete propagules following ingestion and excretion by fungus gnats, shore flies, and snails. Plant Dis. 93:720-726.

Vänninen, I. 2001. Biology of the shore fly *Scatella stagnalis* in rockwool under greenhouse conditions. Entomol. Exp. Appl. 98:317-328.

(Prepared by R. A. Cloyd)

Slugs and Snails

Slugs and snails are not arthropods but rather classified as mollusks. They feed on a wide range of annual bedding plants, especially seedlings, in greenhouses and landscapes. Slugs and snails are particularly problematic when plants are receiving abundant moisture. A number of species may be encountered, depending on geographic location, including the gray fieldslug (*Deroceras reticulatum*), the orange-banded arion or white-soled slug (*Arion fasciatus*), the three-band gardenslug (*Lehmannia valentiana*), the giant gardenslug (*Limax maximus*), the black slug (*Arion ater*), the greenhouse slug (*Milax gagates*), and the brown garden snail (*Cornu aspersum*).

Life Cycle

Slugs and snails are wormlike, legless organisms and vary in length from 19 to 52 mm when full grown, depending on the species. Snails have hardened shells (Fig. 175), whereas slugs, which are often called "naked snails," do not have shells (Fig. 176). Both snails and slugs may be a variety of colors, including yellow, lavender, purple, brown, and black. Some have brown specks or mottled areas on their bodies.

Both slugs and snails hide during the day and emerge at night to feed, leaving a silvery trail of slime. Both reside in moist areas and move around by gliding along a muscular foot that secretes mucus (Fig. 177). Most slugs and snails are bisexual or hermaphroditic, meaning that one individual possesses both sex organs; however, male and female phases may form at different times.

Slugs and snails lay clusters of 20–100 eggs in the cracks and crevices of moist soil or growing medium. In most cases, eggs are laid in masses on the soil surface and covered by a gelatinous shell, which makes them appear milky. Eggs hatch in 10 days or less at temperatures greater than 10°C. Slugs and snails reach maturity in 3 months to 1 year.

Damage

Slugs and snails have chewing mouthparts and cause damage to foliage by creating large, irregularly shaped, ragged holes and sometimes tattered edges on leaves (Fig. 178). In addition, the mucus slime left by slugs and snails may dry into a silvery trail on plant leaves and flowers.

Management

Management of slugs and snails involves implementing a variety of strategies, including conducting stringent inspections of incoming plant shipments, removing all debris from the production area, using barriers around plants (e.g., copper), minimizing the overwatering of plants in both the greenhouse and landscape (i.e., by using a drip or low-volume irrigation system), watering early in the day and not getting the leaves and flowers wet, and applying commercially available molluscicides. In addition, growing plants on benches instead of floors will help to alleviate problems with slugs and snails.

Fig. 175. Snail on a New Guinea impatiens leaf. (Courtesy R. A. Cloyd—© APS)

Fig. 176. Gray field slug (*Deroceras reticulatum*). (Courtesy R. A. Cloyd—© APS)

Fig. 177. Gray field slug (*Deroceras reticulatum*) leaving a trail of mucus secreted by its muscular foot. (Courtesy R. A. Cloyd—© APS)

Fig. 178. Holes in zinnia leaves caused by slug feeding. (Courtesy R. A. Cloyd—© APS)

Selected References

Barker, G. M., ed. 2002. Molluscs as Crop Pests. CABI, Wallingford, Oxfordshire, UK.

Cloyd, R. A. 2012. Insect and mite management in greenhouses. Pages 391-441 in: Greenhouse Operation and Management, 7th ed. P. V. Nelson, ed. Prentice-Hall, Saddle River, NJ.

Cranshaw, W., and Shetlar, D. 2017. Garden Insects of North America, 2nd ed. Princeton University Press, Princeton, NJ.

Dreistadt, S. H. 2001. Integrated Pest Management for Floriculture and Nurseries. Univ. Calif. State. Integr. Pest Manage. Proj. Publ. 3402. Division of Agriculture and Natural Resources, University of California, Oakland.

(Prepared by R. A. Cloyd)

Spotted Cucumber Beetle

The spotted cucumber beetle (*Diabrotica undecimpunctata howardi*) feeds primarily on annual bedding plants in the family Asteraceae, including calendula, chrysanthemum, coreopsis, cosmos, dahlia, gerber daisy, osteospermum, and zinnia, in addition to cucurbits.

Life Cycle

A spotted cucumber beetle adult is 6.3 mm long and yellow-green in color; it has a small, black head and 12 black spots on the abdomen (Fig. 179). The adult feeds mostly on plant pollen.

Fig. 179. Spotted cucumber beetle (*Diabrotica undecimpunctata howardi*) adult. (Courtesy R. A. Cloyd—© APS)

The adult female lays clusters of 25–50 eggs in the soil during the spring, and larvae hatch from the eggs. The larva is 12 mm long and pale colored with a brown head; it resides in the soil, feeding on plant roots, and may cause damage to plants. However, the adult stage causes the most damage to annual bedding plants. There may be one or more generations per year, depending on the geographic location.

Damage

Spotted cucumber beetle adults commonly feed on pollen, but an extensive population may cause substantial damage to certain annual bedding plants in landscapes. Adults feed mainly on the flower petals of annual bedding plants that have light-colored flowers and that bloom in late summer. Adults chew irregularly shaped holes in plant leaves and flowers, especially the petals.

Management

Removing plant debris and weeds will help manage spotted cucumber beetle adults. Placing yellow sticky cards with a pheromone lure within the crop can help attract and kill adults. Contact insecticides can be used to alleviate problems with adult populations, but frequent applications will be required.

Another option is to avoid planting susceptible annual bedding plants when adult spotted cucumber beetles are active and abundant.

Selected References

Cranshaw, W., and Shetlar, D. 2017. Garden Insects of North America, 2nd ed. Princeton University Press, Princeton, NJ.

Westcott, C. 1973. The Gardener's Bug Book, 4th ed. Doubleday, Garden City, NY.

(Prepared by R. A. Cloyd)

Sweetpotato Whitefly

The sweetpotato whitefly (*Bemisia tabaci*) attacks a wide range of annual bedding plants, including basil, begonia, calendula, celosia, chrysanthemum, coleus, dahlia, flowering tobacco, gerber daisy, and salvia.

Life Cycle

A sweetpotato whitefly adult is white to light yellow, narrow, and approximately 1.9–2.8 mm long (Fig. 180). An adult female lays spindle-shaped eggs on the undersides of leaves (both young and mature). Nymphs hatch from the eggs and then move around, looking for feeding locations. The first nymphal

Fig. 180. Sweetpotato whitefly (*Bemisia tabaci*) adults on the underside of a hibiscus leaf. (Courtesy R. A. Cloyd—© APS)

stage grows and molts a second and a third time into a nonfeeding pupal stage (fourth nymphal stage). Adults emerge about 1 month after eggs are laid.

An adult female lives about 4 weeks, and an individual female can lay up to 200 eggs during her life span. The development time from egg to adult is 20–28 days, depending on temperature.

Damage

Sweetpotato whitefly nymphs and adults have piercing–sucking mouthparts, which they use to remove plant fluids. Their feeding results in leaf yellowing (chlorosis) and distortion (curling), as well as plant stunting, wilting, and possibly death (depending on the plant size and the size of the whitefly population).

The sweetpotato whitefly produces honeydew, which serves as a growing medium for certain black sooty molds and can reduce the aesthetic quality of annual bedding plants. The presence of large numbers of adults may be a visual nuisance, reducing the salability of annual bedding plants.

Management

Removing weeds eliminates reservoirs for the sweetpotato whitefly. Growers should avoid overfertilizing annual bedding plants, because the succulent growth that results and the enhanced nutritional quality may increase plants' susceptibility to sweetpotato whitefly adults.

Contact insecticides and insect growth regulators can be effective in suppressing sweetpotato whitefly populations, if frequent applications are made and all plant parts are thoroughly covered. Systemic insecticides can also be effective against the sweetpotato whitefly, especially if applied early in the crop production cycle and before the whitefly population reaches an excessive level. Systemic insecticides can be applied as a drench or as granules to the growing medium or as a spray to plant leaves.

Applications of entomopathogenic fungi, such as *Beauveria bassiana* and *Isaria fumosorosea,* may suppress populations of the sweetpotato whitefly; however, for these applications to be effective, the relative humidity must be greater than 70%. Scouting annual bedding plants and using yellow sticky cards on a regular basis will help time insecticide applications accordingly. Natural enemies, such as the parasitoids *Eretmocerus eremicus* and *E. mundus,* are commercially available for use against sweetpotato whitefly populations and need to be released early in the production cycle. In addition, the predatory mite *Amblyseius swirskii* feeds on sweetpotato whitefly eggs and nymphs; it can be used in conjunction with the parasitoids.

Selected References

Berg Stack, L., Cloyd, R., Dill, J., McAvoy, R., Pundt, L., Raudales, R., Smith, C., and Smith, T. 2015–2016. New England Greenhouse Floriculture Guide: A Management Guide for Insects, Diseases, Weeds, and Growth Regulators. New England Floriculture and New England State Universities.

Dreistadt, S. H. 2001. Integrated Pest Management for Floriculture and Nurseries. Univ. Calif. State. Integr. Pest Manage. Proj. Publ. 3402. Division of Agriculture and Natural Resources, University of California, Oakland.

Powell, C. C., and Lindquist, R. K. 1997. Ball Pest and Disease Manual, 2nd ed. Ball Publishing, Chicago.

(Prepared by R. A. Cloyd)

Tarnished Plant Bug

The tarnished plant bug (*Lygus lineolaris*) feeds on a wide range of annual bedding plants in landscapes, including calendula, chrysanthemum, coreopsis, cosmos, dahlia, gaillardia, impatiens, marigold, osteospermum, verbena, and zinnia.

Fig. 181. Tarnished plant bug (*Lygus lineolaris*) adult on a stem. (Courtesy R. A. Cloyd—© APS)

Life Cycle

A tarnished plant bug adult is 2.5–3.1 mm in length and oval and flat in shape. An adult is mottled brown and has several small yellow, brown, red, or black markings on its body (Fig. 181). A white or black triangle is typically visible on the front of the abdomen.

The adult female deposits eggs into the stems, leaves, and buds of annual bedding plants. The eggs hatch into nymphs, which are round and yellow-green, with black spots on the thorax and abdomen. The antennae and legs are relatively long. Nymphs feed and mature within 1 month.

The life cycle is completed in 21–28 days, depending on the temperature. There may be several generations per year.

Damage

Tarnished plant bugs have piercing–sucking mouthparts, which they insert into the vascular fluids of plants. Feeding by nymphs and adults can cause yellowing and twisting or distortion of terminal growth, and leaves may appear ragged and discolored. In addition, emerging flowers may fail to develop or flower buds may abort.

Management

Removing weeds and plant debris will help manage tarnished plant bug nymphs and adults. Contact insecticides may be effective in suppressing tarnished plant bug populations, as long as the applications are conducted on a frequent basis and all plant parts are thoroughly covered.

Selected References

Cranshaw, W., and Shetlar, D. 2017. Garden Insects of North America, 2nd ed. Princeton University Press, Princeton, NJ.

Johnson, W. T., and Lyon, H. H. 1988. Insects That Feed on Trees and Shrubs, 2nd ed. Cornell University Press, Ithaca, NY.

Westcott, C. 1973. The Gardener's Bug Book, 4th ed. Doubleday, Garden City, NY.

(Prepared by R. A. Cloyd)

Twospotted Spider Mite

The twospotted spider mite (*Tetranychus urticae*) feeds on a wide range of annual bedding plants grown in greenhouses and landscapes, including ageratum, dahlia, fan flower, geranium, gerber daisy, gomphrena, impatiens, marigold, pansy, phlox, primrose, salvia, sweetpotato vine, and viola.

Life Cycle

A twospotted spider mite adult is 0.3–0.4 mm long and oval shaped; it may be orange, green, or yellow and has two dark spots on both sides of the abdomen (Fig. 182). The adult female twospotted spider mite lays eggs in clusters on the undersides of leaves. The female is capable of laying up to 10 eggs per day and can produce more than 100 eggs during her life span. Six-legged larvae emerge from the eggs and develop into eight-legged nymphs and then adults.

During late summer through fall, twospotted spider mites turn orange-red and enter a diapause (resting) stage, as fertilized females residing in soil, on weeds, and in plant debris respond to shorter days and lower temperatures. Growers should use a 10× to 20× hand lens to check the undersides of leaves, especially along the leaf veins, for the presence of twospotted spider mites. Growers can also detect twospotted spider mites on annual bedding plants by gently tapping a plant over a white sheet of paper held underneath the leaves. Twospotted spider mites (which resemble small green, red, or yellow specks, similar in size to grains of pepper) will drop onto the paper and slowly move around.

The development time from egg to adult may take less than 1 week at temperatures of 30–32°C. The twospotted spider mite prefers warm, dry conditions.

Damage

The twospotted spider mite has piercing–sucking mouthparts and generally feeds on the undersides of leaves. The mites use their styletlike mouthparts to pierce and feed on individual plant cells, damaging the spongy mesophyll, palisade parenchyma, and chloroplasts. The damage caused by twospotted spider mite feeding decreases the chlorophyll content and inhibits the plant's ability to manufacture food through the process of photosynthesis.

Damaged leaves may appear bleached and have small white to silver-gray to yellowish speckles (Fig. 183); however, not all plants will exhibit these symptoms. Leaves heavily infested with spider mites may appear yellow to bronze, turn brown, desiccate, and fall off (Fig. 184). In addition, webbing may be present on leaves and plant stems when the twospotted spider mite population is large (Fig. 185). Feeding by twospotted spider mites may also cause flowers to become discolored or faded.

Management

Management of the twospotted spider mite on annual bedding plants involves applications of miticides (acaricides) and implementing a number of cultural practices, such as not overfertilizing plants (especially with nitrogen-based, water-soluble fertilizers), removing old plant material and debris, not allowing plants to become water stressed (which increases susceptibility), removing weeds and heavily infested plants from inside greenhouses and around landscape areas, and applying overhead irrigation in landscapes and greenhouses (which may dislodge or wash off mites from the tops of leaves).

Fig. 182. Twospotted spider mite (*Tetranychus urticae*) adult. (Courtesy R. A. Cloyd—© APS)

Fig. 183. Bleaching of violet leaves caused by the twospotted spider mite (*Tetranychus urticae*). (Courtesy R. A. Cloyd—© APS)

Fig. 184. Speckling and browning of leaves caused by the twospotted spider mite (*Tetranychus urticae*). (Courtesy R. A. Cloyd—© APS)

Fig. 185. Webbing on leaves and stems of sweetpotato vine plant caused by the twospotted spider mite (*Tetranychus urticae*). (Courtesy R. A. Cloyd—© APS)

Miticides can be used to suppress twospotted spider mite populations, but effectiveness requires making frequent applications and obtaining thorough coverage of all plant parts with spray solutions (especially the undersides of leaves). If obtaining this level of coverage is difficult, then using translaminar miticides may be warranted. Translaminar activity refers to absorption by the upper side of the leaf surface so the active ingredient is available even after residues dissipate when mites feed on the leaf underside or untreated leaf surface. The active ingredient penetrates the leaf tissue and forms a reservoir of active ingredient within the leaf that can persist for up to 14 days (although persistence is dependent on the specific translaminar miticide). Growers should rotate miticides with different modes of action to reduce the potential for twospotted spider mite populations to develop resistance.

Preventive releases of predatory mites may be effective in regulating populations of the twospotted spider mite. Commercially available predatory mites include *Amblyseius andersoni, Galendromus occidentalis, Neoseiulus californicus, N. fallacis,* and *Phytoseiulus persimilis*. Predatory mite releases may be especially helpful in regulating twospotted spider mite populations on long-term crops of hanging baskets and long-term combination planters. The larvae of the predatory midge *Feltiella acarisuga* feeds on all the life stages of the twospotted spider mite and may be more effective than predatory mites in finding localized populations. Similarly, the larvae and adult stages of the predatory ladybird beetle *Stethorus punctillum* feed on all the life stages of the twospotted spider mite.

Selected References

Cloyd, R. A. 2011. Twospotted Spider Mite: Management in Greenhouses and Nurseries. Kans. State Univ. Agric. Exp. Stn. Coop. Ext. Serv. Publ. MF-2997. Kansas State University, Manhattan.

Dreistadt, S. H. 2001. Integrated Pest Management for Floriculture and Nurseries. Univ. Calif. State. Integr. Pest Manage. Proj. Publ. 3402. Division of Agriculture and Natural Resources, University of California, Oakland.

Hoy, M. A. 2011. Agricultural Acarology: Introduction to Integrated Mite Management. CRC Press, Boca Raton, FL.

Westcott, C. 1973. The Gardener's Bug Book, 4th ed. Doubleday, Garden City, NY.

Zhi-Quang, Z. 2003. Mites of Greenhouses. CABI, Wallingford, Oxfordshire, UK.

(Prepared by R. A. Cloyd)

Western Flower Thrips

The western flower thrips (*Frankliniella occidentalis*) is a major insect pest of many annual bedding plants grown in greenhouses and incorporated into landscapes, including ageratum, begonia, calendula, calibrachoa, chrysanthemum, dahlia, geranium, gerber daisy, impatiens, marigold, petunia, portulaca, Sweet William, verbena, and zinnia.

Life Cycle

A western flower thrips adult is 2 mm long, and females are commonly larger than males. Females vary in color from yellow to dark brown, whereas males are pale yellow and have a narrow abdomen. The female lives from 30 to 45 days and can lay 150–250 eggs during her lifetime.

The female inserts eggs into leaf tissue, and light-yellow larvae hatch from eggs in 2–4 days. Larvae feed on the leaves and flowers of annual bedding plants until pupation. There are two pupal stages (prepupa and pupa); they do not feed and are located in growing media, leaf debris, and open flowers. Adults emerge from pupae after about 6 days and are attracted to annual bedding plants that have yellow or blue flowers.

The life cycle from egg to adult is completed in 2–3 weeks, but development depends on temperature. Numerous generations may be present simultaneously during the summer months, particularly under greenhouse conditions.

Damage

Feeding by western flower thrips on plant leaves may result in leaf scarring and necrotic spotting, as well as distorted growth and sunken tissue on leaf undersides (Fig. 186). Adult feeding on flowers and unopened buds may cause flower bud abortion or flower deformation (Fig. 187). Feeding on leaf buds before they open may also result in leaf scarring. Damaged flowers and leaves typically have a characteristic white or silvery appearance. In addition, black fecal deposits, or excrement, may be present on the undersides of leaves (Fig. 188).

Western flower thrips can indirectly damage annual bedding plants by transmitting viruses, including *Impatiens necrotic spot virus* and *Tomato spotted wilt virus* (see Part I, the section "Tomato Spotted Wilt, Impatiens Necrotic Spot, and Other Tospovirus Diseases").

Management

Removing weeds from inside greenhouses, around production areas, and in landscapes eliminates refuge sites for the western flower thrips. Moreover, removing the flowers from

Fig. 186. Scarring and spotting of zinnia leaves caused by western flower thrips (*Frankliniella occidentalis*) feeding. (Courtesy R. A. Cloyd—© APS)

Fig. 187. Damage to a phlox flower caused by western flower thrips (*Frankliniella occidentalis*) feeding. (Courtesy R. A. Cloyd—© APS)

Fig. 188. Silvering of a bean leaf caused by western flower thrips (*Frankliniella occidentalis*) feeding. (Courtesy R. A. Cloyd—© APS)

annual bedding plants that are not useable or are senescing can decrease western flower thrips populations.

Contact insecticides may be effective, but they must be applied before the western flower thrips enters terminal or flower buds, because most insecticides are unable to penetrate buds or flowers. Translaminar insecticides (i.e., after application, the material penetrates leaf tissue and new terminal growth, forming a reservoir of active ingredient within the leaf or new growing point) may be more effective given their ability to move into plant tissues, where the western flower thrips feeds. Repeated applications will likely be required, especially if excessive thrips populations are present and overlapping generations have been produced. Growers should rotate insecticides with different modes of action to reduce the potential for western flower thrips populations to develop resistance.

Commercially available natural enemies—including the predatory mites *Amblyseius swirskii* and *Neoseiulus cucumeris* and *Orius* spp. predatory bugs—can be purchased from biological control distributors and suppliers and released to regulate western flower thrips populations. Because of the rapid production time of annual bedding plants in greenhouses, the use of natural enemies may not be feasible. Nevertheless, the use of natural enemies may be an option for bedding plants with longer production times, such as hanging baskets and combination planters. In addition, the use of so-called banker plants, which provide refuge and a food source for *Orius* spp. (e.g., certain cultivars of ornamental pepper, including Purple Flash and Black Pearl), may provide long-term regulation of western flower thrips populations. The soilborne predatory mite *Stratiolaelaps scimitus,* which primarily attacks fungus gnat larvae, may also feed on western flower thrips pupae.

In the greenhouse, scouting or monitoring plants should be done regularly by using yellow or blue sticky cards or by conducting visual inspections. Doing so will determine the population dynamics of the western flower thrips throughout the growing season and help time insecticide applications accordingly.

Selected References

Baker, J. R. 1994. Insect and Related Pests of Flowers and Foliage Plants. N.C. Coop. Ext. Serv. Publ. AG-136. North Carolina State University, Raleigh.

Casey, C., ed. 1990. Integrated Pest Management for Bedding Plants: A Scouting and Pest Management Guide, 2nd ed. Cornell Cooperative Extension, Ithaca, NY.

Cloyd, R. A. 2009. Western flower thrips (*Frankliniella occidentalis*) management on ornamental crops grown in greenhouses: Have we reached an impasse? Pest Technol. 3(1):1-9.

Cloyd, R. A. 2010. Western Flower Thrips: Management on Greenhouse-Grown Crops. Kans. State Univ. Agric. Exp. Stn. Coop. Ext. Serv. Publ. MF-2922. Kansas State University, Manhattan.

Cranshaw, W., and Shetlar, D. 2017. Garden Insects of North America, 2nd ed. Princeton University Press, Princeton, NJ.

Dreistadt, S. H. 2001. Integrated Pest Management for Floriculture and Nurseries. Univ. Calif. State. Integr. Pest Manage. Proj. Publ. 3402. Division of Agriculture and Natural Resources, University of California, Oakland.

Gill, S., and Sanderson, J. 1998. Ball Identification Guide to Greenhouse Pests and Beneficials. Ball Publishing, Chicago.

Parrella, M. P. 1995. IPM—Approaches and prospects. Pages 357-363 in: Thrips Biology and Management. B. L. Parker, M. Skinner, and T. Lewis, eds. Plenum Press, New York.

Reitz, S. R. 2009. Biology and ecology of the western flower thrips (Thysanoptera: Thripidae): The making of a pest. Fla. Entomol. 92(1):7-13.

(Prepared by R. A. Cloyd)

Appendix I

Host Plants

Ageratum houstonianum Mill.—ageratum, floss flower
Amaranthus tricolor L.—amaranth
Anemone L. spp.—anemone, wind flower
Angelonia angustifolia Benth.—angelonia, summer snapdragon
Antirrhinum majus L.—snapdragon
Argyranthemum frutescens (L.) Sch. Blp.—Marguerite daisy, argyranthemum
Aster L. spp.—aster
Begonia × *hiemalis* Fotsch—hiemalis begonia, Rieger begonia
Begonia × *tuberhybrida* Voss.—nonstop begonia
Begonia rex-cultorum—rex begonia hybrids
Begonia semperflorens Link & Otto—wax begonia
Bellis perennis L.—English daisy
Bidens ferulifolia (Jacq.) Sweet—bidens
Brassica oleracea L. (Acephala group)—ornamental kale
Brassica oleracea L. (Capitata group)—ornamental cabbage
Browallia speciosa Hook.—amethyst flower, browallia
Calceolaria crenatiflora Cav.—slipperwort, calceolaria
Calendula officinalis L.—calendula, pot marigold
Calibrachoa hybrids—calibrachoa
Callistephus chinensis (L.) Nees—China aster
Campanula carpatica Jacq.—campanula, bellflower
Capsicum annuum L.—ornamental pepper
Catharanthus roseus (L.) G. Don—annual vinca, Madagascar periwinkle
Celosia argentea L.—celosia
Celosia argentea var. *cristata* (L.) Kuntze—cockscomb
Centaurea cyanus L.—bachelor's button, cornflower
Chaenostoma cordatum (Thunb.) Benth. (syn. *Sutera cordata* (Thunb.) Kuntze)—bacopa
Chrysanthemum maximum L.—Shasta daisy
Chrysanthemum morifolium Ramat. (syn. *Dendranthema* × *grandiflorum*)—chrysanthemum
Clarkia amoena (Lehm.) A. Nelson & J. F. Macbr. (syn. *Godetia grandiflora* Lindl.)—godetia, farewell-to-spring
Cleome houtteana Schltdl. (syn. *Cleome hassleriana* Chodat)—spider flower
Coreopsis auriculata L.—lobed tickseed
Coreopsis grandiflora Hogg ex Sweet—large-flowered tickseed
Coreopsis verticillata L.—whorled tickseed
Cosmos bipinnatus Cav.—garden cosmos, Mexican aster
Cosmos sulphureus Cav.—cosmos
Cuphea ignea A. DC.—cigar flower
Cyclamen persicum Mill.—cyclamen
Dahlia hort. hybrids—dahlia
Dahlia pinnata Cav.—dahlia
Delphinium L. spp.—larkspur
Dianthus barbatus L.—Sweet William
Dianthus caryophyllus—carnation
Dianthus chinensis L.—China pink
Dianthus deltoides L.—maiden pink
Diascia hybrids—twinspur, diascia
Dimorphotheca sinuata DC. (syn. *D. aurantiaca* DC.)—dimorphotheca, African daisy
Echinacea purpurea (L.) Moench—purple coneflower
Erysimum × *cheiri* (L.) Crantz (syn. *Cheiranthus* × *cheiri* (L.))—wallflower
Eustoma grandiflorum (Raf.) Shinners (syn. *E. russellianum* G. Don)—lisianthus
Felicia amelloides (L.) Voss—blue daisy
Fuchsia hybrida hort. ex Siebert & Voss—fuchsia
Gaillardia × *grandiflora* Hort. ex Van Houtte—blanket flower, gaillardia
Gazania linearis (Thunb.) Druce (syn. *G. longiscapa* DC.)—treasure flower, gazania
Gerbera jamesonii Bolus ex Hook. f.—gerber daisy, gerbera daisy, Transvaal daisy
Gomphrena globosa L.—common globe amaranth, gomphrena
Helianthus annuus L.—sunflower
Helichrysum bracteatum (Venten.) Willd.—strawflower
Heliotropium arborescens L.—heliotrope
Iberis sempervirens L.—candytuft
Impatiens balsamina L.—balsam impatiens
Impatiens hawkeri W. Bull—New Guinea impatiens
Impatiens walleriana Hook. f. (syns. *I. holstii* Engl. & Warb.; *I. sultanii* Hook. f.)—garden impatiens
Ipomoea batatas (L.) Lam.—sweetpotato vine
Lavandula angustifolia Mill. (syn. *L. officinalis* Chaix)—English lavender
Linaria maroccana Hook. f.—toadflax
Linum grandiflorum Desf.—scarlet flax
Linum perenne L. and *L. lewisii* Pursh—blue flax
Lobelia erinus L.—garden lobelia
Lobularia maritima (L.) Desv.—sweet alyssum
Lunaria annua L.—honesty
Lupinus albus L. and other *Lupinus* L. spp.—lupine
Matthiola incana (L.) R. Br.—garden stock, ten-weeks stock, hoary stock
Moluccella laevis L.—bells-of-Ireland
Nemesia fruticans Benth.—nemesia
Nicotiana sylvestris Speg. & S. Comes, *N. alata* Link & Otto, and *N. langsdorfii* Schrank—flowering tobacco
Ocimum basilicum L.—basil
Osteospermum ecklonis (DC.) Norl.—Cape Marguerite, osteospermum
Osteospermum fruticosum (L.) Norl.—trailing African daisy, osteospermum
Pelargonium × *hortorum* L. H. Bailey—florist's geranium
Pelargonium domesticum L. H. Bailey—regal geranium, Martha Washington geranium
Pelargonium peltatum (L.) L'Hér.—ivy geranium
Pelargonium zonale (L.) L'Hér. ex Aiton—zonal geranium

Pentas lanceolata (Forssk.) Deflers—star flower
Pericallis hybrida (Regel) B. Nord.—cineraria
Petunia hybrida Vilm.—petunia
Phlox drummondii Hook.—annual phlox
Phlox paniculata L.—garden phlox
Phlox stolonifera Sims—creeping phlox
Platycodon grandiflorus (Jacq.) A. DC.—balloon flower
Plectranthus scutellarioides (L.) R. Br. (syns. *Coleus blumei* Benth.; *Solenostemon scutellarioides* (L.) Codd)—coleus
Portulaca grandiflora Hook.—moss rose, portulaca
Portulaca oleracea L.—purslane, moss rose
Primula × *polyantha* Mill.—false cowslip, primrose
Primula elatior (L.) Hill—true oxlip, primrose
Primula vulgaris Huds.—English primrose, common primrose
Ranunculus asiaticus L.—ranunculus
Rosa hybrids—rose
Rudbeckia hirta L.—black-eyed Susan
Salvia coccinea Buc'hoz ex Etl.—Texas sage
Salvia farinacea Benth.—blue salvia, mealycup sage
Salvia greggii A. Gray—autumn sage
Salvia officinalis L.—common sage
Salvia splendens Sellow ex Schult.—red salvia, red sage
Scaevola aemula R. Br.—fan flower
Senecio cineraria DC.—dusty miller
Senecio cruentus (Masson ex L'Hér.) DC.—cineraria
Tagetes erecta L.—African marigold, American marigold, Mexican marigold, Aztec marigold
Tagetes patula L.—French marigold
Torenia fournieri Linden ex W. Fourn.—torenia, wishbone flower
Tropaeolum majus L.—nasturtium
Tropaeolum officinale R. Br.—nasturtium
Verbena hybrida Groenl. & Rümpler—garden verbena
Viola × *wittrockiana* Gams—pansy
Viola cornuta L.—viola
Zinnia angustifolia Kunth (syn. *Z. linearis* Benth.)—narrowleaf zinnia, Mexican zinnia
Zinnia elegans L.—zinnia
Zinnia violacea Cav. × *Z. angustifolia* Kunth hybrids (syn. *Z. marylandica* D. M. Spooner, Stimart & T. H. Boyle)—zinnia

Appendix II

Diseases of Bedding Plants

Notes:

- Relevant plant hosts are provided in parentheses following pathogen names. A pathogen for which no specific host or hosts are identified has a wide host range.
- The content in this appendix is also available on the website of The American Phytopathological Society in the section "Common Names of Plant Diseases." Any updates that are subsequently made to the taxonomy and nomenclature will be reflected on the website.

BACTERIAL DISEASES

Aster yellows
'Candidatus Phytoplasma asteris,' aster yellows (16SrI-B) group, Lee et al.

Bacterial blight of geranium
Xanthomonas hortorum pv. *pelargonii* (Brown) Vauterin et al. (*Pelargonium* and *Geranium* spp.)
(syn. *Xanthomonas campestris* pv. *pelargonii* (Brown) Dye)

Bacterial blight of *Matthiola* spp.
Xanthomonas campestris pv. *incanae* (Kendrick & Baker) Dye (*Matthiola* spp.)

Bacterial leaf spot and flower spot of zinnia
Xanthomonas campestris (Pammel) Dowson (*Zinnia* spp.)
(syns. *Xanthomonas campestris* pv. *zinniae* (Hopkins & Dowson) Dye; *Xanthomonas nigromaculans* f. sp. *zinniae* Hopkins & Dowson)
[Note: Should be considered *Xanthomonas campestris sensu lato* until further studied.]

Bacterial leaf spot of ranunculus
Xanthomonas campestris (Pammel) Dowson (*Ranunculus* spp.)

Bacterial soft rot
Dickeya chrysanthemi pv. *chrysanthemi* (Burkholder et al.) Samson et al.
(syn. *Erwinia chrysanthemi* pv. *chrysanthemi* Burkholder et al.)
Pectobacterium atrosepticum (van Hall) Gardan et al.
(syn. *Erwinia carotovora* subsp. *atroseptica* (van Hall) Dye)
Pectobacterium carotovorum subsp. *carotovorum* (Jones) Hauben et al. emend Gardan et al.
(syn. *Erwinia carotovora* subsp. *carotovora* (Jones) Bergey et al.)

Crown gall
Agrobacterium tumefaciens (Smith & Townsend) Conn
(syn. *Rhizobium radiobacter* (Beijerinck & van Delden) Young et al.)

Leafy gall (Bacterial fasciation)
Rhodococcus fascians (Tilford) Goodfellow
(syn. *Corynebacterium fascians* (Tilford) Dawson)

Periwinkle wilt
Xylella fastidiosa Wells et al. (*Catharanthus roseus;* many woody plant and weed hosts; some strain specialization)

Pseudomonas leaf spot
Pseudomonas cichorii (Swingle) Stapp
Pseudomonas syringae pv. *antirrhini* (Takimoto) Young et al. (*Antirrhinum* spp.)
Pseudomonas syringae pv. *primulae* (Ark & Gardner) Young et al. (*Primula* spp.)
Pseudomonas syringae pv. *syringae* van Hall
Pseudomonas syringae pv. *tagetis* (Hellmers) Young et al. (*Tagetes* and *Helianthus* spp.)

Southern bacterial wilt
Ralstonia solanacearum (Smith) Yabuuchi et al. (some strain specialization)
(syn. *Pseudomonas solanacearum* (Smith) Smith)

FUNGAL DISEASES

Alternaria leaf spot
Alternaria alternata (Fr.:Fr.) Keissl. (*Catharanthus* spp., *Dahlia* spp., *Gerbera* spp., *Impatiens balsamina,* and *Pelargonium* spp.)
(syn. *Alternaria tenuis* Nees)
Alternaria dianthicola Neerg. (*Dianthus* spp.)
Alternaria ellipsoidea E. G. Simmons (*Dianthus* spp.)
Alternaria nobilis (Vize) E. G. Simmons (*Dianthus* spp.)
(syn. *Alternaria dianthi* F. Stevens & J. G. Hall)
Alternaria saponariae (Peck) Neerg. (*Dianthus* spp.)
Alternaria tagetica S. K. Shome & Mustafee (*Tagetes* spp.)
Alternaria zinniae M. B. Ellis (*Cosmos, Helianthus, Tagetes,* and *Zinnia* spp., among others)

Anthracnose
Colletotrichum gloeosporioides (Penz.) Penz. & Sacc.
(syn. *Glomerella cingulata* (Stoneman) Spauld. & H. Schrenk)
Colletotrichum theobromicola Delacr. (*Cyclamen persicum*)
(syn. *Colletotrichum fragariae* A. N. Brooks)
Cryptocline cyclaminis Sibilia (Arx) (*Cyclamen persicum*)
(syn. *Gloeosporium cyclaminis* Sibilia)

Black root rot
Thielaviopsis basicola (Berk. & Broome) Ferraris (hosts in multiple genera, including *Calibrachoa, Catharanthus, Petunia,* and *Viola* spp.)
(syn. *Chalara elegans* Nag Raj & W. B. Kendr.)

Botrytis blight
Botrytis cinerea Pers.:Fr.
(syn. *Botryotinia fuckeliana* (de Bary) Whetzel)

Cercospora leaf spots
Cercospora apii Fresen. (hosts in multiple genera, including *Moluccella* and *Impatiens* spp.)
Cercospora violae Sacc. (*Viola* spp.)
Choanephora wet rot
Choanephora cucurbitarum (Berk. & Ravenel) Thaxt. (multiple hosts, including *Begonia, Catharanthus, Dahlia, Helianthus, Plectranthus, Tagetes, Tropaeolum,* and *Zinnia* spp.)
Choanephora infundibulifera (Curr.) Sacc. (*Dahlia, Petunia,* and *Zinnia* spp.)
Corynespora leaf spot
Corynespora cassiicola (Berk. & M. A. Curtis) C. T. Wei
(syn. *Helminthosporium cassiicola* Berk. & M. A. Curtis)
Fairy ring leaf spot
Cladosporium echinulatum (Berk.) G. A. de Vries (multiple hosts, including *Dianthus* and *Salvia* spp.)
(syns. *Didymellina dianthi* C. C. Burt; *Heterosporium echinulatum* (Berk.) Cooke; *Mycosphaerella dianthi* (C. C. Burt) Jørst.)
Fusarium root, crown, and stem rots
Fusarium avenaceum (Fr.:Fr.) Sacc. (*Eustoma grandiflorum* and others)
(syn. *Gibberella avenacea* R. J. Cook)
Fusarium oxysporum Schltdl.:Fr.
Fusarium solani (Mart.) Sacc.
Fusarium wilts
Fusarium oxysporum Schltdl.:Fr. (vascular wilts caused by different *formae speciales* on *Cyclamen persicum,* as well as *Dahlia, Eustoma, Gerbera,* and *Ranunculus* spp.)
Itersonilia petal blight
Itersonilia perplexans Derx (*Anemone, Callistephus, Chrysanthemum,* and *Dahlia* spp.; other hosts in Asteraceae and some in Apiaceae)
Itersonilia seedling blight, leaf spots, and root canker
Itersonilia perplexans Derx (*Helianthus* spp. and other hosts in Apiaceae and Asteraceae)
Mycocentrospora leaf spot
Mycocentrospora acerina (R. Hartig) Deighton (*Petunia, Ranunculus,* and *Viola* spp.; other hosts in various families)
(syns. *Centrospora acerina* (R. Hartig) Vienn.-Bourg.; *Cercospora acerina* R. Hartig)
Myrothecium leaf spot and canker
Myrothecium verrucaria (Alb. & Schwein.) Ditmar:Fr. (*Zinnia* spp.)
Paramyrothecium roridum (Tode) L. Lombard & Crous (hosts in multiple genera, including *Impatiens hawkeri* and *Viola* spp.)
(syn. *Myrothecium roridum* Tode)
Phomopsis stem and leaf blight
Diaporthe sp. (*Eustoma* spp. and others)
Phyllosticta leaf spot
Phyllosticta sp. (*Salvia* spp. and others)
Powdery mildews
Erysiphe aquilegiae DC. (*Aquilegia* spp. and primarily other plants in Ranunculaceae)
Erysiphe begoniicola U. Braun & S. Takam. (hosts in Begoniaceae)
(syns. *Microsphaera begoniae* Sivan.; *Oidium begoniae* var. *macrosporum* A. A. Mendonça & Marta Seq.)
Erysiphe cruciferarum Opiz ex L. Junell
(syn. *Erysiphe pisi* var. *cruciferarum* (Opiz ex L. Junell) Ialongo)
Erysiphe pisi var. *pisi* DC. (hosts in Fabaceae and some other families)
Erysiphe polygoni DC. (hosts in Polygonaceae)
Euoidium longipes (Noordel. & Loer.) U. Braun & R. T. A. Cook (hosts in Solanaceae)
(syn. *Oidium longipes* Noordel. & Loer.)
Fibroidium heliotropii-indici (Sawada) U. Braun & R. T. A. Cook (*Heliotropium* spp. and many additional hosts)
Golovinomyces asterum var. *asterum* (Schwein.) U. Braun (hosts in Asteraceae)
Golovinomyces biocellatus (Ehrenb.) Heluta (hosts in Lamiaceae)
Golovinomyces cichoracearum (DC.) Heluta (hosts in Asteraceae and Cucurbitaceae)
(syns. *Erysiphe cichoracearum* DC.; *Erysiphe cichoracearum* var. *cichoracearum* DC.)
Golovinomyces monardae (G. S. Nagy) M. Scholler et al. (hosts in Lamiaceae)
Golovinomyces orontii (Castagne) Heluta
(syns. *Erysiphe orontii* Castagne; *Erysiphe polyphaga* Hammarl.; *Oidium begoniae* Puttemans)
Golovinomyces spadiceus (Berk. & M. A. Curtis) U. Braun (hosts in Asteraceae)
Leveillula taurica (Lév.) G. Arnaud
Neoerysiphe cumminsiana (U. Braun) U. Braun (hosts in Asteraceae)
(syn. *Erysiphe cumminsiana* U. Braun)
Neoerysiphe galeopsidis (DC.) D. Braun (hosts in Acanthaceae and Bignoniaceae; *Althaea, Lamium,* and *Stachys* spp.)
(syn. *Erysiphe cichoracearum* f. *galeopsidis* (DC.) E. S. Salmon)
Podosphaera fuliginea (Schltdl.) U. Braun & S. Takam.
(syns. *Sphaerotheca fuliginea* (Schltdl.) Pollacci; *Sphaerotheca humuli* var. *fuliginea* (Schltdl.) E. S. Salmon)
Podosphaera fusca (Fr.:Fr.) U. Braun & S. Takam.
(syn. *Sphaerotheca fusca* (Fr.:Fr.) S. Blumer)
Podosphaera macularis (Wallr.:Fr.) U. Braun & S. Takam. (*Humulus* spp. and hosts in multiple other genera)
(syns. *Sphaerotheca humuli* (DC.) Burrill; *Sphaerotheca macularis* (Wallr.:Fr.) Magnus)
Podosphaera xanthii (Castagne) U. Braun & Shishkoff
Pseudoidium cyclaminis (Wenzl) U. Braun & R. T. A. Cook (*Cyclamen persicum*)
(syn. *Oidium cyclaminis* Wenzl)
Pseudoidium neolycopersici (L. Kiss) L. Kiss (*Petunia* spp. and other Solanaceae)
(syn. *Oidium neolycopersici* L. Kiss)
Rhizoctonia root and stem rot
Rhizoctonia solani J. G. Kühn
(syn. *Thanatephorus cucumeris* (A. B. Frank) Donk)
Rhizoctonia web blight
Rhizoctonia solani J. G. Kühn
(syn. *Thanatephorus cucumeris* (A. B. Frank) Donk)
Rhizopus blight
Rhizopus stolonifer (Ehrenb.:Fr.) Vuill. (many hosts, including *Catharanthus roseus* and *Gerbera jamesonii*)
(syn. *Rhizopus nigricans* Ehrenb.)
Rusts
Bellis rust
Puccinia lagenophorae Cooke (hosts in Asteraceae and other families)
Brown rust of chrysanthemum
Puccinia chrysanthemi Roze (*Chrysanthemum* spp.)
(syn. *Puccinia tanaceti* DC.)
Chrysanthemum white rust
Puccinia horiana Henn. (*Chrysanthemum* spp.)
Geranium rust
Puccinia pelargonii-zonalis Doidge (*Pelargonium* spp.)
Salvia rust
Puccinia ballotiflora Long (*Salvia* spp.)
(syn. *Puccinia ballotaeflora* Long)

Snapdragon rust

Puccinia antirrhini Dietel & Holw. (*Antirrhinum* and *Cordylanthus* spp.)

Texas sage rust

Puccinia salviicola Dietel & Holw. (*Salvia* spp.)

Sclerotinia blight or white mold

Sclerotinia sclerotiorum (Lib.) de Bary

Southern blight

Athelia rolfsii (Curzi) C. C. Tu & Kimbr.
(syns. *Corticium rolfsii* Curzi; *Pellicularia rolfsii* (Curzi) E. West; *Sclerotium rolfsii* Sacc.)

Verticillium wilt

Verticillium albo-atrum Reinke & Berthold
Verticillium dahliae Kleb.

White smut

Entyloma bellidis Krieg. (*Astranthium* and *Bellis* spp.)
Entyloma calendulae (Oudem.) de Bary (*Calendula* spp.)
Entyloma dahliae Syd. & P. Syd. (*Dahlia* spp.)
(syn. *Entyloma calendulae* f. sp. *dahliae* (Syd. & P. Syd.) Viégas)
Entyloma gaillardianum Vanky (*Gaillardia* spp.)
Entyloma microsporum (Unger) J. Schröt. (*Ranunculus* spp.)
Entyloma polysporum (Peck) Farl. (*Gaillardia* spp. and other Asteraceae)

OOMYCETE DISEASES

Downy Mildews

China aster downy mildew

Basidiophora entospora Roze & Cornu (hosts in Asteraceae, including *Callistephus chinensis*)

Cleome downy mildew

Hyaloperonospora parasitica (Pers.:Fr.) Constant. (*Cleome* sp.)

Coleus downy mildew

Peronospora sp. (*Plectranthus* and *Agastache* spp.)

Gerber daisy downy mildew

Plasmopara halstedii (Farl.) Berl. & de Toni (*Gerbera jamesonii* and other Asteraceae)

Impatiens downy mildew

Plasmopara constantinescui Voglmayr & Thines (*Impatiens* spp.)
(syn. *Bremiella sphaerosperma* Constant.)
Plasmopara obducens (J. Schröt.) J. Schröt. (*Impatiens* spp.)

Lisianthus downy mildew

Peronospora chlorae de Bary (*Eustoma* spp. and other Gentianaceae)

Pansy and viola downy mildew

Peronospora violae de Bary (*Viola wittrockiana*)
Plasmopara megasperma (Berl.) Berl. (*Viola wittrockiana*)
(syn. *Bremiella megasperma* (Berl.) G. W. Wilson)

Primrose downy mildew

Peronospora oerteliana J. G. Kühn (*Primula* spp.)

Salvia spp. downy mildew

Peronospora lamii A. Braun (*Salvia* spp. and other Lamiaceae)
Peronospora swinglei Ellis & Everh. (*Salvia* spp.)

Salvia officinalis downy mildew

Peronospora salviae-officinalis Y. J. Choi et al. (*Salvia officinalis*)

Snapdragon downy mildew

Peronospora antirrhini J. Schröt. (*Antirrhinum majus*)

Stock downy mildew

Hyaloperonospora parasitica (Pers.:Fr.) Constant. (hosts in Brassicaceae)
(syn. *Peronospora parasitica* (Pers.:Fr.) Fr.)
Peronospora matthiolae Gäum. (several *Matthiola* spp.)

Sweet alyssum downy mildew

Hyaloperonospora lobulariae (Ubrizsy & Vörös) Göker et al. (*Lobularia maritima*)
Hyaloperonospora parasitica (Pers.:Fr.) Constant. (*Lobularia maritima* and other Brassicaceae)
(syn. *Peronospora parasitica* (Pers.:Fr.) Fr.)

Verbena downy mildew

Peronospora verbenae U. Braun et al. (*Verbena officinalis*)

Phytophthora Diseases

Late blight

Phytophthora infestans (Mont.) de Bary (hosts in 16 genera, including *Lycopersicon*, *Petunia*, and *Solanum*)

Phytophthora root and crown rot

Phytophthora capsici Leonian (hosts in many genera in many families, including *Calibrachoa* hybrids, *Capsicum annuum*, *Lupinus* spp., and *Nicotiana* spp.)
Phytophthora cryptogea Pethybr. & Laff. (hosts in many genera in many families, including *Antirrhinum majus*, *Begonia* spp., *Bellis perennis*, *Calceolaria crenatiflora*, *Celosia* spp., *Chrysanthemum morifolium*, *Dahlia* hybrids, *Dianthus barbatus*, *Gerbera jamesonii*, *Matthiola incana*, *Osteospermum ecklonis*, *Petunia hybrida*, *Salvia officinalis*, *Senecio cruentus*, *Tagetes erecta*, *Verbena hybrida*, and *Zinnia elegans*)
Phytophthora drechsleri Tucker (hosts in many genera in many families, including *Calibrachoa* hybrids, *Celosia argentea*, *Fuchsia hybrida*, *Gerbera jamesonii*, *Helichrysum bracteatum*, *Salvia splendens*, and *Senecio cruentus*)
Phytophthora nicotianae Breda de Haan (hosts in many genera in many families, including *Begonia* spp., *Catharanthus roseus*, *Gerbera jamesonii*, and *Petunia hybrida*)
(syns. *Phytophthora parasitica* Dastur; *Phytophthora parasitica* var. *nicotianae* (Breda de Haan) Tucker)
Phytophthora tropicalis Aragaki & J. Y. Uchida (hosts in multiple genera, including *Begonia* sp., *Cuphea ignea*, *Cyclamen persicum*, *Gerbera jamesonii*, *Lupinus albus*, and *Senecio cineraria*)

Pythium Root Rots

Globisporangium cryptoirregulare (Garzón et al.) Uzuhashi et al.
(syn. *Pythium cryptoirregulare* (Garzón et al.)
Globisporangium debaryanum (R. Hesse) Uzuhashi et al.
(syn. *Pythium debaryanum* R. Hesse)
Globisporangium intermedium (de Bary) Uzuhashi et al.
(syn. *Pythium intermedium* de Bary)
Globisporangium irregulare (Buisman) Uzuhashi et al.
(syn. *Pythium irregulare* Buisman)
Globisporangium mamillatum (Meurs) Uzuhashi et al.
(syn. *Pythium mamillatum* Meurs)
Globisporangium mastophorum (Drechsler) Uzuhashi et al.
(syn. *Pythium mastophorum* Drechsler)
Globisporangium megalacanthum (de Bary) Uzuhashi et al.
(syn. *Pythium megalacanthum* de Bary)
Globisporangium middletonii (Sparrow) Uzuhashi et al.
(syn. *Pythium middletonii* Sparrow)
Globisporangium paroecandrum (Drechsler) Uzuhashi et al.
(syn. *Pythium paroecandrum* Drechsler)
Globisporangium spinosum (Sawada) Uzuhashi et al.
(syn. *Pythium spinosum* Sawada)
Globisporangium splendens (Hans Braun) Uzuhashi et al.
(syn. *Pythium splendens* Hans Braun)
Globisporangium sylvaticum (W. A. Campb. & F. F. Hendrix) Uzuhashi et al.
(syn. *Pythium sylvaticum* W. A. Campb. & F. F. Hendrix)

Globisporangium ultimum (Trow) Uzuhashi et al. (syn. *Pythium ultimum* Trow)
Phytopythium oedochilum (Drechsler) Abad et al. (syn. *Pythium oedochilum* Drechsler)
Phytopythium vexans (de Bary) Abad et al. (syn. *Pythium vexans* de Bary)
Pythium acanthicum Drechsler
Pythium aphanidermatum (Edson) Fitzp.
Pythium myriotylum Drechsler

White Blister Rusts
Albugo candida (Pers.:Fr.) Kuntze (hosts in Brassicaceae; some reports from other families)
Albugo chardonii W. Weston (*Cleome* spp.)
[Note: Possibly a synonym of *Albugo candida*.]
Pustula centaurii (Hansf.) Thines et al. (hosts in Gentianaceae, including *Eustoma grandiflorum*)
Pustula tragopogonis (Pers.) Thines (hosts in Asteraceae, including *Gerbera, Helianthus,* and *Tragopogon* spp., and several familiar weeds, including *Ambrosia, Antennaria, Artemisia, Cirsium, Matricaria,* and *Parthenium* spp.)
(syn. *Albugo tragopogonis* (Pers.) Gray)

NEMATODES (PARASITIC)

Root-Knot Nematodes
Javanese root-knot nematode
Meloidogyne javanica (Treub) Chitwood
Northern root-knot nematode
Meloidogyne hapla Chitwood
Peanut root-knot nematode
Meloidogyne arenaria (Neal) Chitwood
Southern root-knot nematode
Meloidogyne incognita (Kofoid & White) Chitwood

Foliar Nematodes
Chrysanthemum foliar nematode
Aphelenchoides ritzemabosi (Schwartz) Steiner & Buhrer
Spring crimp nematode (syn. Strawberry crimp nematode)
Aphelenchoides fragariae (Ritzema Bos) Christie
Spring dwarf nematode
Aphelenchoides besseyi Christie

Vector Nematode
American dagger nematode
Xiphinema americanum Cobb

VIRUS AND VIROID DISEASES

Alternanthera mosaic
Alternanthera mosaic virus (AltMV)—genus *Potexvirus;* family *Alphaflexiviridae* (hosts in multiple families, including *Angelonia, Crossandra, Helichrysum, Phlox, Portulaca, Salvia, Scutellaria, Torenia,* and *Zinnia* spp.)
Angelonia flower break
Angelonia flower break virus (AnFBV)—genus *Alphacarmovirus;* family *Tombusviridae* (*Angelonia, Nemesia, Phlox,* and *Verbena* spp.)
Bean yellow mosaic
Bean yellow mosaic virus (BYMV)—genus *Potyvirus;* family *Potyviridae* (hosts in *Fabaceae,* as well as *Eustoma grandiflorum* and various other plants)
Bidens mottle
Bidens mottle virus (BMoV)—genus *Potyvirus;* family *Potyviridae* (many hosts in Asteraceae; other ornamental, vegetable, and weed hosts)
Broad bean wilt
Broad bean wilt virus 1 (BBWV-1)—genus *Fabavirus;* subfamily *Comovirinae;* family *Secoviridae* (many hosts, including *Begonia* and *Petunia* spp.)
Broad bean wilt virus 2 (BBWV-2)—genus *Fabavirus;* subfamily *Comovirinae;* family *Secoviridae* (*Catharanthus roseus* and *Eustoma grandiflorum*)
Calibrachoa mottle
Calibrachoa mottle virus (CbMV)—genus *Carmovirus* (proposed: *Alphacarmovirus*); family *Tombusviridae* (*Calibrachoa* and *Petunia* spp.)
Carnation mottle
Carnation mottle virus (CarMV)—genus *Carmovirus* (proposed: *Alphacarmovirus*); family *Tombusviridae* (*Begonia, Dianthus,* and *Zantedeschia* spp.; a few other hosts)
Chrysanthemum stem necrosis
Chrysanthemum stem necrosis virus (CSNV)—genus *Tospovirus;* family *Bunyaviridae* (*Chrysanthemum, Eustoma,* and *Solanum* spp.)
Clover yellow mosaic
Clover yellow mosaic virus (ClYMV)—genus *Potexvirus;* family *Alphaflexiviridae* (*Antirrhinum* spp., members of Papilionaceae, and several additional weed and cultivated hosts)
Coleus vein necrosis
Coleus vein necrosis virus (CVNV)—genus *Carlavirus;* family *Betaflexiviridae* (*Plectranthus* and *Verbena* spp.)
Cucumber mosaic
Cucumber mosaic virus (CMV)—genus *Cucumovirus;* family *Bromoviridae*
Dahlia mosaic
Dahlia common mosaic virus (DCMV)—possible member of genus *Caulimovirus;* family *Caulimoviridae* (*Dahlia* spp.)
Dahlia mosaic virus (DMV)—genus *Caulimovirus;* family *Caulimoviridae* (*Dahlia* spp.)
(syn. DMV-Portland)
Dahlia variabilis endogenous plant pararetroviral sequence (DvEPRS)—possible member of genus *Petuvirus;* family *Caulimoviridae* (*Dahlia* spp.)
(syn. DMV-D10)
Hibiscus chlorotic ringspot
Hibiscus chlorotic ringspot virus (HCRSV)—genus *Betacarmovirus;* family *Tombusviridae* (*Hibiscus rosa-sinensis* and other Malvaceae, primarily)
Impatiens necrotic spot
Impatiens necrotic spot virus (INSV)—genus *Tospovirus;* family *Bunyaviridae*
Iris yellow spot
Iris yellow spot virus (IYSV)—genus *Tospovirus;* family *Bunyaviridae* (*Alstroemeria, Chrysanthemum, Eustoma, Iris,* and *Portulaca* spp.; select other species in multiple families)
Lettuce mosaic
Lettuce mosaic virus (LMV)—genus *Potyvirus;* family *Potyviridae* (hosts in Asteraceae, including *Gazania* and *Osteospermum* spp.; some members of Brassicaceae, Chenopodiaceae, Cucurbitaceae, Fabaceae, and Solanaceae)
Lisianthus line pattern
Lisianthus line pattern virus (LLPV)—possible member of genus *Ilarvirus;* family *Bromoviridae*
Lisianthus necrosis
Lisianthus necrosis virus (LNV)—possible member of genus *Tombusvirus*; family *Tombusviridae* (*Eustoma grandiflorum*)
[Note: Possibly same as Eggplant mottled crinkle virus (EMCV).]
Lisianthus necrotic ringspot
Lisianthus necrotic ringspot virus (LNRV)—possible member of genus *Tospovirus;* family *Bunyaviridae* (*Eustoma grandiflorum*)
[Note: This disease is possibly caused by *Iris yellow spot virus*.]

Nemesia ring necrosis
Nemesia ring necrosis virus (NeRNV)—genus *Tymovirus;* family *Tymoviridae* (*Alonsoa, Diascia, Lobelia, Nemesia, Sutera,* and *Verbena* spp.)
Pelargonium flower break
Pelargonium flower break virus (PFBV)—genus *Alphacarmovirus;* family *Tombusviridae* (*Pelargonium* spp.; other hosts experimentally)
Pepper veinal mottle
Pepper veinal mottle virus (PVMV)—genus *Potyvirus;* family *Potyviridae* (*Capsicum* and *Solanum* spp.)
Petunia flower mottle
Colombian datura virus (CDV)—genus *Potyvirus;* family *Potyviridae* (*Brugmansia, Petunia,* and *Solanum* spp.)
(syn. *Petunia flower mottle virus* (PetFMV))
Petunia vein banding
Petunia vein banding virus (PetVBV)—genus *Tymovirus;* family *Tymoviridae* (*Petunia hybrida*)
Petunia vein clearing
Petunia vein clearing virus (PVCV)—genus *Petuvirus;* family *Caulimoviridae* (*Petunia hybrida*)
Ranunculus mild mosaic
Ranunculus mild mosaic virus (RanMMV)—genus *Potyvirus;* family *Potyviridae* (*Ranunculus asiaticus*)
Ranunculus mosaic
Ranunculus mosaic virus (RanMV)—genus *Potyvirus;* family *Potyviridae* (*Ranunculus* spp.)
Scrophularia mottle
Scrophularia mottle virus (ScrMV)—genus *Tymovirus;* family *Tymoviridae* (*Antirrhinum, Catharanthus,* and *Scrophularia* spp., plus a few others)
Tobacco etch
Tobacco etch virus (TEV)—genus *Potyvirus;* family *Potyviridae* (many hosts, including *Capsicum, Datura, Linaria, Nicotiana,* and *Solanum* spp.)
Tobacco mosaic
Tobacco mosaic virus (TMV)—genus *Tobamovirus;* family *Virgaviridae*
Tobacco rattle
Tobacco rattle virus (TRV)—genus *Tobravirus;* family *Virgaviridae*
Tobacco ringspot
Tobacco ringspot virus (TRSV)—genus *Nepovirus;* subfamily *Comovirinae;* family *Secoviridae* (many hosts, including *Anemone, Chaenostoma, Pelargonium, Petunia,* and *Portulaca* spp.)
Tobacco streak
Tobacco streak virus (TSV)—genus *Ilarvirus;* family *Bromoviridae* (many hosts, including *Alstroemeria, Dahlia, Eustoma, Helianthus,* and *Impatiens* spp.)
Tomato aspermy
Tomato aspermy virus (TAV)—genus *Cucumovirus;* family *Bromoviridae* (many hosts, including *Canna, Chrysanthemum,* and *Lilium* spp.)
Tomato bushy stunt
Tomato bushy stunt virus (TBSV)—genus *Tombusvirus;* family *Tombusviridae* (relatively restricted host range; includes *Capsicum annuum*)
Tomato chlorotic spot
Tomato chlorotic spot virus (TCSV)—genus *Tospovirus;* family *Bunyaviridae* (host range not well known; includes *Eustoma grandiflorum* and *Impatiens* spp.)
Tomato mosaic
Tomato mosaic virus (ToMV)—genus *Tobamovirus;* family *Virgaviridae* (many hosts in Solanaceae; additional hosts in other families)
Tomato ringspot
Tomato ringspot virus (ToRSV)—genus *Nepovirus;* family *Secoviridae*
Tomato spotted wilt
Tomato spotted wilt virus (TSWV)—genus *Tospovirus;* family *Bunyaviridae*
Turnip mosaic
Turnip mosaic virus (TuMV)—genus *Potyvirus;* family *Potyviridae*

Other Viruses and Viroids

Chrysanthemum chlorotic mottle viroid (CChMVd)—genus *Pelamoviroid;* family *Avsunviroidae* (*Chrysanthemum* spp.)
Chrysanthemum stunt viroid (CSVd)—genus *Pospiviroid;* family *Pospiviroidae* (*Argyranthemum, Chrysanthemum,* and *Petunia* spp.)
Helenium virus S (HVS)—genus *Carlavirus;* family *Betaflexiviridae* (*Helenium amarum, Impatiens holstii*)
Potato spindle tuber viroid (PSTVd)—genus *Pospiviroid;* family *Pospiviroidae* (hosts in Solanaceae)
Potato virus X (PVX)—genus *Potexvirus;* family *Alphaflexiviridae* (many hosts, particularly in Solanaceae)
Potato virus Y (PVY)—genus *Potyvirus;* family *Potyviridae* (many hosts in 31 families)
Primula mosaic virus (PrMV)—possible member of genus *Potyvirus;* family *Potyviridae* (*Primula* spp.)
Primula mottle virus (PrMoV)—possible member of genus *Potyvirus;* family *Potyviridae* (*Primula* spp.)
Verbena virus Y (VVY)—genus *Potyvirus;* family *Potyviridae* (*Verbena* spp.)

Appendix III

Insect, Mite, and Mollusk Pests of Bedding Plants

Aphids
Cotton aphid, melon aphid
Aphis gossypii Glover—Hemiptera, Aphididae
Foxglove aphid
Aulacorthum solani (Kaltenbach)—Hemiptera, Aphididae
Green peach aphid
Myzus persicae (Sulzer)—Hemiptera, Aphididae

Beetles
Japanese beetle
Popillia japonica (Newman)—Coleoptera, Scarabaeidae
Spotted cucumber beetle
Diabrotica undecimpunctata howardi (Barber)—Coleoptera, Chrysomelidae

Caterpillars
Beet armyworm
Spodoptera exigua (Hübner)—Lepidoptera, Noctuidae
Diamondback moth
Plutella xylostella (Linnaeus)—Lepidoptera, Plutellidae
Imported cabbageworm
Pieris rapae (Linnaeus) (syn. *Artogeia rapae* (Linnaeus))—Lepidoptera, Pieridae
Pansyworm, Variegated fritillary
Euptoieta claudia (Cramer)—Lepidoptera, Nymphalidae
Saltmarsh caterpillar
Estigmene acrea (Drury)—Lepidoptera, Erebidae
Tobacco budworm
Heliothis virescens (Fabricius)—Lepidoptera, Noctuidae
Yellowstriped armyworm
Spodoptera ornithogalli (Guenée)—Lepidoptera, Noctuidae

Fungus Gnats
Darkwinged fungus gnats
Bradysia coprophila (Lintner)—Diptera, Sciaridae
Bradysia impatiens (Johannsen)—Diptera, Sciaridae

Leafhoppers
Aster leafhopper
Macrosteles quadrilineatus (Forbes)—Hemiptera, Cicadellidae
Potato leafhopper
Empoasca fabae (Harris)—Hemiptera, Cicadellidae

Leafminer
American serpentine leafminer
Liriomyza trifolii (Burgess)—Diptera, Agromyzidae

Mealybug
Citrus mealybug
Planococcus citri (Risso)—Hemiptera, Pseudococcidae

Mites
Broad mite
Polyphagotarsonemus latus (Banks)—Acari, Tarsonemidae
Cyclamen mite
Phytonemus pallidus (Banks)—Acari, Tarsonemidae
Twospotted spider mite
Tetranychus urticae Koch—Acari, Tetranychidae

Plant Bugs
Fourlined plant bug
Poecilocapsus lineatus (Fabricius)—Hemiptera, Miridae
Tarnished plant bug
Lygus lineolaris (Palisot de Beauvois)—Hemiptera, Miridae

Shore Fly
Shore fly
Scatella stagnalis (Fallen)—Diptera, Ephydridae

Slugs
Black slug
Arion ater (Linnaeus)—Stylommatophora, Arionidae
Giant gardenslug
Limax maximus (Linnaeus)—Stylommatophora, Limacidae
Gray fieldslug
Deroceras reticulatum (O. F. Müller)—Stylommatophora, Agriolimacidae
Greenhouse slug
Milax gagates (Draparnaud)—Stylommatophora, Milacidae
Orange-banded arion, White-soled slug
Arion fasciatus (Nilsson)—Stylommatophora, Arionidae
Three-band gardenslug
Lehmannia valentiana (Férussac)—Stylommatophora, Limacidae
(syn. *Limax valentianus* (Férrusac))

Snail
Brown garden snail
Cornu aspersum (O. F. Müller)—Stylommatophora, Helicidae
(syn. *Helix aspersa* O. F. Müller)

Thrips

Chilli thrips
Scirtothrips dorsalis (Hood)—Thysanoptera, Thripidae

Greenhouse thrips
Heliothrips haemorrhoidalis (Bouché)—Thysanoptera, Thripidae

Western flower thrips
Frankliniella occidentalis (Pergande)—Thysanoptera, Thripidae

Whiteflies

Bandedwinged whitefly
Trialeurodes abutiloneus (Haldeman)—Hemiptera, Aleyrodidae

Greenhouse whitefly
Trialeurodes vaporariorum (Westwood)—Hemiptera, Aleyrodidae

Sweetpotato whitefly
Bemisia tabaci (Gennadius)—Hemiptera, Aleyrodidae

Glossary

A—acre
a.i.—active ingredient
C—Celsius or centigrade
cm—centimeter (1 cm = 0.01 m = 0.3937 in.)
F—Fahrenheit
g—gram (1 g = 0.03527 oz)
gal—gallon (1 gal liquid (U.S.) = 3.785 L)
h—hour
ha—hectare (1 ha = 2.471 acres)
in.—inch (1 in. = 2.540 cm)
kb—kilobase pair
kg—kilogram (1 kg = 2.205 lb)
L—liter (1 L = 1.057 quarts liquid (U.S.))
lb—pound (1 lb = 453.59 g)
m—meter (1 m = 39.37 in.)
mg—milligram (1 mg = 0.001 g)
min—minute
ml—milliliter (1 ml = 0.001 L)
mm—millimeter (1 mm = 0.001 m = 0.03937 in.)
µg—microgram (1 µg = 10^{-6} g)
µm—micrometer (1 µm = 10^{-6} m)
nm—nanometer (1 nm = 10^{-9} m)
oz—ounce (1 oz = 28.35 g); fluid ounce (1 fl oz (U.S.) = 29.57 ml)
ppm—parts per million
sec—second

abaxial—directed away from the stem of a plant; pertaining to the lower surface of a leaf (*see* adaxial)
abiotic—pertaining to the absence of life, as diseases not caused by living organisms
abscise—to separate from a plant, as leaves, flowers, and fruits do when they fall
abscission—the shedding of leaves or other plant parts as the result of physical weakness in a specialized layer of cells (the abscission layer), which develops at the base of the structure
acaricide—a pesticide used to kill mites (syn. miticide)
accession—a uniquely identified and catalogued member of a plant collection
acervulus (pl. acervuli)—an erumpent, cushionlike fruiting body bearing conidiophores, conidia, and sometimes setae
acicular—needle shaped
acid—having a pH less than 7
acid precipitation—precipitation of low pH because of the presence of nitric and sulfuric acid formed by the combination of air pollutants (NO_x and SO_2) with water
acid rain—*see* acid precipitation
acropetal—upward from the base to the apex of a shoot of a plant (*see* basipetal); in fungi, producing spores in succession in the direction of the apex, so that the apical spore is the youngest
acute—with reference to symptoms, developing suddenly (*see* chronic)
adaxial—directed toward the stem of a plant; pertaining to the upper surface of a leaf (*see* abaxial)
adventitious—arising from other than the usual place, as roots from a stem rather than branches of a root
aeciospore—a dikaryotic spore produced in the aecium of a rust fungus; in heteroecious rust fungi, the spore stage that infects the alternate host
aecium (pl. aecia; adj. aecial)—the fruiting body of a rust fungus in which the first dikaryotic spores (aeciospores) are produced
aerial—occurring in air
aerobic—living only in the presence of oxygen
aflatoxin—a chemical by-product of *Aspergillus flavus* and *A. parasiticus* that is harmful to humans and other animals
agar—a jellylike material derived from algae and used to solidify liquid culture media; a medium containing agar
alate—winged (*see* apterous)
albino (n. albinism)—white or light colored; having a marked deficiency in pigmentation
aleuriospore—a thick-walled spore that is released into the soil only after the hypha of a fungus disintegrates; a nondeciduous chlamydospore
alkaline—having basic (nonacidic) properties; having a pH greater than 7
alkaloid—any of various nitrogen-containing ring compounds produced by plants and having physiological effects on animals
allele—any of one or more alternative forms of a gene
alpha conidium—one type of conidium formed by *Diaporthe* spp.; single celled and oval
alternate host—one of two kinds of plants on which a parasitic fungus (e.g., a rust fungus) must develop to complete its life cycle
alternative host—a plant other than the main host that can be colonized by a parasite but that is not required for completion of the developmental cycle of the parasite
ambisense—a single-stranded genome that has both positive-sense and negative-sense sections
amphigynous—(describing a type of antheridium) surrounding the oogonium
anaerobic—living in the absence of oxygen
anamorph (adj. anamorphic)—the imperfect state or asexual form in the life cycle of certain fungi, producing asexual spores (e.g., conidia) or no spores (*see* holomorph; teleomorph)
anastomosis (pl. anastomoses)—the fusion of branches of the same or different structures (e.g., hyphae) to make a network
angiosperms—the flowering plants
annual—a plant that completes its life cycle and dies within one year (*see* biennial; perennial)
antagonist—an organism or substance that counteracts or limits the action of another organism or substance
anterior—situated toward the front or head (*see* posterior)
anther—the pollen-bearing portion of a flower
antheridium (pl. antheridia)—the male sexual organ (male gametangium) of certain fungi
anthesis—the period of the opening of a flower, during which pollination can occur
anthracnose—disease caused by acervuli-forming fungi (order Melanconiales); characterized by sunken lesions and necrosis
antibiotic—any of various chemical compounds produced by microorganisms and killing or inhibiting the growth of other living organisms
antibody—a protein formed in the blood of warm-blooded animals in response to the presence of an antigen
antigen—any foreign chemical (normally, a protein) that induces antibody formation in warm-blooded animals
antiserum (pl. antisera)—blood serum containing an antibody
apex (pl. apices; adj. apical)—the tip of a root or shoot; contains the apical meristem
aphid—any of numerous species of small, sucking insects of the family Aphididae (order Hemiptera) that produce honeydew and injure plants when present in large populations
aplerotic—with reference to oospores, not filling the oogonium

apothecium (pl. apothecia)—open, cuplike or saucerlike, ascus-bearing fruiting body (ascocarp) of ascomycetous fungi; often supported on a stalk
appressed—closely flattened down or pressed against a surface (syn. adpressed)
appressorium (pl. appressoria)—a swollen, flattened portion of a fungal filament that adheres to the surface of a higher plant, providing anchorage for invasion by the fungus
apterous—wingless (*see* alate)
arborescent—treelike
arthropod—any member of the phylum Arthropoda, which consists of animals with articulated bodies and limbs, including insects, arachnids, and crustaceans
ascocarp—a sexual fruiting body (e.g., apothecium, ascostroma, cleistothecium, perithecium, pseudothecium) of ascomycetous fungi, producing asci and ascospores (syn. ascoma)
ascogenous—pertaining to ascus-producing hyphae
ascogonium (pl. ascogonia)—a specialized cell that gives rise to hyphae that produce asci
ascoma (pl. ascomata)—*see* ascocarp
ascomycetes (adj. ascomycetous)—a group of fungi in the phylum Ascomycota (kingdom Fungi) characterized by the production of sexual spores (ascospores) within an ascus
ascospore—a sexual spore borne in an ascus
ascostroma (pl. ascostromata)—a fruiting body of ascomycetous fungi containing bitunicate (double-walled) asci in locules (cavities); usually dark and containing multiple locules but sometimes containing only a single locule (*see* pseudothecium)
ascus (pl. asci)—a saclike structure containing ascospores (typically eight) and usually borne in a fruiting body
aseptate—lacking cross-walls (septa); nonseptate; coenocytic
asexual—vegetative; lacking sex organs, gametes, or sexual spores (as in the imperfect or anamorphic stage of a fungus)
asexual reproduction—any type of reproduction not involving the union of gametes and meiosis
asexual stage—the imperfect (anamorphic) stage of certain fungi
attenuate—to narrow; to weaken; to decrease in virulence or pathogenicity
autoecious—with reference to rust fungi, producing all spore forms on one species of host plant (*see* heteroecious)
avirulent—nonpathogenic; unable to cause disease (*see* virulent)
axenic—culture in the absence of living bacteria or other organisms; pure culture
axil—the angle formed by a leaf petiole and the stem
axillary—pertaining to or placed within an axil

bacilliform—shaped like a short rod with rounded ends
backcross—to cross (mate) an offspring with one of its parents
bactericide—a chemical or physical agent that kills bacteria
bacterium (pl. bacteria)—a prokaryotic, microscopic, single-celled organism having a cell wall and increasing by binary fission
ballistospores—forcibly expelled spores
basidiocarp—the sexual fruiting body of fungi in the group basidiomycetes (syn. basidioma)
basidiomycetes (adj. basidiomycetous)—a group of fungi in the phylum Basidiomycota (kingdom Fungi) characterized by the formation of external basidiospores on basidia
basidiospore—a haploid (1*n*) sexual spore produced on a basidium
basidium (pl. basidia; adj. basidial)—a structure on which basidiospores are produced externally
basipetal—situated down from the apex toward the base of a shoot of a plant; developing in the direction of the base, so that the apical part is oldest (*see* acropetal)
beta conidium—one type of conidium formed by *Diaporthe* spp.; single celled but filiform and sometimes curved
biennial—a plant that produces seed and dies at the end of its second year of growth (*see* annual; perennial)
biguttulate—(describing spores) containing two oil droplets
binucleate—having two nuclei per cell (*see* multinucleate; uninucleate)
bioassay—any test (assay) using a living organism
biocide—a compound toxic to all forms of life
biological control—exploitation of the natural competition, parasitism, or antagonism of organisms for the management of pests and pathogens (syn. biocontrol)
biotic—pertaining to living organisms; with reference to disease, caused by a living organism
biotroph—an organism that can live and multiply only on another living organism (syn. obligate parasite) (*see* necrotroph)
biotype—a subdivision of a species, subspecies, or race based on some identifiable physiological trait, such as a particular virulence pattern
biovar—(of a prokaryote) a strain that is biologically different
bipartite genome—(of a virus) nucleic acid in two separate molecules
bitunicate—having two walls
blight—the sudden, severe, and extensive spotting, discoloration, wilting, or destruction of leaves, flowers, stems, or entire plants
blotch—an irregularly shaped, usually superficial spot or blot
botryose—shaped like a bunch of grapes
bract—a reduced leaf associated with a flower or inflorescence; a modified leaf from the axil of which a flower arises
breeding line—a plant strain used in a plant breeding program and usually containing one or more desirable agronomic or breeding characteristics
broadcast application—application of fertilizer by spreading or scattering it on the soil surface
bursa—an extension or flap of cuticle at the side of the male nematode sex organ; used for orienting the body during mating

caducous—(describing sporangia) easily detached
calcareous—rich in calcium carbonate (lime)
callus—specialized tissue that forms over a wound or cut in a plant (e.g., cork cambium may form, and the cells produced will gradually seal the wound)
calyx (pl. calyces)—the collection of sepals surrounding the petals of a flower
canker—a plant disease characterized (in woody plants) by the death of cambium tissue and malformation or loss of bark and (in nonwoody plants) by the formation of sharply delineated, dry, necrotic, localized lesions on the stem; a lesion caused by such a disease, particularly in woody plants
canopy—the expanded leafy top of a plant or plants
capsid—the protective layer of protein surrounding the nucleic acid core of a virus; the protein molecules that make up this layer (syn. coat protein)
carbohydrate—any of various chemical compounds composed of carbon, hydrogen, and oxygen, such as sugars, starches, and cellulose
carpel—the ovule-bearing structure of a flower in angiosperms
catenulate—with reference to the spores of certain fungi, formed in a chain
cation—a positively charged ion
causal agent—the organism or agent that produces a given disease
chasmothecium (pl. chasmothecia)—a spherical ascocarp found in powdery mildew fungi (syn. cleistothecium)
chelated iron—iron complexed so as to increase its availability as a plant nutrient
chemotropism—growth oriented toward a chemical attractant
chimera, chimaera—a plant or organ consisting of two or more genetically different tissues
chlamydospore—a thick-walled or double-walled, asexual resting spore formed from hyphal cells (terminal or intercalary) or by transformation of conidial cells; functions as an overwintering stage
chlorophyll (adj. chlorophyllous)—any of a group of green pigments found in chloroplasts and critical to photosynthesis
chloroplast—a disklike structure containing chlorophyll in which photosynthesis occurs in green plants
chlorosis (adj. chlorotic)—the failure of chlorophyll development, caused by disease or a nutritional disturbance; the fading of green plant parts to light green, yellow, or white
chromosome—the structure that contains the genes of an organism (in eukaryotes, chromosomes are in the nucleus and can be visualized with an optical microscope as threads or rods during meiosis and mitosis; in bacteria, the chromosome is usually a single circle of DNA and cannot be visualized with an optical microscope)
chronic—with reference to symptoms, slow-developing, persistent, or recurring (*see* acute)
circulative—with reference to viruses, passing through the gut and circulating in the body of an insect vector before being transmitted to a host

circulative transmission—virus transmission in which the virus must accumulate within or pass through the lymphatic system of an insect vector before it can be transmitted to a plant
cirrhus (pl. cirrhi), cirrus (pl. cirri)—a curled, tendril-like mass of exuded spores held together by a slimy matrix
clamp connection—a bridge- or buckle-like hyphal protrusion in basidiomycetous fungi, formed at cell division and connecting the newly divided cells
clavate—club shaped (syn. claviform)
cleistothecium (pl. cleistothecia)—a spherical ascocarp that is closed at maturity
coalesce—to grow together into one body, as enlarging lesions may join together to form a single spot
coenocytic—having multiple nuclei embedded in cytoplasm without cross-walls; nonseptate
colonize—with reference to plant pathogens, to infect and ramify through plant tissue
colony—a microorganism growing in mass, especially as a pure culture
color break, color breaking— a symptom of virus infection in which the normal petal coloration is disfigured by discolored or bleached areas (syn. flower color break, flower color breaking)
conidiogenous—producing and bearing conidia
conidioma (pl. conidiomata)—a fungal fruiting structure bearing conidia
conidiophore—a simple or branched hypha on which conidia are produced
conidium (pl. conidia)—an asexual, nonmotile fungal spore that develops externally or is liberated from the cell that formed it
coralloid—corallike in form
cortex (adj. cortical)—the region of parenchyma tissue between the epidermis and phloem in stems and roots; the region beneath the rind of a sclerotium
cotyledon—the primary embryonic leaf within the seed, in which nutrients for the new plant are stored (monocots have one cotyledon; dicots have two)
crown—the upper dome of a tree, bearing leaves, flowers, and fruits; the junction of the root and stem of a plant, usually at the soil line; in grafted woody plants, the rootstock portion of the plant near the soil surface
crown rot—the infection of a plant at the soil line, often involving both stem and root tissue
crucifer (adj. cruciferous)—a member of the cabbage family; the Brassicaceae (syn. Cruciferae)
culm—the stem of grasses, cereals, and bamboos
cultivar (abbr. cv.)—a cultivated variety of a plant species, resulting from deliberate genetic manipulation and having recognizable characteristics, such as color; shape of flowers, fruits, and seeds; height; and form (syn. variety)
cultural practices—the methods by which plants are grown (e.g., application of nutrients, irrigation, cultivation, and so forth)
culture—the growth and propagation of microorganisms on nutrient media; the growth and propagation of living plants
culture indexing—a lab technique for clean stock development in which slices from the bases of cuttings are cultured to detect the presence of vascular pathogens (bacterial or fungal)
cuticle (adj. cuticular)—the noncellular outer layer of an insect or nematode; the water-repellent, waxy layer of epidermal cells of plant parts, such as leaves, stems, and fruits
cyst—in fungi, a resting structure in a protective membrane or shell-like enclosure; in nematodes, the egg-laden carcass of a female nematode; in bacteria, a specialized cell enclosed in a thick wall; often dormant and resistant to environmental conditions
cytoplasm—the living protoplasm in a cell, not including the nucleus

damping-off—the death of a seedling before or shortly after emergence caused by decomposition of the root or lower stem (it is common to distinguish between preemergence and postemergence damping-off)
defoliation—the loss of leaves from a plant, whether normal or premature
dehiscent—opening by breaking into parts
dematiaceous—(describing fungi) pigmented; having dark (brown, black) hyphae
demicyclic—pertaining to the life cycle of rust fungi (e.g., many species of *Gymnosporangium*) that lack the urediniospore (repeating) stage (*see* macrocyclic; microcyclic)
desiccate—to dry out
determinate—(describing a sporangiophore) ceasing growth after the production of a sporangium, thus having a predetermined length
diagnostic (n. diagnosis)—pertaining to a distinguishing characteristic important for the identification of a disease or disorder
dichotomous—branching, often successively, into two more or less equal arms
dicot—a dicotyledon, or plant having two cotyledons (*see* monocot)
dieback (v. die back)—the progressive death of shoots, leaves, or roots, beginning at the tips
digitate—having lobes radiating from a common center
dikaryon (adj. dikaryotic)—an organism having two sexually compatible haploid nuclei per cell that divide simultaneously in the reproductive phase called dikaryophase
dimorphic—having two distinct shapes or forms
diploid—having two complete sets of chromosomes ($2n$ chromosomes) (*see* haploid; polyploid)
disease—the abnormal functioning of an organism
disease cycle—the succession of events and interactions among a host, a parasite, and the environment, from initial infection through pathogenesis to overseasoning, until another infection occurs
disinfest—to kill pathogens that have not yet initiated disease or other contaminating microorganisms that occur in or on inanimate objects (e.g., soil or tools) or on the surfaces of plant parts (e.g., seed)
dispersal—the spread of infectious material (inoculum) from diseased to healthy plants (syn. dissemination)
dissemination—*see* dispersal
distal—situated away from the point of attachment or origin or away from the main body (*see* proximal)
diurnal—daily
DNA—deoxyribonucleic acid; the double-stranded, helical molecule that contains genetic code information; consists of repeating units, or nucleotides, each of which is composed of deoxyribose (a sugar), a phosphate group, and a purine (adenine or guanine) or a pyrimidine (thymine or cytosine) base
dolipore septum—a type of cross-wall in fungi in the group basidiomycetes characterized by distinctive swellings and membranes in association with the septal pore
dominant—pertaining to a phenotypic trait that is expressed in hybrid progeny of diploid organisms even if contributed by only one of the parents (*see* recessive)
dormancy (adj. dormant)—a condition of suspended growth and reduced metabolism of an organism; generally induced by internal factors or environmental conditions as a mechanism of survival
dormant—resting; living in a state of reduced physiological activity
dorsal—situated toward the back or top of the body (ant. ventral)
dwarfing—underdevelopment of a plant or plant organs caused by disease, inadequate nutrition, or unfavorable environmental conditions

ebb and flood—an irrigation system in which potted crops grown in holding trays on benches are shallowly flooded with water or nutrient solution for a short period to allow uptake by capillary action before the bench is drained
echinulate—having small spines projecting from the cell wall
ecology—the study of the interactions between individual organisms, groups of organisms, and organisms and their environments
ectomycorrhiza (pl. ectomycorrhizae)—a symbiotic relationship between a fungus and the roots of a vascular plant in which the fungus forms a coating (called a "mantle") on the outsides of the root tips and hyphae penetrate the roots
ectoparasite—a parasite that feeds from the exterior of its host (*see* endoparasite)
ectotrophic—pertaining to fungal development primarily over the root surface
edema, oedema—blisters produced on leaves or other plant parts under conditions of high moisture and restricted transpiration (syn. intumescence)
effuse (adj.)—stretched out, especially with reference to a filmlike growth
egg mass—a group of eggs held together by a gelatinous matrix

electron microscope—a high-resolution microscope that uses a beam of electrons, rather than light, to illuminate the subject; suitable for viewing virus particles

ELISA—enzyme-linked immunosorbent assay; a serological test in which the sensitivity of the reaction is increased by attaching an enzyme that produces a colored product in response to one of the reactants

ellipsoidal—elliptical in the plane section

embryo—an organism in the early stages of development, such as a young plant in the seed or a nematode before hatching from the egg

emergence—the growth of a seedling shoot through the surface of the soil

enation—an abnormal outgrowth from the surface of a stem or leaf

encapsidate—to enclose viral nucleic acid in a protein coat

encyst—to form a cyst or protective covering

endemic—native to a particular place; pertaining to a low and steady level of natural disease occurrence

endocarp—the inner layer of a fruit wall (*see* pericarp)

endoconidium (pl. endoconidia)—a conidium produced inside a hypha or conidiophore

endodermis—the layer of cells between the vascular tissue and the cortex of roots

endogenous—arising from inside; (describing the nucleic acid of a virus) incorporated into the host DNA (*see* exogenous)

endomycorrhiza (pl. endomycorrhizae)—a symbiotic relationship between a fungus and the roots of a vascular plant in which the fungus penetrates the root cortex; also known as "arbuscular mycorrhiza"

endoparasite—a parasite that lives and feeds inside its host (*see* **ectoparasite**)

endophyte—a plant developing inside another organism; any of the various endoparasitic fungi associated with grass species

endosperm—the nutritive tissue formed within the embryo sac of seed plants

entomophagous—feeding on insects

enzyme—a protein that catalyzes a specific biochemical reaction

enzyme-linked immunosorbent assay (ELISA)—a serological test in which the sensitivity of the reaction is increased by attaching an enzyme that produces a colored product in response to one of the reactants

epicotyl—the portion of the stem of a plant embryo or seedling above the node where the cotyledons are attached (*see* hypocotyl)

epidemic—an increase of disease in a population; a general and serious outbreak of disease (*see* epiphytotic)

epidemiology (adj. epidemiological)—the study of factors influencing the initiation, development, and spread of infectious disease; the study of disease in populations

epidermis (adj. epidermal)—the surface layer of cells of leaves and other plant parts

epinasty—the abnormal downward curling of a leaf, leaf part, or stem

epiphyllous—located on the upper surface of a leaf

epiphyte—an organism growing on a plant surface but not as a parasite

epiphytotic—pertaining to the widespread and destructive outbreak of a plant disease

episomal—(describing the nucleic acid of a virus) free in the cytoplasm and capable of replication; as opposed to endogenous nucleic acid, which is incorporated into the host DNA and is detectable but not active; *Petunia vein clearing virus* (PVCV) is an example of a virus that may be either episomal or endogenous

epitope—the part of an antigen at which an antibody attaches

eradicant—a chemical used to eliminate a pathogen from a host or environment

eradication—the control of plant disease by eliminating the pathogen after it has become established or by eliminating the plants that carry the pathogen

erumpent—bursting or erupting through the surface of a substrate

ethylene—a plant hormone influencing vegetative growth, fruit ripening, abscission of plant parts, and senescence of flowers

etiolation—the elongation of stems caused by low levels of light intensity

etiology—the study of the causes of diseases

eukaryote (adj. eukaryotic)—any of various organisms whose cells contain a membrane-bound nucleus and other organelles, including all higher plants, animals, fungi, and protists (*see* prokaryote)

exogenous—originating from the outside (*see* endogenous)

extracellular—not occurring or located within a cell

exudate—a liquid excreted by or discharged from diseased tissues, roots, leaves, and fungi

f. sp. (pl. ff. spp.)—*forma specialis* (pl. *formae speciales*); a taxonomic group within a pathogenic species, defined by its host range (members of different *formae speciales* infect different groups of plants)

facultative—capable of changing lifestyle (e.g., from saprophytic to parasitic or the reverse)

facultative parasite—an organism that is normally saprophytic but capable of living as a parasite

facultative saprophyte—an organism that is normally parasitic but capable of living as a saprophyte

falcate—sickle shaped

fallow—pertaining to cultivated land kept free from a crop or weeds during the normal growing season

fasciation—the malformation of shoots or floral organs manifested as enlargement and flattening, as if several parts were fused

fascicle (adj. fasciculate)—a small group, bundle, or cluster

fastidious—with reference to prokaryotic organisms, having special requirements for growth and nutrition

feeder root—a fine root that absorbs water and dissolved nutrients

filament (adj. filamentous)—a thin, flexible, threadlike structure

filamentous—threadlike (syn. filiform)

filiform—long, needlelike (syn. filamentous)

fingerprinting—using DNA sequences to profile organisms for identification purposes

flaccid—wilted; lacking turgor

flagellum (pl. flagella)—a hairlike, whiplike, or tinsel-like appendage of a motile cell, bacterium, or zoospore that provides locomotion

fleck—a minute, discolored spot in green tissue

flexuous—having turns or windings; capable of bending

flower color break, flower color breaking—a symptom of virus infection in which the normal petal coloration is disfigured by discolored or bleached areas (syns. color break, color breaking)

focus (pl. foci)—a small area of diseased plants within a larger plot or field

foliar—pertaining to leaves

forma specialis (pl. *formae speciales*)—a taxonomic group within a pathogenic species, defined by its host range (members of different *formae speciales* infect different groups of plants); abbr. f. sp. (pl. ff. spp.)

FRAC group—a group of fungicides with the same mode of action, as determined by the Fungicide Resistance Action Committee (FRAC)

fructification—any of the various spore-bearing organs formed by macro- and microfungi

fruiting body—any of the various complex, spore-bearing structures of fungi

fumigant (v. fumigate)—a gas or volatile substance used to kill or inhibit the growth of microorganisms or other pests

Fungi Imperfecti—a group of fungi lacking a sexual stage; a group comprising the asexual stages of fungi in the groups ascomycetes and basidiomycetes (syn. deuteromycetes)

fungicide (adj. fungicidal)—a chemical or physical agent that kills or inhibits the growth of fungi

fungistat (adj. fungistatic)—a compound that inhibits fungal growth or sporulation but does not kill fungi

fusiform—spindle shaped; tapering at each end

gall—an abnormal swelling or localized outgrowth, often roughly spherical, produced by a plant as a result of attack by a fungus, bacterium, nematode, insect, or other organism (syn. tumor)

gametangium (pl. gametangia)—a cell containing gametes or nuclei that act as gametes

gamete—sex cell

gelatinous—resembling gelatin or jelly

gene—a unit located on a chromosome and controlling the transmission of a heritable characteristic

genetic—pertaining to heredity or heritable characteristics

genetically modified organism (GMO)—an organism possessing a gene from another species; having been the subject of genetic engineering

geniculate—bent, like a knee

genome—the complete genetic information of an organism or virus
genotype—the genetic constitution of an individual or group; a class or group of individuals sharing a specific genetic makeup (*see* phenotype)
genus (pl. genera)—a taxonomic category that consists of species that are closely related structurally or phylogenetically (the genus or generic name is the first name in a Latin binomial)
germ tube—a hypha resulting from an outgrowth of the spore wall and cytoplasm after germination
germinate (n. germination)—to begin growth, as of a seed, spore, sclerotium, or other reproductive body
germplasm—the bearer of heredity material (often loosely applied to cultivars and breeding lines)
giant cell—an enlarged, multinucleate cell formed in a root by repeated nuclear division without cell division; induced by secretions of certain sedentary plant-parasitic nematodes
girdle—to circle and cut through a stem or the bark and outer few rings of wood, disrupting the phloem and xylem
globose—nearly spherical
glycoprotein—a peptide chain with carbohydrate attached; one of the components of a virus capsid that is important for host and vector recognition
graft transmission—the transmission of a pathogen from one host plant to another through fusion of living tissue from the diseased host with living tissue of a healthy host
gram-negative bacteria—bacteria that are stained red or pink after treatment with Gram's stain
gram-positive bacteria—bacteria that are stained violet or purple after treatment with Gram's stain
gravid—containing an egg or eggs; capable of depositing eggs
guttation—the exudation of watery, sticky liquid from hydathodes, especially along leaf margins
guttulate—(describing a spore) containing one or more oil droplets

haploid—having a single complete set of chromosomes (*see* diploid; polyploid)
hardiness—the ability to withstand stress
haustorium (pl. haustoria)—a specialized branch of a parasite formed inside a host cell to absorb nutrients
hemocoel—the body cavity of the majority of invertebrates, including insects; contains the circulatory fluid (hemolymph)
hemolymph—the fluid of the circulatory systems of insects and most other invertebrates; contained in the hemocoel
herbaceous—pertaining to primary, soft, nonwoody tissue of a plant or plant part; having the characteristics of an herb
herbarium—a collection of plant or fungal specimens carefully organized and preserved for botanical research
herbicide—any of various chemicals that kill plants or inhibit their growth (e.g., a weed or grass killer)
hermaphrodite (adj. hermaphroditic)—having both male and female reproductive organs in a single individual
heteroecious—with reference to rust fungi, passing different stages of the life cycle on alternate hosts (*see* autoecious)
heterokaryon (adj. heterokaryotic)—a cell with genetically different nuclei
heterothallism (adj. heterothallic)—a fungal life cycle in which sexual reproduction can occur only in the presence of genetically different mycelia (*see* homothallism)
heterotroph—an organism that absorbs or consumes food from various organic sources because it cannot produce its own food
heterozygous—having the same form (allele) of a gene on different chromosomes (*see* homozygous)
hilum—a scar left on a seed or spore after detachment
holomorph—a fungus having sexual and asexual forms in its life cycle, considered in all its forms (*see* anamorph; teleomorph)
homothallism (adj. homothallic)—a fungal life cycle in which sexual reproduction occurs in a single, self-fertile thallus (*see* heterothallism)
homozygous—having the same form (allele) of a gene on homologous chromosomes (*see* heterozygous)
honeydew—a sugary ooze or exudate often secreted by aphids; a characteristic symptom of ergot
host plant—a living plant attacked by or harboring a parasite or pathogen, from which the invader obtains part or all of its nourishment
host range—the range of plants on which an organism (particularly a parasite) feeds
hull—the outer coat of a seed
hyaline—transparent or nearly so; translucent; colorless
hybrid (v. hybridize)—the offspring of two individuals of different genotypes
hydathode—an epidermal leaf structure specialized for the secretion or exudation of water; the opening at the terminus of a leaf vein
hygroscopic—absorbing moisture readily
hymenium—the continuous, spore-bearing layer of a fungal fruiting body
hyperplasia (adj. hyperplastic)—an abnormal increase in the number of cells, often resulting in the formation of galls or tumors
hypersensitive—extremely or excessively sensitive, often with reference to an extreme reaction to a pathogen
hypertrophy (adj. hypertrophic)—an abnormal increase in the size of cells in a tissue or organ, often resulting in the formation of galls or tumors
hypha (pl. hyphae; adj. hyphal)—a tubular filament of a fungal thallus or mycelium; the basic structural unit of a fungus
hyphomycete—a fungus that produces spores on a conidiophore that is not enclosed in a protective structure; a form classification for the "imperfect" spore stage of fungi that are primarily in the Ascomycota
hypocotyl—the portion of the stem below the cotyledons and above the root (*see* epicotyl)

icosahedral (n. icosahedron)—having 20 faces, like polyhedral virus particles (*see* isometric)
immune—not capable of being infected by a given pathogen
immunogenic—able to induce the production of antibodies
imperfect state—the anamorph, or asexual form in the life cycle of certain fungi, in which asexual spores (e.g., conidia) or no spores are produced (*see* perfect state)
in vitro—in glass, on an artificial medium, or in an artificial environment; not within a living host
in vivo—within a living organism
inclusion body—a structure developed within a plant cell as a result of infection by a virus; often useful in identifying the virus
indeterminate—(describing a sporangiophore) continuing to extend growth after a sporangium has been produced and thus not having a predetermined length
indicator plant—a plant in which specific or distinctive symptoms develop in reaction to a pathogen or certain environmental conditions; used to detect or identify the pathogen or determine the effects of the environmental conditions
infection—the process in which an organism enters, invades, or penetrates and establishes a parasitic relationship with a host plant
infection court—a site in or on a host plant where infection can occur
infection cushion—an organized mass of hyphae formed on the surface of a plant and producing numerous infective hyphae
infection peg—the specialized, narrow hyphal strand located on the underside of an appressorium and penetrating host cells (syn. penetration peg)
infectious—with reference to disease, capable of spreading from plant to plant
infective—able to attack a host and cause infection; with reference to vectors, carrying or containing a pathogen and able to transfer it to a host plant
infest (n. infestation)—to attack as a pest (used especially for insects and nematodes); to contaminate (as with microorganisms); to be present in large numbers
inflorescence—a flower or flower cluster
initial inoculum—inoculum (usually from an overwintering source) that initiates disease in the field, as opposed to inoculum that spreads disease during the growing season (syn. primary inoculum)
injury—damage caused by transitory interaction with an agent such as an insect, chemical, or unfavorable environmental condition
inoculate (n. inoculation)—to place inoculum in an infection court; to insert a pathogen into healthy tissue
inoculum (pl. inocula)—a pathogen or its parts capable of causing infection when transferred to a favorable location
inoculum density—a measure of the number of propagules of a pathogenic organism per unit of area or volume
insecticide—a pesticide used to kill insects
Insecticide Resistance Action Committee (IRAC)—a group that maintains a database on arthropod pesticide resistance

integrated pest management (IPM)—a broad-based system for management of diseases, insects, and other pests that makes use of sanitation practices and resistant crop cultivars, as well as carefully timed cultural, biological and chemical controls, with the goal of effective control with minimal environmental impact
intercalary—inserted within (e.g., located along a hypha, as opposed to located at the end of a hypha)
intercellular—between or among cells
internode (adj. internodal)—the portion of a stem between two successive nodes
interveinal—between veins (of a leaf)
intracellular—through or within cells
intumescence—blisters produced on leaves or other plant parts under conditions of high moisture and restricted transpiration (syn. edema)
isolate—(n.) a culture or subpopulation of a microorganism separated from its parent population and maintained under controlled conditions; (v.) to remove from soil or host material and grow in pure culture
isometric—with reference to virus particles, having an icosahedral structure and thus appearing approximately round
ITS—internal transcribed spacer region (of DNA); used in fungi identification

juvenile—an immature form that appears similar to the adult stage but is usually smaller and is not sexually mature, as in nematodes or insect species that undergo gradual metamorphosis

knot—a gall or localized abnormal swelling

lamina—the expanded part of a leaf (*see* petiole)
larva (pl. larvae)—the immature or caterpillar stage in the life cycle of an insect that undergoes complete metamorphosis
latent—present but not manifested or visible, as a symptomless infection
latent infection—an infection that does not have visible symptoms
latent period—the time elapsed between infection and the appearance of symptoms or production of new inoculum (sometimes synonymous with incubation period); the time elapsed after the acquisition of a pathogen by a vector but before the pathogen can be transmitted
leaf spot—a lesion typically restricted in development after reaching a characteristic size
leafhopper—any of various species of mobile insects with sucking mouthparts in the order Hemiptera
leaflet—one of the separate blades or divisions of a compound leaf
legume—a simple, dry, dehiscent fruit developing from a simple pistil and splitting at maturity along two seams; any plant in the family Fabaceae (formerly Leguminosae)
lenticel—a natural opening in the surface of a stem, tuber, fruit, or root, permitting gas exchange
lenticular—lens shaped (convex on both faces)
lesion—a localized diseased area or wound
life cycle—the cyclical stages in the growth and development of an organism
lignin—a complex organic substance or group of substances that impregnate the cell walls of xylem vessels and certain other plant cells; constituting wood
limoniform—lemon shaped
lobate—having lobes
local lesion—a small, restricted lesion; often the characteristic reaction of a differential cultivar to a specific pathogen, especially in response to mechanical inoculation with a virus
lodge—with reference to hay or grain crops, to fall over
lumen (pl. lumina)—the central cavity of a cell or other structure
lunate—crescent shaped

macerate—to cause disintegration of tissues by separation of cells; to soften by soaking
macroconidium (pl. macroconidia)—the larger of two kinds of conidia formed by certain fungi (*see* microconidium)
macrocyclic—pertaining to the life cycle of rust fungi that typically exhibit all five spore stages (pycniospores, aeciospores, urediniospores, teliospores, and basidiospores) in the course of their development (*see* demicyclic; microcyclic)
macroelements—the nutrients needed by plants in relatively large quantities for metabolism and growth, including calcium, magnesium, nitrogen, phosphorous, potassium, and sulfur, as well as carbon, hydrogen, and oxygen
manual transmission—the spread or introduction of inoculum to an infection court by hand
mating types—compatible strains (usually designated + and – or A and B), both of which are necessary for sexual reproduction in heterothallic fungi
mechanical injury—injury of a plant part by abrasion, mutilation, or wounding
mechanical transmission—the spread or introduction of inoculum to an infection court by human manipulation, accompanied by physical disruption of host tissues (wounding)
medium (pl. media)—a mixture of organic or inorganic chemical compounds and water providing the nutrients needed for the growth of a microorganism in vitro; with reference to higher plants, a mixture of fertilizers and other components in which plants are grown
medulla—the loosely arranged hyphae at the centers of some sclerotia
meiosis—nuclear division in which the number of chromosomes per nucleus is halved (i.e., the diploid state is converted to the haploid state) (*see* mitosis)
melanin (adj. melanoid)—a brown-black pigment
meristem (adj. meristematic)—plant tissue characterized by frequent cell division, producing cells that become differentiated into specialized tissues
meristem culture—an aseptic culture of a plant or plant part from a portion of the meristem
mesophyll—the central, internal, nonvascular tissue of leaves, consisting of the palisade and spongy mesophyll
messenger RNA (abbr. mRNA)—a form of RNA that carries information to direct the synthesis of protein
metabasidium—the part of the basidium in which meiosis occurs
metabolite—any chemical participating in metabolism; a nutrient
microclimate—weather conditions on a small scale, as at the surface of a plant or within a field
microconidium (pl. microconidia)—the smaller of two kinds of conidia formed by certain fungi (*see* macroconidium)
microcyclic—pertaining to the life cycle of rust fungi that produce only teliospores and basidiospores (*see* demicyclic; macrocyclic)
microelements—the nutrients needed by plants as trace elements for metabolism and growth, including boron, chlorine, copper, iron, manganese, molybdenum, nickel, and zinc
microflora—the combination of all of the microorganisms in a particular environment
microorganism—an organism small enough that it can be seen only with the aid of a microscope (syn. microbe)
microsclerotium (pl. microsclerotia)—a microscopic, dense aggregate of darkly pigmented, thick-walled hyphal cells
middle lamella—the layer between the walls of adjacent plant cells; consists largely of pectic substances
midrib—the central, thickened vein of a leaf
migratory—moving from place to place on a plant or from plant to plant when feeding (*see* sedentary)
mildew—a thin coating of mycelial growth and spores on the surfaces of infected plant parts
miticide—a pesticide used to kill mites (syn. acaricide)
mitochondrion (pl. mitochondria)—a cellular organelle outside the nucleus; functions in respiration
mitosis—nuclear division in which the number of chromosomes per nucleus remains the same (*see* meiosis)
MLO—mycoplasmalike organism (*see* phytoplasma)
mold—any microfungus with conspicuous, profuse, or woolly superficial growth (mycelium or spore masses) on various substrates; commonly grow on damp or decaying matter and on the surfaces of plant tissues, especially with reference to economically important saprobes
mollicute—any of a group of prokaryotic organisms bounded by a flexuous membrane and lacking a cell wall (*see* phytoplasma; spiroplasma)
molt—to shed a cuticle or body encasement during a phase of growth
monilioid—in a chain of individual units with regular constrictions; having a form like beads in a necklace
monocot—a monocotyledon, or a plant having only one cotyledon, such as species of grasses, including grain crops and corn (*see* dicot)

monoculture—the cultivation of plants of the same species in close proximity and with few or no other plant species present

monoecious—having male and female reproductive organs on a single individual

monogenic—determined by a single gene (*see* polygenic)

monopodial—a form of branching in which there is a single stem with subordinate lateral branches

morphology (adj. morphological)—the study of the form of organisms; the form and structure of organisms

mosaic—a disease symptom characterized by nonuniform coloration, with intermingled normal, light-green and yellowish patches; usually caused by a virus (often used interchangeably with mottle)

mother block—a collection of clean stock of a cultivar from which all cuttings are derived

motile—capable of self-propulsion by means of flagella, cilia, or amoeboid movement

mottle—a disease symptom characterized by light and dark areas in an irregular pattern; usually caused by a virus (often used interchangeably with mosaic)

mucilaginous—viscous; slimy

mulch—a layer of material, such as organic matter or plastic film, applied to the surface of the soil for purposes such as retention of water and inhibition of weeds

multinucleate—having more than one nucleus per cell (*see* binucleate; uninucleate)

multiseptate—having many septa (cross-walls)

muriform—having transverse and longitudinal septa

mutation—an abrupt heritable or genetic change in a gene or individual as a result of an alteration in a gene or chromosome or an increase in chromosome number

mycelium (pl. mycelia; adj. mycelial)—the mass of hyphae constituting the body (thallus) of a fungus

mycoparasite—a fungus that attacks another fungus

mycophagous—feeding on fungi

mycoplasmalike organism (abbr. MLO)—*see* phytoplasma

mycorrhiza (pl. mycorrhizae; adj. mycorrhizal)—a symbiotic association between a nonpathogenic or weakly pathogenic fungus and the roots of a plant

mycotoxin—any poisonous compound produced by a fungus

necrosis (adj. necrotic)—the death of cells or tissue; usually accompanied by blackening or browning

necrotroph—a parasite that typically kills and obtains its energy from dead host cells (*see* biotroph)

negative sense—describes a single-stranded RNA virus that cannot be translated directly into protein but must first be transcribed into a positive-sense RNA

nematicide—an agent that kills nematodes; usually a chemical

nematode—any of various unsegmented roundworms (animals) that is parasitic on plants or animals or free living in soil or water

node (adj. nodal)—the enlarged portion of a shoot at which leaves or buds arise

nodule—a small knot or irregular, rounded lump; with reference to leguminous plants, a structure located on the root and containing nitrogen-fixing bacteria

noninfectious disease—a disease caused by an abiotic agent and not capable of being transmitted from plant to plant

nonpersistent transmission—virus transmission in which a virus is acquired and transmitted by a vector after a short feeding time and retained by the vector for only a short time (syn. stylet-borne transmission)

nonseptate—lacking cross-walls; coenocytic

nymph—the juvenile stage in the life cycle of an insect with incomplete metamorphosis; superficially resembles the adult

obclavate—shaped like an upside-down club

obligate parasite—an organism that can grow only as a parasite in association with its host plant and cannot be grown in artificial culture media (syn. biotroph)

obovoid—shaped like an upside-down egg

obtuse—rounded or blunt; pertaining to an angle greater than 90°

oedema—*see* edema

oncogenes—genes coding for transformation of normal cells into cells that overgrow (in plants, forming a gall)

oogonium (pl. oogonia)—a female gametangium of fungi in the group oomycetes containing one or more gametes

oomycetes (adj. oomycetous)—a group of funguslike organisms in the phylum Oomycota (kingdom Chromista); typically has nonseptate mycelium, asexual sporangia and zoospores, and sexual oospores

oospore—the thick-walled, sexually derived resting spore of organisms in the group oomycetes

ooze—a mass of bacterial cells mixed with host fluids

organelle—any of various membrane-bound structures contained within cells and having a specialized function (e.g., mitochondria and chloroplasts)

ostiole (adj. ostiolate)—a pore; the opening in the papilla or neck of a perithecium, pseudothecium, or pycnidium through which spores are released

ovary—the female reproductive structure of an organism; in plants, the enlarged basal portion of the pistil, containing the ovules and developing into the fruit

overseason—to survive or persist from one planting season to the next

oversummer—to survive or persist through the summer

overwinter—to survive or persist through the winter

oviposit—to deposit or lay eggs with an ovipositor

ovule—an enclosed structure that after fertilization becomes a seed; an egg contained within an ovary

ozone—a highly reactive form of oxygen (O_3) injurious to plants

palisade parenchyma—the tissue located beneath the upper epidermis of leaves; composed of elongate, tubular cells arranged upright in the manner of posts in a palisade fortification

papilla (pl. papillae; adj. papillate)—a small, blunt projection

paracrystalline—having an ordered lattice but lacking the long-range ordering typical of crystals, in at least one direction

paraphysis (pl. paraphyses)—a hairlike cell within a fungal fruiting structure

pararetrovirus—a reverse-transcribing virus that uses an RNA intermediate for replication; endogenous pararetroviruses (EPRVs) may be integrated into plant genomes

parasexual—pertaining to the recombination of genetic characters without a sexual process

parasite (adj. parasitic)—an organism that lives in intimate association with another organism, on which it depends for its nutrition; not necessarily a pathogen

parasitoid—an insect that parasitizes and kills another insect host; immature stages are inside or outside the host, but adults are free living

parenchyma (adj. parenchymatous)—the soft tissue of living plant cells with undifferentiated, thin, cellulose walls

parthenogenesis (adj. parthenogenetic)—reproduction by the development of an unfertilized egg

pasteurization—the process by which a material (usually a liquid) is treated with heat to eliminate selected harmful microorganisms

pathogen (adj. pathogenic)—a disease-producing organism or agent

pathogenesis—the initiation and development of disease

pathogenicity—the ability to cause disease

pathology—the study of diseases

pathotype—a subdivision of a pathogenic species characterized by its pattern of virulence or avirulence to differential host varieties

pathovar (abbr. pv.)—a subdivision of a plant-pathogenic bacterial species, defined by its host range (a pathovar of a bacterial species is analogous to a *forma specialis* of a fungal species)

PCR—polymerase chain reaction; used to amplify a DNA segment starting from one or a few copies

pedicel—a small, slender stalk; a stalk bearing an individual flower, inflorescence, or spore

peduncle—the stalk or main stem of an inflorescence; part of an inflorescence or fructification

penetration—the initial invasion of a host by a pathogen

penetration peg—the specialized, narrow hyphal strand located on the underside of an appressorium and penetrating host cells (syn. infection peg)

perennial—a plant that survives for several to many years (*see* annual; biennial)

perfect—sexual; capable of sexual reproduction

perfect flower—a flower possessing both stamens and pistils

perfect state—the teleomorph, or sexual form in the life cycle of certain fungi (*see* imperfect state)
pericarp—the outer layer of a seed or fruit (*see* endocarp)
periphysis (pl. periphyses)—a hairlike cell within a fungal fruiting structure
perithecium (pl. perithecia)—the flask-shaped or subglobose, thin-walled fruiting body (ascocarp) of an ascomycetous fungus, containing asci and ascospores and having a pore (ostiole) at its apex, through which ascospores are expelled or released
peritrichous—having hairs or flagella distributed over the whole surface
persistent transmission—virus transmission in which a virus is acquired and transmitted by a vector after a relatively long feeding time and remains transmissible for a prolonged period while in association with the vector (syn. circulative transmission)
pest—any organism that damages plants or plant products
pesticide—any of various chemicals used to control pests
petiole—the stalk portion of a leaf (*see* lamina)
pH—the negative logarithm of the effective hydrogen ion concentration; a measure of acidity (pH 7 is neutral; values less than pH 7 are acidic, and values greater than pH 7 are alkaline)
phenotype—the composite of observable physical qualities of an organism produced by the interaction of its genotype with the environment
phialide—the end cell of a conidiophore with one or more open ends, through which a basipetal succession of conidia develops
phialospore—a conidium produced on a phialide
phloem—the food-conducting, food-storing tissue in the vascular system of roots, stems, and leaves
photochemical oxidant—any of various highly reactive compounds formed by the action of sunlight on less toxic precursors
photosynthate—a chemical product of photosynthesis
photosynthesis—the manufacture of carbohydrates from carbon dioxide and water in the presence of chlorophylls; uses light energy and releases oxygen
phyllody—a disorder in which floral organs are transformed into leaflike structures
phylogeny (adj. phylogenetic)—the evolutionary development of an organism
phylotype—a plant or other organism classified by its appearance, rather than its genetic relationship
phytoalexin—any of various substances produced by higher plants in response to chemical, physical, or biological stimuli and inhibiting the growth of certain microorganisms
phytopathology—the study of plant diseases (syn. plant pathology)
phytoplasma—any of various plant-parasitic pleomorphic mollicutes (prokaryotes lacking cell walls) found in phloem tissue and not capable of growth on artificial nutrient media (previously called mycoplasmalike organisms (MLOs))
phytotoxic—harmful to plants; usually with reference to chemicals
pigment—a colored compound, such as chlorophyll, in the cells of plants or fungi
pinnate—featherlike; having parts arranged along two sides of an axis
pionnote—a colony form (of *Fusarium*) having no discrete sporodochia but forming a continuous, slimy spore layer; usually lacking aerial mycelium
pistil—the ovule-bearing organ of a plant, consisting of the ovary and its appendages (e.g., the style and stigma)
pith—parenchymatous tissue occupying the center of a stem
plant pathology—the study of plant diseases (syn. phytopathology)
plasmodium (pl. plasmodia)—a naked multinucleate mass of protoplasm moving and feeding in amoeboid fashion
pleomorphic—able to assume various shapes and perhaps sizes; having a life cycle characterized by a succession of distinctly different forms
polar—located at one end (pole) of the cell
pollen—male sex cells produced by the anthers of flowering plants or the cones of seed plants
pollination—the transfer of pollen from anther to stigma or from a staminate cone to an ovulate cone
polyclonal antibodies—a collection of antibodies that recognizes multiple epitopes on an antigen
polygenic—pertaining to or governed by many genes (*see* monogenic)
polyploid—having three or more complete sets of chromosomes (*see* haploid; diploid)
positive-sense RNA—RNA that can serve directly as messenger RNA
posterior—situated toward the back or tail (*see* anterior)
predator—an organism (an insect or mite, in this compendium) that attacks and feeds on another organism
predispose (n. predisposition)—to make prone to infection and disease
prepupa—a nonfeeding, inactive stage between the larval and pupal stages of an insect
primary inoculum—inoculum (usually from an overwintering source) that initiates disease in the field, as opposed to inoculum that spreads disease during the season (syn. initial inoculum)
primary leaf—the first true leaf that emerges from a plant following the cotyledons
primary root—a root that develops directly from the radicle of an embryo rather than from a crown or node
primer—a short sequence of nuclei acid (RNA) used in synthesis of DNA
prokaryote (adj. prokaryotic)—any of various organisms lacking internal membrane-bound organelles and a distinct nucleus, such as bacteria and mollicutes (*see* eukaryote)
promycelium (pl. promycelia)—in rust and smut fungi, a germ tube issuing from a teliospore and bearing basidiospores
propagative virus—a virus that multiplies within its arthropod vector
propagule—any part of an organism capable of independent growth
protectant—an agent (usually a chemical) applied to a plant surface in advance of exposure to a pathogen for the purpose of preventing infection
protein—any of numerous nitrogen-containing organic compounds containing combinations of amino acids
protoplasm—the living contents of a cell
protoplast—the living contents of a cell, exclusive of the cell wall
proximal—situated toward or near the point of attachment or main body (*see* distal)
pseudoseptate—appearing septate but lacking true septa
pseudothecium (pl. pseudothecia)—a perithecium-like fruiting body containing asci and ascospores dispersed rather than in an organized hymenium; an ascostroma with a single locule (cavity) containing bitunicate asci
pupa (pl. pupae; v. pupate)—the quiescent stage between the larval and adult stages of certain insects that undergoes complete metamorphosis
pustule—a small, blisterlike elevation of the plant epidermis that forms as spores emerge
pv.—pathovar; a subdivision of a plant-pathogenic bacterial species defined by its host range (a pathovar of a bacterial species is analogous to a *forma specialis* of a fungal species)
pycnidiospore—a spore (conidium) produced in a pycnidium
pycnidium (pl. pycnidia)—an asexual, globose or flask-shaped fruiting body of certain imperfect fungi, producing conidia
pycniospore—a haploid, sexually derived spore formed in the pycnium of a rust fungus (syn. spermatium)
pycnium (pl. pycnia)—the globose or flask-shaped haploid fruiting body of rust fungi, bearing receptive hyphae and pycniospores (syn. spermagonium)
pyriform—pear shaped

quarantine—an enforced isolation imposed to prevent the spread of disease; pertaining to plant disease, control of the transport of plants or plant parts to prevent the spread of pests or pathogens
quiescent—dormant; inactive

race—a subgroup or biotype within a species or variety that is distinguished from other races by virulence, symptom expression, or host range but not by morphology
rachis—the elongated main axis of an inflorescence
radicle—the part of the plant embryo that develops into the primary root
receptacle—the structure of a flower that bears the reproductive organs
receptive hypha—in rust fungi, the part of a pycnium (spermagonium) that receives the nucleus of a pycniospore (spermatium)
reniform—kidney shaped
resinosis—the exudation of resin
resistance management—a practice using effective rates of pesticides and rotating among active ingredients with different modes of action

to slow the target pest or pathogen population's development of resistance to a chemical control (an insecticide or fungicide, primarily)
resistant (n. resistance)—possessing properties that prevent or impede disease development (*see* susceptible)
respiration—a series of chemical reactions that make energy available through oxidation of carbohydrates and fat
resting spore—a spore (often thick walled) that can remain alive in a dormant state for some time, germinating later and capable of initiating infection
reticulate—having netlike markings
RFLP—restriction fragment length polymorphism; a lab procedure in which enzymes are used to separate and identify desired fragments of DNA
Rhizobium—a genus of bacteria that live symbiotically with roots of leguminous plants, converting atmospheric nitrogen into a form useable by the plants
***Rhizobium* nodules**—root galls caused by *Rhizobium* spp.
rhizoid—a rootlike filament that anchors a fungus or plant to its substrate
rhizome—a mostly horizontal, jointed, fleshy, often elongated, usually underground stem
rhizomorph—a macroscopic ropelike strand of compacted tissue formed by certain fungi
rhizosphere—the microenvironment in the soil immediately around a plant's roots
ribosome—a subcellular protoplasmic particle made up of one or more RNA molecules and several proteins; involved in protein synthesis
rind—the outer layer of some sclerotia
ring spot—a disease symptom characterized by yellowish or necrotic rings enclosing green tissue, as in some plant diseases caused by viruses
RNA—ribonucleic acid; any of several nucleic acids composed of repeating units of ribose (a sugar), a phosphate group, and a purine (adenine or guanine) or a pyrimidine (uracil or cytosine) base, transcribed from DNA and involved in translation to proteins
rogue—to remove and destroy individual plants that are diseased, infested by insects, or otherwise undesirable
root cap—a group of cells that protects the growing tip of a root
root hair—a threadlike, single-celled outgrowth from a root epidermal cell
rosette—a disease symptom characterized by short, bunchy growth habit caused by a shortening of internodes with no comparable reduction in leaf size
rot—softening, discoloration, and often disintegration of plant tissue as a result of fungal or bacterial infection
rotation—the practice of growing different kinds of crops in succession in the same field; also the practice of applying pesticides with different modes of action in a sequence to slow the development of resistance in the target pest or pathogen population
rugose—wrinkled; roughened
runner—a slender, horizontal stem that grows close to the soil surface (syn. stolon)
russet—a brownish, roughened area on the skin of fruit, resulting from cork formation
rust—any of several diseases caused by specialized fungi in the group basidiomycetes, some of the spores of which are a rusty color

sanitation—the destruction or removal of infected or infested plants or plant parts; the decontamination of tools, equipment, containers, work space, hands, and so on
saprobe—*see* saprophyte
saprophyte (adj. saprophytic)—an organism that obtains nourishment from nonliving organic matter (syn. saprobe)
satellite virus—a virus that accompanies another virus and depends on it for its multiplication
scab—a roughened, crustlike diseased area on the surface of a plant organ
scald—a necrotic condition in which tissue is usually bleached and has the appearance of having been exposed to high temperature
sclerenchyma (adj. sclerenchymatous)—tissue made up of thick-walled plant cells
sclerotium (pl. sclerotia)—a vegetative, resting body of a fungus composed of a compact mass of hyphae with or without included host tissue; usually has a darkened rind
scorch—a symptom that suggests the action of flame or fire on the affected tissue, often at the margins of leaves
secondary infection—an infection resulting from the spread of infectious material from a primary or secondary infection without an intervening inactive period
secondary inoculum—inoculum produced by an infection established earlier in the same growing season
secondary organism—an organism that multiplies in already diseased tissue but is not the primary pathogen
secondary root—a branch of a primary root
sedentary—remaining in a fixed location (*see* migratory)
seed—a ripened ovule consisting of an embryo and stored food enclosed by a seed coat
seed treatment—the application of a biological agent, chemical substance, or physical treatment to seed to protect the seed or resulting plant from pathogens or to stimulate germination or plant growth
seedborne—carried on or in seed
selective medium—a culture medium containing substances that inhibit or prevent the growth of certain species of microorganisms
semipersistent transmission—a type of virus transmission in which a virus is acquired and transmitted by a vector after a short feeding time and retained by the vector for hours or days
senesce (adj. senescent; n. senescence)—to decline, as with maturation, age, or disease stress
sepal—one of the modified leaves composing a calyx
septate—having septa (cross-walls)
septum (pl. septa; adj. septate)—a dividing wall; in fungi, a cross-wall
sequevar—a strain (of a bacterium or fungus) classified according to a genetic sequence
serology (adj. serological)—a method using the specificity of the antigen–antibody reaction for the detection and identification of antigenic substances and the organisms that carry them
sessile—with reference to leaves, leaflets, flowers, florets, fruits, ascocarps, basidiocarps, and so on, lacking a stalk, petiole, pedicel, stipe, or stem; with reference to nematodes, permanently attached and not capable of moving about
seta (pl. setae)—a bristle- or hairlike structure, usually deep yellow or brown and thick walled
sexual spore—a spore produced during the sexual cycle of a fungus
sharpshooters—leafhoppers in the tribes Proconiini and Cicadellini within the Cicadellidae
sheath—the lower part of a grass leaf, which clasps the culm; a membranous cover
shot hole—a symptom in which small lesions on a leaf fall out, giving the leaf the appearance of being hit by buckshot
sign—an indication of disease from direct observation of a pathogen or its parts on a host plant (*see* symptom)
sinuous—having curves, bends, or turns
slime molds—saprophytic organisms in the class Myxomycetes forming vegetative amoeboid plasmodia and spores
slow-sand filtration—a method of water purification in which water is passed slowly through a quantity of sand to remove pathogens
smut—any of several diseases caused by fungi in the group basidiomycetes that typically release masses of black, dusty teliospores at maturity
soft rot—a mushy disintegration of fleshy fruit or vegetables as a result of bacterial or fungal infection
soil drench—the application of a solution or suspension of a chemical to the soil, especially a pesticide to control soilborne pathogens
soilborne—carried on or beneath the soil surface
solanaceous—referring to members of the family Solanaceae, which includes petunia and tomato
solarization—a disease management practice in which soil is covered with polyethylene sheeting to increase heating by sunlight and thus control soilborne plant pathogens
sooty mold—black, nonparasitic, superficial fungal growth; often develops on honeydew produced by aphids and other phloem-feeding insects
sorus (pl. sori)—a compact fruiting structure, especially with reference to spore masses of rust and smut fungi
sp. (pl. spp.)—species (the singular and plural abbreviations are used only after genus names; the singular form is used to refer to an undetermined species, and the plural form is used to refer to two or more species without naming them individually)

spermagonium (pl. spermagonia)—a structure in which male reproductive cells are produced; in rust fungi, the pycnium, a globose or flask-shaped haploid fruiting body composed of receptive hyphae and spermatia (pycniospores)

spermatium (pl. spermatia)—a male sex cell; a nonmotile male gamete; a haploid male gamete; in rust fungi, a pycniospore

spicule—the copulatory organ of a male nematode

spiroplasma—any of various spiral-shaped, plant-pathogenic mollicutes (prokaryotes lacking cell walls)

sporangiolum (pl. sporangiola)—a small sporangium that contains only a few spores

sporangiophore—the sporangium-bearing body of a fungus

sporangiospore—a nonmotile, asexual spore borne in a sporangium

sporangium (pl. sporangia)—a saclike fungal structure, the entire contents of which are converted into an indefinite number of asexual spores

spore—the reproductive structure of fungi and some other organisms, containing one or more cells; a bacterial cell modified to survive an adverse environment

sporidium (pl. sporidia)—the basidiospore of rust fungi, smut fungi, and other members of the group basidiomycetes

sporocarp—a spore-bearing fruiting body

sporodochium (pl. sporodochia)—a superficial, cushion-shaped, asexual fruiting body consisting of a cluster of conidiophores

sporulate—to produce spores

sporulating—producing and (often) liberating spores

spot—a small necrotic area occurring as a disease symptom on a leaf, flower, or stem

spreader–sticker—a material used to improve control from a pesticide spray through its surfactant (spreader) and adhesion (sticker) properties

stamen (adj. staminal)—the male structure of a flower, composed of a pollen-bearing anther and a filament, or stalk

staminate flower—a male flower

stele—the central cylinder of vascular tissue, especially with reference to roots

stem pitting—a viral disease symptom characterized by depressions on the stem

sterigma (pl. sterigmata)—a small, usually pointed projection that supports a spore

sterile—unable to reproduce sexually; free of living microorganisms

sterile fungus—a fungus that is not known to produce any kind of spores

sterilization (adj. sterilized)—the total destruction of living organisms by various means, including heat, chemicals, and irradiation

sticky card, sticky trap—a rectangular card (often yellow) that is coated with sticky glue for the purpose of catching insect pests to monitor their presence and population size (usually, within a greenhouse)

stigma—the portion of a flower that receives pollen and on which pollen germinates

stipe—a stalk

stippling—a pattern of small dots or speckles on leaves where chlorophyll is absent

stipule—a small, leaflike appendage at the base of a leaf petiole; usually occurs as one of a pair

stock plants—plants maintained for purposes of propagation by seed or cuttings

stolon—a slender, horizontal stem that grows close to the soil surface (syn. runner); in fungi, a hypha that grows horizontally along a surface

stoma, stomate (pl. stomata; adj. stomatal)—a structure composed of two guard cells and the opening between them in the epidermis of a leaf or stem; functions in gas exchange

strain—a distinct form of an organism within a species or virus, differing from other forms of the species biologically, physically, or chemically

striate (n. striations)—to mark with delicate lines, grooves, or ridges

stroma (pl. stromata)—a compact mass of mycelium (with or without host tissue) that supports fruiting bodies or in which fruiting bodies are embedded

stunting—a reduction in the height of plants resulting from a progressive reduction in the lengths of successive internodes or a decrease in their number

style—the slender part of many pistils, located between the stigma and the ovary and through which the pollen tube grows

stylet—the stiff, slender, hollow feeding organ of plant-parasitic nematodes or sap-sucking insects, such as aphids and leafhoppers

stylet-borne transmission—virus transmission in which a virus is acquired and transmitted by a vector after a short feeding time and is retained by the vector for only a short time (syn. nonpersistent transmission)

subepidermal—located or occurring just below the epidermis

subgenomic RNA—a piece of viral RNA, shorter than the entire genome of the virus, found in cells infected by the virus and sometimes encapsidated

substrate—the substance on which an organism lives or from which it obtains nutrients; a chemical substance acted upon, often by an enzyme

sunscald—the injury of plant tissues burned or scorched by direct sunlight

suscept—a susceptible plant

susceptible (n. susceptibility)—prone to develop disease when infected by a pathogen (*see* resistant)

symbiosis (adj. symbiotic)—a mutually beneficial association of two different kinds of organisms

symbiont—either of the organisms in a symbiotic association

symptom—an indication of disease by a reaction of the host (e.g., canker, leaf spot, wilt) (*see* sign)

symptomatology—the study of disease symptoms

symptomless carrier—a plant that is infected with a pathogen (usually a virus) but has no obvious symptoms

syncytium (pl. syncytia)—a multinucleate structure in root tissue formed by the dissolution of common cell walls induced by secretions of certain sedentary plant-parasitic nematodes (e.g., cyst nematodes)

synergism (adj. synergistic)—an interaction in which the total effect of the interacting factors is greater than the additive effects of the factors acting individually

synnema (pl. synnemata)—compact or fused, generally upright conidiophores, with branches and spores that form a headlike cluster (syn. coremium)

systemic—pertaining to a disease in which the pathogen (or a single infection) spreads generally throughout a plant; pertaining to pesticides that spread throughout a plant by moving through the vascular system

taproot—a primary root that grows vertically downward and from which smaller lateral roots branch

taxonomy (adj. taxonomic)—the science of naming and classifying organisms

teleomorph—the perfect state, or sexual form in the life cycle of certain fungi (*see* anamorph; holomorph)

teliospore—a thick-walled resting spore produced by some fungi (notably rust and smut fungi) from which a basidium is produced

telium (pl. telia)—the fruiting body (sorus) that produces teliospores of a rust fungus

testa (pl. testae)—seed coat

thallus—the vegetative body of a fungus

tiller—a lateral shoot, culm, or stalk arising from a crown bud; common in grasses

tissue—a group of cells, usually of similar structure, that perform the same or related functions

tissue culture—an in vitro method of propagating healthy cells from plant tissues

titer—a measure of the concentration of a substance in a solution

tolerance (adj. tolerant)—the ability of a plant to endure an infectious or noninfectious disease, adverse conditions, or chemical injury without serious damage or yield loss; with reference to pesticides, the amount of chemical residue legally permitted on an agricultural product entering commercial channels, usually measured in parts per million (ppm)

tomentose—covered with a dense mat of hair

toxicity—the capacity of a substance to interfere with the vital processes of an organism

toxin—any poisonous substance of biological origin

transgenic—possessing a gene from another species; having been the subject of genetic engineering

translocation—the movement of water, nutrients, chemicals, or food materials within a plant
translucent—clear enough to allow the transmission of light
transmit (n. transmission)—to spread or transfer, as in spreading an infectious pathogen from plant to plant or from one generation of plants to another
transovarial passage—the passage of a virus through the eggs or offspring of its vector and then to the next generation of host plants
transpiration—the loss of water by evaporation from leaf surfaces and through stomata
trichome—a plant epidermal hair, of which several types exist
truncate—to end abruptly, as though an end has been cut off
tuber—an underground stem adapted for storage; typically produced at the end of a stolon
turgid—swollen or inflated; plump or swollen as a result of internal water pressure
turgidity—the state of being rigid or swollen as a result of internal water pressure
tylosis (pl. tyloses)—a balloonlike extrusion of parenchyma cells into the lumina of contiguous vessels, partially or completely blocking them

ultrastructure—the submicroscopic structure of a macromolecule, cell, or tissue
unicellular—one celled (*see* multicellular)
uniflagellate—having one flagellum
uninucleate—having one nucleus per cell (*see* binucleate; multinucleate)
urediniospore, **urediospore**, **uredospore**—the asexual, dikaryotic, often rust-colored spore of a rust fungus, produced in a uredinium; the repeating stage of a heteroecious rust fungus, in which it is capable of infecting the host plant on which it was produced
uredinium (pl. uredinia), uredium (pl. uredia)—the fruiting body (sorus) in which urediniospores of a rust fungus are produced

vacuole—a generally spherical organelle within a plant cell, bound by a membrane and containing dissolved materials, such as metabolic precursors, storage materials, and waste products
variegation—a pattern of two or more colors in a plant part, as in a green and white leaf
variety (adj. varietal)—a plant type within a species, resulting from deliberate genetic manipulation and having recognizable characteristics, such as color; shapes of flowers, fruit, and seeds; height; and form (syn. cultivar)
vascular—pertaining to fluid-conducting tissues (xylem and phloem) in plants
vascular bundle—a strand of conductive tissue; usually composed of xylem and phloem (in leaves, small bundles are called *veins*)
vascular wilt disease—a xylem disease that disrupts the normal uptake of water and minerals, resulting in wilting and yellowing of foliage
vector—a living organism (e.g., insect, mite, bird, higher animal, nematode, parasitic plant, human) that is able to carry and transmit a pathogen; in genetic engineering, a cloning vehicle, or self-replicating DNA molecule, such as a plasmid or virus, used to introduce a fragment of foreign DNA into a host cell
vegetative—pertaining to the somatic or asexual parts of a plant, which are not involved in sexual reproduction
vegetative propagation—asexual reproduction; in plants, the use of cuttings, bulbs, tubers, or other vegetative parts to grow new plants
vein—a small vascular bundle in a leaf
vein banding—a symptom of virus disease in which the regions along veins are darker green than the tissue between veins
vein clearing—disappearance of green color in or around leaf veins
vermiform—worm shaped
verrucose—covered with wartlike bumps
vesicle—the thin sac in which zoospores are differentiated and from which they are released; the bulbous head terminating the conidiophores of *Aspergillus* spp.; a structure formed by an endomycorrhizal fungus within living cells of the root
vessel—the water-conducting structure of xylem tissue, with pit openings in end walls
viable (n. viability)— alive; capable of germination (e.g., of seeds, fungus spores, sclerotia, etc.); capable of growth
virescence—a state or condition in which normally white or colored tissues (e.g., flower petals) become green
virion—a complete virus particle
viroid—any of various small, unencapsidated (naked), circular, single-stranded RNA molecules capable of causing disease in plants
viroplasm—an inclusion body within a cell where viruses are replicated
virulence—the degree or measure of pathogenicity; the relative capacity to cause disease
virulent—pathogenic; having the capacity to cause disease (*see* avirulent)
viruliferous—virus laden, usually with reference to insects or nematodes as vectors
virus—any of numerous submicroscopic, intracellular, obligate parasites consisting of a core of infectious nucleic acid (either RNA or DNA) usually surrounded by a protein coat
volunteer—a self-set plant; seeded by chance

water soaking—a disease symptom in which plant tissues or lesions appear wet, dark, and usually sunken and translucent
web blight—an attack on the aboveground parts of plants by a fungus or oomycete; webs of hyphae are sometimes visible under highly humid conditions
whorl—a circular arrangement of like parts
wild type—the phenotype characteristic of most individuals of a species under natural conditions
wilt—the drooping of leaves and stems caused by a lack of water (inadequate water supply or excessive transpiration); a vascular disease in which normal water uptake is interrupted
witches'-broom—a disease symptom in which many weak shoots arise at or close to the same point, producing abnormal massed, brushlike growth

xylem—the water- and mineral-conducting, food-storing, supporting tissue of a plant

yellows—a plant disease characterized by chlorosis and stunting

zonate—pertaining to the targetlike development of a tree canker, characterized by successive perennial rings of callus; pertaining to any symptom appearing in concentric rings
zoosporangium—a sporangium (spore case) bearing zoospores
zoospore—a fungal spore with flagella; capable of locomotion in water
zygomycetes (adj. zygomycetous)—a group of fungi in the phylum Zygomycota (kingdom Fungi) that produce resistant, spherical sexual spores following fusion of hyphal gametangia
zygospore—the sexual resting spore formed from the union of gametangia of fungi in the group zygomycetes

Index